Digestible Quantum Field Theory

Andrei Smilga

Digestible Quantum Field Theory

 Springer

Andrei Smilga
SUBATECH
University of Nantes
Nantes
France

ISBN 978-3-319-86735-9   ISBN 978-3-319-59922-9   (eBook)
DOI 10.1007/978-3-319-59922-9

Printed on acid-free paper

This Springer imprint is published by Springer Nature
The registered company is Springer International Publishing AG
The registered company address is: Gewerbestrasse 11, 6330 Cham, Switzerland

To the memory of Elizaveta Glinka

Menu

Part III   Chef's Secrets

Part IV   Entrées

Part I
Aperitif

Chapter 1
Introduction

Half a century ago, when the author of this book was somewhat younger than he is now, he was interested in science, especially in physics. One thing that I wanted to understand at that time was what this famous *theory of relativity* was about. Not yet prepared to learn it from the corresponding Landau and Lifshitz volume, I had to look for the answer in popular books.

The first book that I happened to read on this subject was *Relativity for the Million* by Martin Gardner (in Russian translation, of course). This book was written in a way adapted for a general public and did not contain a single formula. The author tried to explain the main relativistic effects (time dilation and twin paradox, length contraction, mass-energy equivalence and so on) using only words. Well, I must say that it did not work in my case. I did not understand much.

My second try was, however, much more successful. It was *Readable Relativity* by Clement Durell. That book did involve some simple algebraic formulas. They failed to confuse me because I had just read the high-school manual for maths and already had some idea about what x, y and even $\sqrt{z}$ might mean. The book also involved transparent and pedagogical explanations how the time dilation and length contraction mentioned above can be derived from Einstein's two postulates of special relativity. I understood it.[1]

Later I read many other good popular articles and popular books. There were books by Yakov Perelman: *Physics can be Fun, Mathematics can be Fun, Astronomy can be Fun* and so on). I remember the article by Murray Gell-Mann about the *Eightfold Way* (the quark model and the Mendeleev-like classification of "elementary" particles). Still later, I enjoyed reading *The First Three Minutes* by Steven Weinberg about the origin of the Universe — what happened during the first three minutes of its existence. At that time I was already at graduate school. But I maintain that this book

---

[1] Recalling my positive emotions from reading my first book on a serious subject in serious physics, I have chosen the title of my own book as a variant of Durell's title.

© Springer International Publishing AG 2017
A. Smilga, *Digestible Quantum Field Theory*,
DOI 10.1007/978-3-319-59922-9_1

is understandable for a bright high-school student, in spite of (I would say, *due to*) the presence of a number of simple algebraic relations.

Like it or not, physics is an exact science. And *exact* means that it is expressed in the language of mathematics. Without maths one just cannot understand it. For sure, there are different levels of maths. Going back to special relativity, it can well be expressed in the language of square roots. To understand it *completely* (better than Einstein understood it back in 1905 when writing his famous original papers), one should also be familiar with vectorial and tensorial analysis and with the basics of group theory.

The reader has probably already guessed that *this* book devoted to quantum field theory will not be deprived of mathematics. Moreover, the necessary level of maths to grasp this subject is essentially higher than for special relativity. Thus, this book is hopefully going to be readable, but not by a 12-year-old boy. Even a bright one. But it is written to be accessible to any person who studied physics at University at the *undergraduate* level.

This was actually my main motivation to write the book. I know many engineers, mathematicians, chemists, colleagues working in other branches of physics, who want to learn what *quantum chromodynamics* or the *Standard Model* are, who *are* capable of learning it at some level, not spending the large amount of time required to acquire a professional understanding, but have not been able to do so till now. The first two parts of this book (the first two courses in the dinner I am suggesting you enjoy) are for them.

The prerequisites in maths are not only elementary algebra, but also elementary analysis and linear algebra (ordinary and partial derivatives, integrals, differential equations, matrices). The prerequisites in physics are a university course of general physics including quantum mechanics. The best such course is due to Feynman. However, an acquaintance with the *Feynman Lectures in Physics* is a sufficient, but not necessary condition to read the first two parts of this book.

They have a popular character. In a certain sense, the beginning of the book is *more popular* than the Durell book on special relativity mentioned above. Whereas after reading Durell's book one would obtain a rather clear and comprehensive understanding of the subject, we cannot promise the same to a person who limits his acquaintance with our book to its first two parts.

Unfortunately, quantum field theory is a more complicated subject than special relativity, and the prerequisites listed above are not sufficient to fully understand what quantum field theory is. At this preliminary stage we will only *announce* the results and give whenever possible their heuristic explanations, without being able to *derive* them.

One can make a general comment in this respect. Even though physical laws are formulated in the language of mathematics, there is a substantial difference between physics and mathematics and between the way a student learns them. Different branches of mathematics are related between themselves, but not so tightly. An expert in, say, functional analysis can have a rather superficial knowledge of group theory or number theory. And vice versa. Group theory and functional analysis are different logical systems based on different sets of axioms. They can be studied separately. A broad mathematical culture, knowing well several such logical systems and not only the one a mathematician is working on at a given moment, brings him perspective and can facilitate his insights, but it is not absolutely necessary.

But physics is different. It has basically only one subject of study: the world around us. Different physical phenomena are very much intertwined and also their mathematical descriptions have very much in common; The same wave equation, $\Box f = 0$, describes light, sound and surf. That is why it is next to impossible to study physics branch by branch. One simply cannot study optics at a deep level having no idea about mechanics (including analytical mechanics), about Maxwell's equations, etc.

A physics student is traditionally invited to be treated in a way similar to the way a piece of work is treated by a lathe — in a sequence of revolutions bringing it finally to the required shape. *(i)* At (a good) high school, s/he acquires the knowledge of the whole of physics at a superficial level and learns to calculate the trajectory of a canon ball and the capacity of a condenser. This is the first revolution. *(ii)* The second revolution is a course of general physics in college. Among many other things, a student learns that light is nothing but an electromagnetic wave representing a solution to Maxwell's equations. *(iii)* The third revolution is a course on theoretical physics.

And our book also involves several such revolutions. The first one is Part II. In Chaps. 3 and 4 I give a general bird's eye overview of what the Universe looks like and how physicists understand and describe its structure. In Chap. 5, still hovering in the skies, we go down a little to focus better on the main subject of the book, quantum field theory, and give a superficial and mostly descriptive (with many words and only few formulas) synopsis of the modern theories of the electromagnetic, strong and weak interactions.

Chapter 5 closes the *Hors d'Oeuvres* part of the book, and the reader who feels that his hunger is already somewhat satisfied and decides that s/he now has more urgent things to do, may skip all the other parts except, probably, Part V, the *Trou Normand*,[2] where I do not discuss physics, but tell a few human stories about physicists and impose on the reader's attention with some personal recollections.

If the reader wants to undergo the second revolution and to learn quantum field theory at a more profound level, s/he is invited to deal with four *Entrées* that I serve up in Part IV of the book. In Chaps. 11 and 12, the two ingredients of modern quantum field theory (it carries the low-profile name "Standard Model"), the theory of the strong interactions and the unified theory of the electromagnetic and the weak interactions, are described in detail.

But to understand these chapters, one needs to learn mathematics, classical and quantum mechanics at a deeper level— beyond the standard university courses. All the necessary extras are given in the *Chef's Secrets* part.

---

[2]The "trou normand" is a special course in a traditional French dinner — a glass of calvados or something similar — which helps a banqueter to digest the meals already consumed and prepares him for the dessert.

Chapter 6 is a crash course in group theory, which is indispensable to understand the mathematical formulation of field theories, given in Chaps. 10–12. I also introduce there the notion of Grassmann numbers needed to describe fermion fields, which we do in Chap. 9.

In Sect. 7.1 I briefly describe analytical mechanics, the Lagrangian and Hamiltonian methods. This should be familiar to our target reader. But the other sections of Chap. 7 involve material that is not always taught at universities. Please have a look at it.

Chapter 8 is devoted to scattering theory in quantum mechanics and to the diagrammatic representation of the elastic scattering amplitude. We derive the latter quite accurately, in contrast to the relativistic diagram technique, where we have chosen to be somewhat sloppy.

I must say here that though parts III and IV do not exactly represent what one calls a popular reading, they are still popular in some sense. I tried to give there only the *minimal* technical details that are necessary to understand how things work at some level, which is not, however, the professional level. For example, I will discuss Feynman diagrams, will try to explain how they appear and what kind of analytical expressions for the scattering amplitudes they encode. But I will neither derive the Feynman rules accurately, nor teach the reader how to use them in exact calculations. I do not explain complicated theoretical issues (like quantum anomalies). I acquaint the reader with the notion of path integral, but do not go into detail.

A student who wants to learn all this must consult more serious textbooks. Among them, I first want to mention the brilliant book of Anthony Zee *Quantum Field Theory in a Nutshell*, written with a lot of pedagogical explanations at a level close to professional. And then there are, of course, comprehensive manuals, like *An Introduction to Quantum Field Theory* by Michael Peskin and Daniel Schroeder.

For the *Dessert*, I mostly serve some sweet speculations. Chapter 14 is devoted to supersymmetry. Supersymmetry is a very elegant idea about the degeneracy between bosons (particles with integer spin) and fermions (particles with half-integer spin). It may or may not be realized in Nature. We give some arguments in its favour at the end of Chap. 12 and in Chap. 14. Irrespective of its relevance, it is so beautiful that I could not refrain from explaining what it is.

The main subject of the book is quantum field theory. But there is one particular field theory that is known to us today only in its classical form and which successfully resists our attempts to quantize it. I mean *gravity*. We will discuss the classical theory of gravity, *general relativity*, in Chap. 15.

The last chapter is devoted to the mystery of quantum gravity, to string theory (the mainstream candidate for the role of unified Theory of Everything) and to my own heretical ideas on this subject. This chapter and, to a lesser extent, Chap. 14 are more complicated than the others. They are addressed to a person who already knows the Standard Model fairly well — from our book or from other sources — and wants to meditate together with the author on what is beyond.

Our book has one feature that I would like to mention here. There are frequent cross-references both to the previous and to the subsequent chapters and sections. These references are indispensable: after all, there is a lot of material, and one cannot

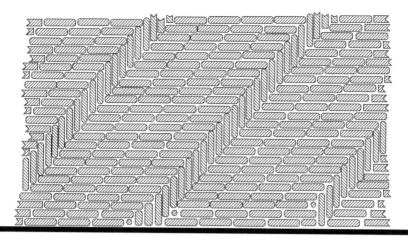

Fig. 1.1 Fishbone brickwork with cross-references.

expect the reader to keep in memory everything that he had read in Chap. 5, while reading Chaps. 11 and 12. S/he needs reminding. And for the benefit of those who are not satisfied with the heuristic hand-waving arguments in Chaps. 4 and 5, we gave there references to the later parts of the book where the same subjects are treated with more rigour.

In so doing, we took our inspiration from the great architect of the 15th century Filippo Brunelleschi, who put some bricks vertically in the fishbone (*spina pesce*) pattern, when building the magnificent Florentine Dome. The vertical bricks kept the inclined wall of the Dome together and prevented it from collapsing during construction. Hopefully, our cross-references will do a similar job (Fig. 1.1).

Acknowledgements I am not a native English speaker. My English is good enough for writing scientific papers, but barbarisms inevitably creep in when I try to write a text of a more general scope. Understanding that, I hesitated until recently to realize my longtime project and to write such a semi-popular book.

Maybe this book would never have been written (at least, in English) if Hugh Jones from Imperial College in London had not make a generous proposal to *edit and correct* my style throughout the whole book. Knowing well what quantum field theory is about, he also made many valuable remarks on the content. My gratitude to him is difficult to overstate.

I am indebted to Jean-Paul Blaizot, Masud Chaichian, Igor Klebanov, Heinrich Leutwyler, Alexei Morozov, Mikhail Shaposhnikov, Anca Tureanu, Arkady Vainshtein and Andrei Varlamov, to whom I showed the manuscript of this book, for useful comments. It is also my pleasure to thank Tatiana Eletsky for drawing the pictures in Figs. 4.2 and 15.1, Krystyna Haertle for her artistic contributions in Figs. 8.1 and 8.2, Albane Smilga for creating the Universe in Fig. 16.7, and Oleg Kancheli for providing me with the photo in Fig. 13.2 from his archive.

Finally, I would like to thank the whole Springer team, and especially Dr. Angela Lahee for her encouragement and support.

Chapter 2
Units □ Fundamental Constants □ Conventions

· We are using in this book different units and unit systems. Quite often (though not always) we give numerical values for the masses and distances in kilograms and meters, as is prescribed by the International System (*SI*). However, although it is good for practical engineering purposes, SI is a kind of devil's tool as far as *teaching* physics is concerned. This applies not so much to meters and kilograms, but to its messy electric units.

In SI the ampere is an independent unit. As a result, the Coulomb law

$$F^{SI} = \frac{1}{4\pi\epsilon_0} \frac{q_1 q_2 \boldsymbol{r}}{r^3} . \tag{2.1}$$

involves an esoteric factor $1/(4\pi\epsilon_0)$, which has absolutely no physical meaning and distracts students' attention from what is relevant. (It is possible, of course, to learn physics using SI, but ϵ_0 simply makes this task somewhat more difficult.) Even the brilliant *Feynman Lectures* were written using amperes, which is a pity. The reason is, of course, that Feynman's book was written on the basis of a real lecture course that Feynman gave in Caltech. And he had to use the official unit system imposed by the Ministries of Education all around the world including the USA. Our book is not supervised by any ministry and we can make our own choices.

Thus, (2.1) is the only formula in this book involving ϵ_0 or μ_0. Whenever we need to write an explicit formula describing the physics of the electromagnetic interactions, we will often use the CGS unit system. The basic units there are centimeter, gram and second, all other units, including the electric charge, magnetic field and so on, being expressed in terms of them. For example, the CGS unit for the electric charge q is 1 esu[1] $= 1\,g^{1/2} \cdot cm^{3/2} \cdot s^{-1}$, as follows from the Coulomb law written in the form

---

[1] *esu* means *electrostatic unit* of charge.

© Springer International Publishing AG 2017
A. Smilga, *Digestible Quantum Field Theory*,
DOI 10.1007/978-3-319-59922-9_2

$$F = \frac{q_1 q_2 \boldsymbol{r}}{r^3} \tag{2.2}$$

without extra factors.

CGSE/M is better pedagogically than SI, but it is not the best. The best (not for all purposes, but for many) is the so-called *theorists'* or *natural* unit system. The idea is simple. There are several basic fundamental constants in physics. In particular, the Planck constant $\hbar$ and the speed of light c. These constants enter all formulas. We now may choose the unit system where $\hbar = c = 1$. This allows us to get rid of these constants and simplify a lot of the formulas, so that their *relevant* structure is much more clearly displayed.

The consequences are rather radical and may not at first seem desirable. To begin with, due to the condition $c = 1$, time and length are now measured in the same units. Instead of centimeters or meters, lengths are measured in "light-seconds". Well, actually, the readers of science fiction and popular astronomy books, familiar with the light-year measure, might not be too surprised and shocked here. In the theory of relativity time and length have a similar nature, which is better displayed if they are measured in the same units.

For those who are still feeling uneasy, I propose to imagine an almost flat world with very strong gravity, such that the movements in the vertical direction are much more restricted than in our world. Would not the third z spatial axis be then perceived completely differently compared to the x and y axes? And would it not be natural and convenient in everyday life to measure z in different units compared to x, y ?[2] Imagine now a physicist in this world, an indigenous Einstein, who discovered that the height actually has the same nature as two other direction. And he then proposes to measure them in the same units to clarify the underlying space structure. I believe that he would have his reasons.

In this system speed is dimensionless, it is measured in fractions of c. The energy and momentum have the same dimension as mass and are measured in the same units.

The condition $\hbar = 1$ brings about still more drastic changes. Given any mass m and knowing $\hbar$ and c, one can cook up a quantity of dimension of length:

$$l_C = \frac{\hbar}{mc}. \tag{2.3}$$

For a particle of mass m, this quantity is called its reduced *Compton wavelength* (The ordinary Compton wavelength involves an extra factor 2π). The meaning is the following. In quantum mechanics, there is a dualism between particles and waves. An electron with momentum p can be described by a de Broglie wave of wavelength $\lambda = 2\pi\hbar/p$. The wavelength $2\pi l_C$ corresponds to the momentum $p = mc$ when the

---

[2]In fact, something like that takes place even in our world with our moderate gravity. Some nomadic peoples have a traditional measure of length: the distance covered over a day by horse. But, obviously, this is only a measure of distances in x and y direction, not in z direction. Different measures for x, y and for z are also traditionally used in England: the altitudes are usually measured in feet while the distances between the points with the same gravitational potential—in yards and miles.

speed of the particle is relativistic:

$$v = \frac{pc^2}{E} = \frac{pc^2}{\sqrt{p^2c^2 + m^2c^4}} = \frac{c}{\sqrt{2}}. \tag{2.4}$$

One can say that the Compton wavelength is an approximate limit beyond which (at smaller distances) the usual non-relativistic quantum mechanics does not apply.

In the system $\hbar = c = 1$, the Compton wavelength of a particle is just its inverse mass. In this system one measures distances not in meters and not in (light-) seconds, but in inverse (quantum-light-) kilograms!

To be more precise, usually not in kilograms. In the micro-world, a kilogram (or a gram) is too large and inconvenient a mass unit. The units one traditionally uses to measure masses and energies of elementary particles are *electron-volts*. One electron-volt (denoted eV) is the energy that an electron acquires when going across a condenser with the potential difference 1 V. In other words,

$$1\,\text{eV} = |e| \cdot (1V) \approx 1.6 \cdot 10^{-19} J \,, \tag{2.5}$$

where e is the electron charge. In practice, one also uses kilo-electron-volts (1 keV $= 10^3$ eV), mega-electron-volts (1 MeV $= 10^6$ eV), giga-electron-volts (1 GeV $= 10^9$ eV), and tera-electron-volts (1 TeV $= 10^{12}$ eV).

Bearing in mind the convention $c = 1$, these units are also used to measure momenta and masses. For example, the mass of electron is $m_e = 511$ keV, the mass of the proton is $m_p = 938$ MeV and so on.

Also for distances, one uses not the meter, but its small fractions. The popular units are the *angstrom* (1 Å $= 10^{-10}$ m), the characteristic size of the atom, and the *fermi* (1 fm $= 10^{-15}$ m), the characteristic size of the atomic nucleus. As was discussed before, the distances can also be measured in time units and in inverse mass units. Thus, 1 fm $\approx 3.3 \cdot 10^{-24}$ s $\approx (200\,\text{MeV})^{-1}$.

The information about different units and the values of the most important fundamental constants is assembled in the table below (Table 2.1).

We draw special attention to the value of Newton's gravitational constant. In natural units,

$$G_N = \frac{1}{m_P^2} \,, \qquad m_P \approx 1.22 \cdot 10^{19}\,\text{GeV} \approx 2.2 \cdot 10^{-5}\,\text{g} \,. \tag{2.6}$$

The mass m_P is called the Planck mass. In usual units,

$$m_P = \sqrt{\frac{\hbar c}{G_N}} \,. \tag{2.7}$$

One can in principle use a "supernatural" system where all masses are dimensionless, being measured in units of the Planck mass. In such a system, the electron mass is

Table 2.1 Units and fundamental constants.

	Name	Symbol	Value
Units	Angstrom	Å	10^{-10} m
	Fermi	fm	10^{-15} m $\approx$ $(200\,\text{MeV})^{-1}$
	Electron-volt	eV	$1.6 \cdot 10^{-19}$ J $\approx$ $11600\,K$
	Degree kelvin	K	$1.38 \cdot 10^{-23}$ J
Constants	Speed of light	c	$3 \cdot 10^8$ m/s
	Planck constant	$\hbar$	$1.05 \cdot 10^{-34}$ J· s
	Gravitational constant	G_N	$6.67 \cdot 10^{-11}$ (SI) $\approx$ $(1.22 \cdot 10^{19}\,\text{GeV})^{-2}$
	Electron charge	e	$-4.8 \cdot 10^{-10}$ esu $\approx$ $-1.6 \cdot 10^{-19}$ C
	Fine-structure constant	$\alpha = \frac{e^2}{\hbar c}$	$1/137$
	Electron mass	m_e	$9.1 \cdot 10^{-31}$ kg $\approx$ 511 keV
	Proton mass	m_p	$1.67 \cdot 10^{-27}$ kg $\approx$ 938 MeV

$$m_e \approx 4.2 \cdot 10^{-23} \,, \tag{2.8}$$

which is very, very small. Thus, the masses of all elementary particles are very small compared to the Planck mass. One can also say that the Planck mass, being of the same order as the weight of the *Amoeba Proteus*, is very large compared to the mass of elementary particles.

Besides the Planck fundamental mass, one can also define the Planck fundamental length and time. In usual units,

$$l_P = \sqrt{\frac{\hbar G_N}{c^3}} \approx 1.6 \cdot 10^{-35}\ \text{m}, \qquad t_P = \sqrt{\frac{\hbar G_N}{c^5}} \approx 5.4 \cdot 10^{-44}\ \text{s}. \tag{2.9}$$

When theorists work in their system, they usually give it an ultimate polishing touch and use the *Heaviside* definition of the electric charge, $q_{\text{Heaviside}} = \sqrt{4\pi}\, q_{\text{CGS}}$. Then the Coulomb law acquires the factor 4π downstairs, but Maxwell's equations (4.31) and the Lagrangian, where they follow from, have no π.

In Heaviside natural units, which we will mostly use, the fine-structure constant is expressed as

$$\alpha = \frac{e^2}{4\pi}\,. \tag{2.10}$$

We will use the Minkowski metric $\eta_{\mu\nu} = \text{diag}(1, -1, -1, -1)$ and will carefully distinguish between covariant and contravariant vectors and tensors (we recall this formalism in Sect. 6.1.2) throughout the book. In most cases this distinction is irrelevant, but sometimes it is. Whenever we care to distinguish time and space coordinates, our convention is $x^\mu = (ct, \boldsymbol{x})$ and $\partial_\mu = \left(\frac{1}{c}\frac{\partial}{\partial t}, \frac{\partial}{\partial \boldsymbol{x}}\right)$. For the other 4-vectors we adopt the convention $V^\mu = (V_0, \boldsymbol{V})$ and hence $V_\mu = (V_0, -\boldsymbol{V})$. We will also meet matrix-valued 4-vectors: two different 2×2 matrices

$$\sigma^\mu = (\mathbb{1}, \boldsymbol{\sigma}), \qquad \bar{\sigma}^\mu = (\mathbb{1}, -\boldsymbol{\sigma}), \tag{2.11}$$

where $\boldsymbol{\sigma}$ are the Pauli matrices;[3] and the Dirac 4×4 γ matrices

$$\gamma^\mu = \begin{pmatrix} 0 & \bar{\sigma}^\mu \\ \sigma^\mu & 0 \end{pmatrix}, \tag{2.12}$$

which satisfy the Clifford algebra $\gamma^\mu \gamma^\nu + \gamma^\nu \gamma^\mu = 2\eta^{\mu\nu} \mathbb{1}$.

The last remark is about our conjugation symbols policy. The star * will denote complex conjugation of ordinary numbers. The dagger † will be used for complex conjugation of the Grassmann numbers and for Hermitian conjugation of operators. The bar $^-$ is reserved for Dirac conjugation of bispinors [see Eq. (9.45)] and will also mark fermion antiparticles (p stands for proton, while $\bar{p}$ for antiproton, etc.).

---

[3] One can observe that $\sigma^\mu \bar{\sigma}_\mu = 4 \cdot \mathbb{1}$.

Part II
Hors d'Oeuvres

Chapter 3
The Universe as We Know It

> *... To myself, I seem to have been only like a boy playing*
> *on the seashore, and diverting myself in now and then*
> *finding a smoother pebble or a prettier shell than ordinary*
> *whilst the great ocean of truth lay all undiscovered before me.*
>
> I. Newton

When Newton wrote this, people had just started to study Nature, and Newton's comparison was only a little hyperbolic. At that time the only known fundamental law was Newton's law of universal gravitational attraction. People were ignorant about the atomic structure of ordinary matter, had no idea why the Sun burns, what sunlight is and so on.

Since then, more than 300 years have passed. The situation has changed drastically. We do know now up to a point what matter consists of. We know what ignites the Sun. Basically (with some notorious exceptions to be discussed later), we understand how the world around us functions from the microscopic scale $\sim 10^{-18}$ m (eight orders of magnitude smaller than the size of the atom) up to the scale $\sim 10^{26}$ m (the size of the observable part of the Universe). The ocean of truth is now mostly discovered and charted.

The Universe was created about 13 billion years ago. The act of its creation is called the *Big Bang*. At the very moment of the Big Bang, the Universe had zero size and infinite temperature. Then it started to expand and to cool down rapidly. It is still expanding: distant galaxies run away from us at high speed. The more distant they are, the greater is the speed.

Frankly speaking, we do not know now what happened at the *very moment* of the Big Bang. Neither are we sure that such a moment existed. And definitely, we cannot answer the question: *what happened before the Big Bang?* Well, people often give *an* answer to that (you can find it in popular and scientific books) that Time together with Space were created just during Big Bang, and hence the question "*what was before*" has no meaning. But this answer simply reflects the limits of our knowledge. It is true that, in Friedmann's solution of the *classical* general relativity equations,

© Springer International Publishing AG 2017
A. Smilga, *Digestible Quantum Field Theory*,
DOI 10.1007/978-3-319-59922-9_3

time and space originate in the moment of the Big Bang. The problem is, however, that the classical theory simply does not apply to this very moment. Quantum effects are essential there. And we do not know today what quantum theory of gravity *is*…

However, we know for sure that there was a time when the Universe was very small and very hot. We begin to understand the dynamics of the expanding Universe reasonably well from the moment $t \approx 10^{-10}$ s after the Big Bang when the Universe had already cooled down to the temperature $T \approx 100\,\text{GeV} \approx 10^{15}$ K.[1]

3.1   Elementary Particles

The Universe is filled with *Matter*. There are many different types of matter. Sometimes, physicists call these types *fields* and sometimes *particles*, the latter being elementary excitations (quanta) of the former. The temperature of the Universe we have just talked about roughly coincides (up to a numerical coefficient) with the mean energy of an individual particle.

All the known elementary particles are listed in Table 3.1.

Let us discuss this table.

There are four basic types of particles represented there. First of all, there are *gauge bosons*[2] — mediators of four types of widely known *fundamental interactions*: the strong, weak, electromagnetic, and gravitational. Thus, the mediator of the electromagnetic interaction is the photon. The reader has seen it. The mediator of the gravitational interaction is called the *graviton*. The existence of gravitational waves (representing coherent flows of a large number of gravitons) was predicted by Einstein a hundred years ago, and very recently they were detected in experiment — we will discuss this in detail in Chap. 15. The mediators of the weak interaction are the so-called *intermediate bosons*, $W^{\pm}$ and Z. Their existence was also first predicted theoretically and then they were discovered in accelerator experiments. The mediator of the strong interaction is called the *gluon*, In contrast to the photon, graviton, $W^{\pm}$ and Z; gluons have not been *directly* observed. Such an observation is in fact impossible for a rather peculiar reason. It turns out that an individual gluon (like an individual quark) just cannot exist in separation from other quarks and gluons due to *confinement*. We will discuss this at length later.

The next group are the *leptons*, their most known representative being the electron. The word "lepton" comes from Greek λεπτός, which means light. Indeed, the electron is light, lighter than most of the other particles. Neutrinos are even lighter — we do not even exactly know now how light they are; only the upper limits for their masses

---

[1] Today its temperature (the temperature of the cosmic microwave background radiation) is around 2.7 K.

[2] The reader probably understands the word *boson* — a particle with integer spin which likes to stay in the company of its comrades. On the other hand, he probably does not understand the word *gauge* yet. By the middle of this book, he will.

Table 3.1 Elementary particles. In the last column, "G" stands for *gravitational*, "EM" for *electromagnetic*, "W" for *weak* and "S" for *strong*.

Name		Mass	Electric charge (in units of $|e|$)	Spin	Participation in interactions
Gauge bosons	Graviton	0	0	2	G
	Photon (γ)	0	0	1	G, EM
	Gluon (g)	0	0	1	G, S
	$W^\pm$	80 GeV	± 1	1	G, EM, W
	Z	91 GeV	0	1	G, W
Leptons	Electron (e^-)	511 keV	-1	1/2	G, EM, W
	Electron neutrino (ν_e)	<0.12 eV	0	1/2	G, W
	Muon (μ^-)	106 MeV	-1	1/2	G, EM, W
	Muon neutrino (ν_μ)	<0.12 eV	0	1/2	G, W
	Tau lepton(τ^-)	1.78 GeV	-1	1/2	G, EM, W
	Tau neutrino (ν_τ)	<0.12 eV	0	1/2	G, W
Quarks	Up (u)	3 MeV	2/3	1/2	G, EM, W, S
	Down (d)	6 MeV	$-1/3$	1/2	G, EM, W, S
	Strange (s)	150 MeV	$-1/3$	1/2	G, EM, W, S
	Charmed (c)	1.35 GeV	2/3	1/2	G, EM, W, S
	Beautiful (b)	4.8 GeV	$-1/3$	1/2	G, EM, W, S
	Top (t)	170 GeV	2/3	1/2	G, EM, W, S
Higgs boson (H)		125 GeV	0	0	G, W

were established by direct measurements.[3] But the tau-lepton is not particularly light. It is almost 2 times heavier than the proton. In fact, leptons are distinguished not by their lightness, but by the fact that they (in contrast to quarks) do not interact with gluons, do not participate in the strong interactions.

There are 6 types (physicists say —*flavours*) of *quarks*. Some of them are rather light (in particular, the u and d quarks), and some are heavy. Quarks participate in all four types of interaction.

The name "quark" was invented by Gell-Mann. He took it from *Finnegans Wake* by James Joyce, where this word had some mysterious and rather obscure meaning. But in German *der Quark* means cottage cheese.

Quarks and leptons form three families called *generations*. Each generation involves a pair of leptons and a pair of quarks. The pairs (or *doublets*) (ν_e, e) and (u, d)

---

[3] We know, however, that neutrinos are not completely massless. This follows from the analysis of the so-called *neutrino oscillations* — see Sect. 12.6.

constitute the first generation. The pairs (ν_μ, μ) and (c, s) — the second generation, and the pairs (ν_τ, τ) and (t, b) — the third generation.[4]

The names *up* and *down* for the quarks of the first generation just indicate their position in the doublet represented as

$$\begin{pmatrix} u \\ d \end{pmatrix}.$$

The names *strange, charmed* and *beautiful* reflect a poetic imagination of modern physicists. The latter was to a considerable extent exhausted when the need to christen the heaviest quark flavour came. "Top" means again "the upper doublet component". Matching this, also the *b* quark has got another often-used low-profile name: *bottom* instead of "beautiful".

Both leptons and quarks carry spin $1/2$.[5] They are fermions and have very bad social manners. They are so bad that two identical fermions just cannot stand staying together in the same quantum state (the so-called *Pauli exclusion principle*).

In addition to the fermions listed in the Table, there are also antifermions. For example, the anti-electron is called the *positron*. It has the same mass, but carries positive electric charge. The antiparticle for the *u* quark is the quark $\bar{u}$ with electric charge $-2/3$. As for the bosons, most of them (the graviton, photon, Z and Higgs bosons, in a certain sense also the gluon) coincide with their antiparticles. The antiparticle for W^+ is W^-.

Let us say right now that the ordinary matter, atoms and nuclei, is made of the light fermions of the first generation: *u* quarks, *d* quarks and electrons. But, as was already mentioned, the quarks are not observed directly due to confinement. What we see are the composite objects: the proton made of two *u* quarks and one *d* quark and the neutron made of two *d* quarks and one *u* quark.

The fourth group of elementary particles has only one representative — the Higgs particle. It is described by a scalar field. Hence it has zero spin and hence (I will explain this further in the book) it is a boson and likes socializing. It participates only in two interactions out of the four — weak and gravitational.

However, the Higgs boson also participates in two other less known types of interaction not mentioned in the Table: *(i)* the *Higgs self-interaction*, which endows this particle with a mass and *(ii)* the *Yukawa interaction*, giving mass to quarks and leptons. These interactions do not have gauge mediators. The Yukawa interaction and, especially, the Higgs self-interaction are important to make consistent the theoretical

---

[4]In fact, things are somewhat more complicated: there is mixing between generations. But it is not yet time for us to discuss this.

[5]The reader probably knows this, but I recall that the spin of a particle is its intrinsic angular momentum. Spin is measured in units of $\hbar$. Thus, saying that the electron has spin $1/2$ means that its angular momentum is equal to $\hbar/2$.

description of Nature, but they do not immediately display themselves when one looks around.[6]

On the other hand, two out of the four gauge interactions, gravitational and electromagnetic, act at large distances and their effects are seen in everyday life without any devices. (Apples fall down; this is the gravitational interaction. We see the sunlight; this is the electromagnetic interaction.) The strong and weak interactions act at small distances, one fermi and less. They are responsible for the microscopic structure of matter.

Let us now talk about their role in the Universe in more detail.

3.2 Gravitational Interaction

At the microscale, it is very weak, the force of gravitational attraction between two electrons being 43 orders of magnitude weaker than the force of their electric repulsion.[7] However, it is the gravitational interaction which determines the structure of the Universe at the cosmic scale. This is due to *universality* of gravitational attraction — the more matter you have, the larger its mass and the larger the gravitational force. The structure of the solar system (Kepler's laws and so on) is determined by the gravitational interaction — that was already known to Newton.

Newton's law

$$F = -\frac{G_N m_1 m_2 r}{r^3} \tag{3.1}$$

represents a non-relativistic approximation of Einstein's general relativity. The latter brings about nontrivial effects.

One such effect is the influence of the gravitational field on time measurements. Clocks tick faster on satellites than on the Earth's surface. For the satellites used in the Global Positioning System and placed in a high orbit at an altitude of about 20000 km from the ground, the clocks advance by about 38 μs every day.[8] 38 μs seems to be nothing, but it is not so. It was absolutely necessary to take this effect in consideration while designing the GPS. Otherwise, it simply would not work.

The most spectacular prediction of general relativity is the existence of *black holes*. Black holes are massive and dense cosmic objects. Being very massive and usually very dense, they have such a strong gravitational field that nothing, including light, can escape it.

---

[6]Even if one does not just look, but performs ingenious experiments with special devices. The most sophisticated such device is the Large Hadronic Collider near Geneva, the length of its ring being about 30 km.

[7]The ratio 10^{-43} is nothing but m_e^2/α where m_e is expressed in Planck mass units, see (2.8).

[8]There are actually two effects. The satellite clocks tick *slower*, falling behind the clocks on the ground by ≈ 7 μs per day because the satellites are moving; this is a special relativity effect. But they also get ahead of the clocks on the ground by ≈ 45 μs per day because of the difference of gravitational potentials; this is a general relativity effect.

In fact, to understand this phenomenon, one does not need general relativity. Everybody knows that stones thrown by hand have the tendency to fall down on the ground. To launch a spaceship which would leave the Earth, one needs to endow it with a considerable velocity ≈ 11.2 km/s. On the Sun, where gravity is stronger, the escape velocity is larger: ≈ 620 km/s. The black hole is an object where the escape velocity exceeds the maximal possible speed existing in the Universe — the speed of light. An elementary high school calculation in the framework of Newton's theory gives an estimate for the size of a black hole of mass M:

$$r_g \approx \frac{2 G_N M}{c^2} \tag{3.2}$$

Surprisingly, this *exactly* coincides with the result (15.25) of the accurate general relativity calculation. The quantity (3.2) is called the *gravitational radius* of a body of a given mass. For Earth, the gravitational radius is around 1 cm. For the Sun it is larger, about 3 km. It is still much smaller than the actual size of the Sun, meaning that the Sun is *not* a black hole.

A black hole with a mass of about a million solar masses sits in the middle of our Milky Way galaxy. Its gravitational radius, and hence its size is $\sim 3 \cdot 10^6$ km. The Milky Way is not an exception; massive black holes are found in the centres of all the sufficiently large galaxies. In deep outer space, there are black holes with masses up to several billion solar masses (they are called *quasars*). Their size is comparable to the size of our Solar System.

We would like now to make a remark concerning the black hole density. As is seen from (3.2), it decreases with increasing mass, $\rho \sim 1/M^2$. For a solar mass black hole, the density is of the order

$$\rho_{\text{BH}}(M_\odot) \sim 10^{16} \frac{g}{\text{cm}^3} . \tag{3.3}$$

But for a supermassive quasar, it is just $\sim 10^{-2}$ g/cm^3, less than the density of water!

The Sun is not a black hole, and it does not have a chance to become one, even when the stock of nuclear fuel which ignites it is exhausted. But stars several times more massive than the Sun can well collapse into a black hole at the end of their evolution. A natural question is *why*. What is the nature of this solar mass scale beyond which a star may collapse? The answer is that the stellar masses are determined by an interplay between gravitational and strong forces. We will evaluate their characteristic value in this chapter, but a little bit later.

3.3   Electromagnetic Interactions

The Coulomb force (2.2) is proportional to the product of the two electric charges $q_1 q_2$. In contrast to the mass, which is always positive, the charges, and hence their products, can be both positive and negative. Thus, the force can be attractive (when $q_1 q_2 < 0$) or repulsive (when $q_1 q_2 > 0$). In a large piece of matter (a brick or a planet), positive and negative charges almost perfectly balance each other, so that its net electric charge is close to zero. As a result, the Coulomb attraction or repulsion practically does not display itself at the macroscale.

I said *practically* because one can, of course, charge a cat by combing it with an amber comb, but this would remove from the cat only a tiny fraction of its electrons.[9] For a metal ball charged by a Van de Graaff generator, this fraction is somewhat larger than for the cat; it may be as large as $\sim 10^{-17}$. Technologies based on macroscopic charges and currents are indispensable now in everyday life. Still, on the cosmic scale, Coulomb forces play no role.

On the other hand, they play a crucial role in determining the microscopic structure of matter. Atoms are made of positively charged nuclei and negatively charged electrons. The characteristic size of the atom is

$$l_{at} \sim \frac{1}{m_e \alpha} \qquad (3.4)$$

in natural units. This makes roughly one angstrom.

All other forces one meets with in everyday life — the forces of pressure, of support reaction (which does not allow the reader to fall down *through* his chair), of friction and so on — have an electromagnetic nature; they come from the interaction of large sets of electrons in contacting bodies.

We all know that the density of water is $\approx 1 \, g/cm^3$ and that all other solids and liquids have comparable densities. An approximate estimate for characteristic matter densities can be derived from first principles. Indeed, in solids and liquids, atoms are packed rather tightly, such that the characteristic distance between them is of the order of their size. For hydrogen, the mass of its atom is close to $m_p \gg m_e$. Thus, a *rough* estimate for the density of liquid hydrogen is

$$\rho_H = \frac{m_p}{l_{at}^3} \sim m_p (m_e \alpha)^3 . \qquad (3.5)$$

In usual units this gives $\rho_H \approx 11.2 \, g/cm^3$, which is, of course, an overestimate, the actual density being ~ 160 times less. The latter circumstance is due to different factors:

- Eq. (3.4) is a characteristic Bohr radius of the hydrogen atom. Its diameter is twice as large.
- Liquid and solid hydrogen are found in nature in molecular form, and the characteristic distance between the atoms there is ≈ 3 times larger than the atomic size.

Still, the estimate (3.5) gives a correct order of magnitude for the density of ordinary matter. To perform the actual *calculation* and to determine with reasonable accuracy the numerical coefficient by which the estimate (3.5) should be multiplied

---

[9]Incidentally, the word *electron* comes from the Greek root meaning *amber*.

for hydrogen, for water or, which is probably easier, for a simple ionic crystal like Na-Cl, is a very difficult technical task. One needs to solve a very complicated quantum mechanical many-body problem. To the best of my knowledge,[10] this problem is not solved yet. Which is a pity.

Equation (3.5) is an estimate for the density of ordinary matter existing at low temperatures and pressures. When squeezed, the matter can become much more dense. In the middle of the Sun, the density is $\sim 150\,\mathrm{g/cm}^3$. In the interior of neutron stars, it can reach nuclear density and more, up to $8 \cdot 10^{14}\,\mathrm{g/cm}^3$.

Going back to ordinary matter, we can estimate not only its density, but its characteristic stiffness and strength. The quantitative measure of the former is Young's modulus E. The latter is characterized by the so-called *ultimate strength* — the maximal stress that a material can withstand without being broken. The ultimate strength is measured in the same units as Young's modulus, but for certain reasons unimportant for us it is typically several hundred times smaller.

The physical dimension of Young's modulus is the same as of the pressure: force divided by area or energy divided by volume. It can be estimated as a characteristic energy of the outer atomic shells $\sim m_e \alpha^2$ divided by l_{at}^3. This gives

$$E_{\mathrm{char}} \sim \frac{m_e^4 e^{10}}{\hbar^8} \approx 3 \cdot 10^{13}\,\mathrm{Pa} \tag{3.6}$$

in usual units.

Again, this is an overestimate. In fact, for steel, $E \sim 2 \cdot 10^{11}\,\mathrm{Pa}$; for diamond, it is $\sim 10^{12}\,\mathrm{Pa}$. However, for the so-called *carbyne nanoparticles*,[11] Young's modulus reaches the estimate (3.6). Anyway, the essential thing is that the atomic and molecular structure of ordinary matter is determined by the Coulomb forces between electrons and nuclei and depends on m_e, m_p, the electron charge and the Planck constant.

We talked up to now only about matter, but there is also radiation. The light, the microwaves in your kitchen and radio all have an electromagnetic nature.

3.4 Strong Interactions

As the electromagnetic interactions are responsible for the structure of atoms, so the strong interactions are responsible for the structure of nuclei. An atomic nucleus is made out of protons and neutrons; that was understood shortly after the discovery of the neutron in 1932. Protons and neutrons (or *nucleons*, a common name for these two particles — they have the same properties with respect to the strong interactions and differ only by their electric charge: the proton is positively charged, while the

---

[10]I am not an expert in quantum chemistry, but I asked experts, and nobody could answer me and give a reference to a scientific paper where such a model-independent calculation would be done.

[11]Do not wonder what they are. I do not know myself!

neutron is, of course, neutral) attract each other at small distances ~ 1 fm. The scale ~ 1 fm also gives the characteristic size of nucleons.

In contrast to electromagnetic forces, which act both at small and large distances, strong forces are short-ranged and fall off exponentially fast at large distances. We are not yet ready to discuss the *nature* of the strong interactions, the only remark we want to make right now is that, as is seen from what was just said and from Table 2.1, the size of the nucleon expressed in natural units is of the same order as its inverse mass (several times larger).

One can actually call m_p a *fundamental constant* of the strong interactions. In contrast to the electromagnetic fundamental constant, the dimensionless (in natural units) electron charge, the strong fundamental constant carries dimension.

We can now pay our debt and explain what determines the mass of the Sun and other stars.

When the nuclear fuel stored in a star runs low, the pressure of radiation can no longer cope with the forces of gravitational attraction, so that the star shrinks down and becomes very dense. Some stars (including our Sun) spend (or will spend) the final stage of their life in the form of a gradually cooling *white dwarf*. White dwarfs are small, much smaller than usual stars; the size of a white dwarf of solar mass is of the same order as the size of the Earth.

More massive stars undergo a tremendous *supernova* explosion, sweeping away their outer layers. An extremely dense hard core is left. It is observed as a *pulsar*. Another name for the pulsar is *neutron star*. In fact, it is nothing but a single giant nucleus. The characteristic size of a neutron star of solar mass is ~ 10 km.

At this stage the star may be stabilized such that it shrinks no more. And maybe not. In the latter case gravitational forces overcome the internal pressure and the star collapses, forming a black hole. This happens when the binding gravitational energy exceeds the internal energy of the star (i.e. its mass). In other words, the collapse occurs when

$$\frac{G_N M^2}{R} \gtrsim M \,, \tag{3.7}$$

where M is the mass of the star and R is its radius.

The radius of the star and its mass are related as $M \sim \rho R^3$, where ρ is the density. It is clear from the previous discussion that the characteristic density of nuclear matter is $\rho \sim m_p^4$. Substituting this in (3.7) and bearing in mind (2.6), we finally derive the upper mass limit beyond which the collapse is imminent:

$$M^* = \frac{m_{\text{Planck}}^3}{m_{\text{proton}}^2} \approx 3.7 \cdot 10^{30} \, \text{kg} \,, \tag{3.8}$$

which is roughly twice the solar mass, $M_\odot \approx 2 \cdot 10^{30}$ kg.

Surprisingly, this crude estimate turns out to be rather precise. A more sophisticated treatment based on the available theoretical models of nuclear matter gives the so-called *Landau–Oppenheimer–Volkoff limit*, $M^* \approx 3 M_\odot$.

Please do not confuse this LOV limit with the better known *Chandrasekhar limit*, $M_C \approx 1.4 M_\odot$. The Chandrasekhar limit is the condition for the stability of white dwarfs made not of nuclear matter, but of so-called electron-degenerate matter. Well, we need not go into further details here. We only ask you to note that the LOV limit slightly exceeds the Chandrasekhar limit. The masses of all pulsars lie in the interval between these two limits:

$$1.4 M_\odot < M_{\text{pulsar}} < 3 M_\odot. \tag{3.9}$$

In other words, a white dwarf of mass lying in this interval eventually collapses into a pulsar, while a white dwarf of a larger mass collapses into a black hole.

What is remarkable is not the astrophysical details, but the existence of the relation (3.8), expressing the characteristic stellar mass[12] via the fundamental constants of the strong and gravitational interactions.

Going down to our small cold planet, strong interactions show up here, as was mentioned, as the forces keeping the protons and neutrons in nuclei together. However, the numbers of protons and neutrons that can form a nucleus are not arbitrary. For example, a proton and a neutron can be bound together by the attractive nuclear force (forming a nucleus of deuterium, the heavy hydrogen isotope), but two protons do not form a bound state: first, for a special reason associated with the Pauli exclusion principle, their nuclear attraction is weaker than the attraction between a proton and a neutron and, second, there is an extra Coulomb repulsion. The system of two protons and one neutron is stable (this is the nucleus of ^{3}He), but the system of three protons and one neutron is not.

The strong interactions are short-ranged. That means that a nucleon feels only the presence of its immediate neighbours. The total strong binding energy is thus proportional to the total number of nucleons A. On the other hand, the electromagnetic interactions are long-ranged and the associated *positive* contribution to the nucleus energy grows as αZ^2, where α is the fine-structure constant and Z is the number of protons (alias the nucleus charge; alias the atomic number). For the nucleus to be stable, a condition like

$$\alpha Z^2 - A < 0 \tag{3.10}$$

should be satisfied.[13] In other words, for the nucleus to be stable, the proportion of protons there should decrease with the size of the nucleus. Heavy nuclei involving a significant fraction of protons are torn apart by Coulomb repulsion.

At this stage, we do not understand yet why nuclei containing only neutrons — two, a hundred or a billion of them — do not exist in nature. The explanation will be given in a moment.

---

[12]The stellar masses *are* of this order. When the mass of an object is too small, smaller than $\sim 0.1 M_\odot$, it has too weak gravitation. It is not squeezed enough to ignite nuclear reactions. It is a planet, not a star. On the other hand, when the mass of a star exceeds, say, ten solar masses, it burns out very rapidly, after which collapse to a black hole is imminent.

[13]This is rather a rough idea of such a condition. The actual one is somewhat more complicated.

3.5   Weak Interactions

Soon after the discovery of radioactivity by Becquerel in 1896, three different types of radiation were distinguished: α, β and γ. We now know that gamma radioactivity is the emission of energetic photons by excited nuclei. Obviously, it is due to electromagnetic interactions.

Alpha radioactivity is the emission of α-particles, each particle representing a tightly-packed system of 2 protons and 2 neutrons, a nucleus of ^{4}He. Both electromagnetic and strong interactions play a role in this process.

And beta radioactivity is the emission of electrons by nuclei. At the dawn of nuclear physics people concluded that, before being emitted, these electrons were hidden inside the parent nucleus. We now know that this is not the case, the electrons do not participate in the nuclear structure. They cannot — the Heisenberg uncertainty principle dictates that a light electron localized within the nuclear volume $\sim 1\,\text{fm}^3$ must have a very large ultrarelativistic momentum. It would immediately pop out of the nucleus.

The electrons are *created* during emission. The basic underlying process is neutron decay

$$n \; \rightarrow \; p + e^- + \bar{\nu}_e . \tag{3.11}$$

In words: a neutron is converted into a proton with the emission of an electron and an antineutrino.[14]

The neutrino and antineutrino interact very *weakly* with the medium, and it is very difficult to observe them. The antineutrinos that accompany beta emission were first detected only in 1956, half a century after β-radioactivity was discovered.

Now, it is the β-radioactivity which makes nuclei with too large a proportion of neutrons in their nucleon content unstable.

One can ask at this point why the *ordinary* nuclei are stable. Why do their constituent neutrons not all decay? The explanation lies in energy conservation. The neutron is heavier than the proton by 1.3 MeV. The electron weighs 0.5 MeV, so that for a *free* neutron, the process is energetically possible. But the neutrons in nuclei are not free. The mass of a nucleus is less than the sum of the masses of its constituent nucleons. There is a binding energy.

For example, for ^{4}He, the total binding energy is 28 MeV. On the other hand, the total binding energy of ^{3}He is only 8 MeV. Thus, a hypothetical decay process

$$^4\text{He} \; \rightarrow \; ^3\text{He} + p + e^- + \bar{\nu}_e \tag{3.12}$$

is not allowed by energy conservation, and the nucleus ^{4}He is stable.

---

[14]We are gradually approaching our theme; neutron decay is a genuine quantum field theory phenomenon. In the standard university courses of quantum mechanics, only photons are sometimes radiated and absorbed, but not other particles.

On the other hand, for nuclei with rich enough neutron content, beta decay is energetically possible.[15]

In other words, the existence of a finite number of distinct stable nuclei and hence of a distinct number of chemical elements, their rich chemistry, biochemistry and the existence of life are all due to the concerted efforts of the strong, electromagnetic and weak interactions. And the existence of our cosy planet, which is warmed by our friendly Sun in such a way that the development of life has become possible, is also helped a lot by gravity. It seems that Voltaire was not quite correct when he made a mockery of Pangloss and Leibnitz for their conviction that we live in the best of all possible worlds. They were right after all: even a slight change of fundamental constants would make the world unhabitable.

The Universe also employs weak interactions to get rid of the heavy particles from the second and third generations. For example, the muon, a heavy counterpart of the electron from the second generation decays following a similar pattern to the neutron:

$$\mu^- \ \to \ e^- + \bar{\nu}_e + \nu_\mu \tag{3.13}$$

The tau lepton and the quarks s, c, b, t do the same. The bosons $W^\pm, Z, H$ are also unstable, but they decay into two rather than three particles.

When I was myself a student and read popular articles about particle physics, I was confused by the fact that the basic manifestation of the weak interactions were *decays*. After all, in the colloquial use, the word "interaction" implies the presence of at least two interacting agents.

Well, at the heuristic level, one can say that the decaying neutron or muon interact with the products of their decay. More detailed explanations of what a quantum field theorist understands by "interaction" will be given later in this book. One can, however, mention right now that, besides the decay processes like (3.11) and (3.13), there are also associated scattering processes (physicists call them *crossed channels*)

$$\bar{\nu}_e + p \ \to \ n + e^+ , \qquad \nu_\mu + e^- \ \to \ \nu_e + \mu^- \tag{3.14}$$

and so on. Here the interaction is manifest.

---

[15] Knowing the basics of quantum mechanics, it is easy to understand why. Consider as a limiting case a *neutronium*, a proton-free nucleus. The Pauli exclusion principle says that two neutrons cannot occupy the same quantum state. As a result, the neutrons occupy a chain of states from the bottom of the potential well up to some level. It is the upmost neutron that decays. On the other hand, the proton that appears after its conversion falls to the bottom of the well and its potential energy is much less than the energy of the parent neutron. This decay is energetically even more favourable than for a free neutron.

A watchful reader can now ask — how come neutron stars are stable? Why are neutrons not converted into protons there? A heuristic answer is that, when one discusses the stability of a nucleus, one neglects gravity. The electrons created in beta decay leave the nucleus. But in a neutron star, they would be kept by gravity and "make a nuisance" (increase pressure, the enthalpy and Gibbs free energy). Under extreme conditions of very high pressure and density, the star tries to get rid of the electrons and to force them into protons to make neutrons.

The weakness of the weak interactions means that the probability of these processes is very small compared, for example, with the probability for an electron passing through matter to be scattered. And going back to the decays (3.11) and (3.13), the weakness of the interaction shows up in the small probability of these decays and, hence, in large lifetimes. The average muon lifetime is 2.2 μs. This is many orders of magnitude more than the characteristic lifetime of strongly interacting and strongly decaying unstable particles, $\tau_{\text{strong}} \sim 10^{-23}$ s. And the neutron (which can interact strongly, but decays due to weak interactions) lives ≈ 17 min. Compared to 10^{-23} s, that is just an eternity!

Chapter 4
The Cubic Edifice of Physical Theories

Physics is the science studying the basic laws of Nature. In the previous chapter we talked very briefly about different facets of Nature — what the world around us is made of and what it looks like.

The primary source of knowledge in physics is observations and experiments. But to systematise and really *understand* the observed facts and phenomena is the job of a theorist. In this chapter, we will talk about the methods a theorist uses to process the experimental observations and deduce the laws formulated in a solid mathematical language which allows one to make verifiable predictions.

4.1 Classical Mechanics

One usually starts a systematic study of physics at university or at a good high school with Newton's laws of classical mechanics. Indeed, the latter represent a cornerstone of the whole building of theoretical physics. And the hard core of this cornerstone is Newton's first two laws.

The first law postulates the insensitivity of all other laws of physics under the choice of frame as long as the frame stays inertial. In other words, it says that these laws are *invariant* under *Galilei transformations* relating different inertial frames.

The second law

$$\boldsymbol{F} = m\ddot{\boldsymbol{x}} \qquad (4.1)$$

is the *equation of motion*. The left-hand side of this wonderful fundamental relation is the force acting on a body we are interested in. The force can be calculated from the other laws describing the interactions of our body with its surroundings. And the right-hand side is proportional to the acceleration. Having evaluated the latter and knowing the initial conditions, the position and the velocity of the body at $t = 0$, we

© Springer International Publishing AG 2017
A. Smilga, *Digestible Quantum Field Theory*,
DOI 10.1007/978-3-319-59922-9_4

can solve what mathematicians call the *Cauchy problem* and determine the trajectory of the body — its position at later times.

For more complicated classical systems involving several pointlike masses or several solids, the story is a little more complicated, but the principle is the same: knowing the interactions of our system, we write down the equations of motion, a set of ordinary differential equations involving time derivatives, for which we formulate the Cauchy problem. This allows us to determine the time evolution of the system.

There are two main realms of application of classical mechanics. The first is mechanics in its historical meaning — the science describing the behaviour of different mechanisms. The second is celestial mechanics, allowing one to analyse and predict the movement of celestial bodies.

Very well known is the story about the discovery of Neptune in 1846. People studied the motion of Uranus and observed some deviations of its trajectory from the calculated one. They suggested that these deviations could be explained by the existence of some hitherto unknown planet which attracted Uranus and perturbed its motion. On the basis of this hypothesis, Urbain Le Verrier calculated precisely the position and the mass of the new planet. He asked Johann Galle from Berlin Observatory to look in the sky and, lo and behold, Neptune was there![1]

4.2 Relativistic Mechanics

This describes the motion of bodies when their speed is comparable to the speed of light. It is also based on two of Newton's laws, but these laws are modified.

- One extends Galilei's principle of relativity by saying that not only do the results of any *mechanical* experiment performed on the ground and the same experiment performed in a closed cabin of a ship sailing on a smooth sea coincide, but it is also true for any *optical* experiment, so that it is impossible to detect the motion of this ship with respect to the nonexistent *aether*.
- The Eq. (4.1) is modified to

$$F = \frac{d\boldsymbol{p}}{dt} = \frac{d}{dt}\left(\frac{m\boldsymbol{v}}{\sqrt{1 - \boldsymbol{v}^2/c^2}}\right). \tag{4.2}$$

Einstein's principle of relativity formulated above entails nontrivial *Lorentz transformations* relating the spatial and time coordinates (x, t) in the rest frame and the

---

[1] As Neptune was not known to the Greeks and Romans, it played no role in classical astrology. Incidentally, the same is true of Uranus, discovered only in the 18th century. Modern astrologists managed to incorporate these planets in their schemes, but call them *transcendental*. Having written that, I looked in Wikipedia just out of curiosity and found out that Pluto, discovered in 1930, also plays now some role in horoscopes. But a little bit more massive *Eris*, discovered in 2005 — not yet.

same coordinates (x', t') in the moving frame. When the latter moves with velocity v in the direction of the positive x-axis, the Lorentz transformations read

$$t' = \frac{t - vx/c^2}{\sqrt{1 - v^2/c^2}},$$

$$x' = \frac{x - vt}{\sqrt{1 - v^2/c^2}},$$

$$y' = y,$$

$$z' = z. \tag{4.3}$$

When the velocities of the bodies whose motion we are interested in are small, the Lorentz transformations reduce to the Galilei transformations

$$t' = t,$$

$$x' = x - vt,$$

$$y' = y,$$

$$z' = z, \tag{4.4}$$

keeping time unchanged. Relativistic effects represent only small corrections.

But when the velocities are comparable to c, the kinematics is quite different. We do not observe around us macroscopic bodies moving so fast. But for microscopic bodies, the elementary particles, it is not uncommon for them to be relativistic. Relativistic mechanics describes the motion of particles in accelerators, their subsequent scattering and decays. Also cosmic rays — the flow of particles coming to Earth from the depths of space — are relativistic and ultrarelativistic.

Accelerators are specially made experimental devices. Cosmic rays are produced by different natural astrophysical accelerators.[2] One may ask whether particles with nonzero rest mass can have relativistic speed without being specially accelerated. Are they present in ordinary matter, for example, in a sofa where our dear reader may repose now?[3]

The answer is positive. The characteristic velocity of the inner atomic electrons is $\sim(Z\alpha)c$ where Z is the charge of the nucleus and α is the fine structure constant. For hydrogen, that gives one hundredth of the speed of light, which is nonrelativistic. But the sofa is not made of hydrogen. For iron ($Z = 26$), the speed of the inner electrons is roughly one fifth of the speed of light. And for tungsten or gold (the reader may carry a wedding ring), it is more than half of c...

---

[2] Thus, extremely relativistic cosmic rays are probably created during supernovae explosions.

[3] When a philosopher wants to give a reference to a material object, he usually says: "for example, a table", the table being the first thing he sees when writing his/her essay. But I am willing to perform the discussion, taking into account the *reader's* perspective.

4.3 Quantum Mechanics

For microscopic particles, the notion of a classical trajectory loses any meaning. The reason is the famous Heisenberg uncertainty principle

$$\Delta x \Delta p \geq \frac{\hbar}{2}, \tag{4.5}$$

where Δx is the uncertainty in measuring the position of a particle, and Δp the uncertainty in measuring its momentum. One cannot simultaneously measure both with an infinite precision.

Let us recall why. To determine precisely the position of a particle (say, an electron), one needs to *look* at it. And what does *looking* mean? We illuminate the object we are looking at by a lamp or by sunlight and observe the scattered light. The inaccuracy in determining the position of the object is of the order of the wavelength of light. Light consists of photons, however. The energy of a photon is[4]

$$E = pc = \hbar\omega. \tag{4.6}$$

The photon hits the electron and transfers to it a considerable fraction of its large momentum. The information about the electron momentum before the collision is thus lost.

In physics, things that cannot be measured do not exist. Thus, one introduces, instead of the notion of a classical electron trajectory $x(t)$, the notion of the electron wave function $\Psi(x, t)$. $\Psi(x, t)$ is complex, but the quantity $|\Psi(x, t)|^2$ is, naturally, real. It represents the density of probability to find our electron at the point x. The integral

$$P = \int |\Psi(x, t)|^2 \, d^3x \tag{4.7}$$

is the probability to find our electron *somewhere*. If the electron is stable, is still there at any time and no new electron pops up out of the blue to confuse us (these are the basic assumptions of standard non-relativistic quantum mechanics), this probability is always unity.

The three paragraphs above were a basic course in quantum mechanics, *level one*. We are now proceeding to *level two*.

The basic dynamical equation in quantum mechanics describes not the evolution of the position of the particle, but the evolution of its wave function:

---

[4]People say that the following story once happened at some university whose exact name and geographical location is now difficult to establish.

A student was taking the oral exam in quantum mechanics. Professor saw that the student knew practically nothing and, wanting to ask him a very simple basic question, showed him Eq. (4.6) and asked to explain the meaning of the different quantities entering there. "For example, what is ω?" "The constant of the plank!" answers the student. "And what is $\hbar$ then?" asks the surprised professor. "The height of this plank!".

$$i\hbar\frac{\partial\Psi}{\partial t} = \hat{H}\Psi, \tag{4.8}$$

where $\hat{H}$ is a certain differential operator called the *Hamiltonian*.[5]
For a particle moving in an external potential field $V(x)$,

$$\hat{H} = -\frac{\hbar^2}{2m}\left(\frac{\partial^2}{\partial x^2} + \frac{\partial^2}{\partial y^2} + \frac{\partial^2}{\partial z^2}\right) + V(x). \tag{4.9}$$

Equation (4.8) is a time-dependent *Schrödinger equation*. Again, one can pose a Cauchy problem: if the wave function is known at the initial moment, one can determine $\Psi(x, t)$ at all later times.

For practical purposes, the formulation of the problem given above is inconvenient and even unphysical. Indeed, it is impossible to follow the time evolution of the wave function at all spatial points. How would you measure it?

On the basis of (4.8), one can, however, formulate two other problems which are quite physical.

1. The first is the *stationary state* problem. Let us suppose that our wave function represents the product of a certain function of the spatial coordinates and a certain function of time:

$$\Psi(x, t) = \psi(x)\chi(t). \tag{4.10}$$

Substituting this into (4.8), we obtain the system of two equations:

$$i\hbar\frac{d\chi}{dt} = E\chi(t), \tag{4.11}$$

$$\hat{H}\psi(x) = E\psi(x), \tag{4.12}$$

where E is a constant.
The solution of (4.11) is trivial,

$$\chi(t) = C\exp\left\{-\frac{iEt}{\hbar}\right\}. \tag{4.13}$$

The second equation (called the *stationary Schrödinger equation*) can be solved if we impose certain boundary conditions on $\psi(x)$. For example, bearing in mind the finite (equal to unity) normalization integral (4.7), we may require for $\psi(x)$ to vanish at spatial infinity. It then turns out that the solutions of (4.12) exist only for certain values of the constant E (which is, of course, nothing but the energy of our system). In other words, the possible values of the energy of a quantum

---

[5]This word, obviously, comes from the name *Hamilton*. William Hamilton died in 1865 and knew nothing about quantum mechanics. He knew, however, a lot about *classical* mechanics. The notion of classical Hamiltonian that he introduced will be discussed at length in Chap. 7. We will also briefly elucidate there the relationship between classical and quantum Hamiltonians.

system are *quantized*. The set of admissible energies is called the *spectrum* of the Hamiltonian.

There is nothing mysterious in such quantization. The mathematical reason is the same as that for the presence of a discrete set of frequencies with which a guitar string can vibrate or a drum resound.

For the Hamiltonian (4.9) with a real and not too wild potential $V(x)$, the spectrum is real. And it is very nice that it is. Indeed, were E to be complex, the norm of the factor $\chi(t)$ in (4.13) would grow or decrease with time. We could not then normalize the total probability (4.7) to unity and the basic assumption of this section (that the particles are neither created, nor do they disappear) would not be satisfied.

A qualified reader who followed a university course of quantum mechanics may protest at this point. "How come the energies are always quantized? The author seems to have forgotten about the continuum spectrum!" "For example," says the irritated reader, "for a free particle when $V(x) = 0$, the solutions of (4.12) read

$$\psi(x) = C \exp\left\{\frac{i\,p \cdot x}{\hbar}\right\}\tag{4.14}$$

giving an arbitrary positive energy

$$E = p^2/2m\,.\tag{4.15}$$

,,

No, the author has not forgotten. Note, however, that the solution (4.14) does not satisfy our normalization condition $P = 1$. The only way to impose the latter is not to allow a particle to move in the whole space, but rather to put it in a *fictitious* finite spatial box of size L and to impose some boundary condition at its border. Mathematically, the most simple choice would be periodic boundary conditions

$$\psi(x + L, y, z) = \psi(x, y + L, z) = \psi(x, y, z + L) = \psi(x, y, z)\,.\tag{4.16}$$

But then the admissible momenta p *are* quantized:

$$p = \frac{2\pi\hbar n}{L}\,,\tag{4.17}$$

where n is an integer-valued vector. So are the energies (4.15). In the physically interesting limit $L \to \infty$, the spacings between the energy levels become very small, but for any finite L, however large, the spectrum is discrete.

If one cares to keep the normalization $P = 1$, the factor C in (4.14) should be chosen as $C = 1/\sqrt{L^3}$. Given that L has no physical meaning, people often forget about it, choosing $C = 1$. But one should keep in mind that one can always discretize the spectrum by imposing the boundary conditions (4.16). This gives a mathematical justification for working with the wave functions of continuum spectrum in physics.

2. The second important problem that is treated in standard quantum mechanics is the *scattering* problem.

   We consider now the time-dependent Schrödinger equation (4.8) with the potential localized in a finite region of space in the vicinity of the origin, $V(|x| \to \infty) = 0$. But instead of following the evolution of the wave function at all times, we are interested only in what happened in the distant past, $t \sim -\infty$ and in what will happen in the distant future, $t \sim \infty$. We assume further that our particle did not hang all that time around the origin in the region with potential, but was far from it at $t = -\infty$ and will go far away at $t = \infty$. In both those cases, the particle does not feel the presence of the potential and can be considered as free. It is described by the plane wave solutions (4.14) with some *ingoing* momentum p_{in} at $t = -\infty$ and an *outgoing* momentum p_{out} at $t = \infty$. Due to the fact that the particle, while passing near the origin, was affected by the action of the potential, these momenta need not be the same. One can ask then what is the probability (or rather, the complex amplitude of probability) for the particle to scatter — to change its momentum.

   We will not pursue this discussion now, but will come back to it in Chap. 8. Once we understand what this *scattering amplitude* is and how is it calculated, it will be much easier for us to understand how the scattering problem is formulated and solved in quantum field theory.

When the scalar product of the characteristic momentum p_{char} and the characteristic distance x_{char} become large compared to $\hbar$ (this product has the dimension of *action* and actually *is* the characteristic action of the particle S_{char}), the wave function (4.14) oscillates quite rapidly. In that case, one can consider a wave function localized in a small spatial region (a *wave packet*) and be convinced that the trajectory of the motion of its centre, $x_{centre\ of\ wave\ packet}(t)$ is described by the classical equation just coinciding with second Newton's law.

This is the famous *correspondence principle*: the quantum dynamics goes over to the classical dynamics when the characteristic action is large compared to $\hbar$. The same correspondence relates wave optics to geometrical optics. When the optical path (or a relevant difference of optical paths) is large compared to the wavelength, one can forget about the wave nature of light and use the laws of geometrical optics.

4.4 Relativistic Quantum Mechanics

The logic of our discussion in this chapter is the following. First, we recalled the very basics of physics, Newtonian mechanics. And then we are showing how all other branches of physics grow out of it, as branches and leaves of a tree grow from its body. We discussed relativistic mechanics — what happens when the parameter

$\kappa_{rel} = v_{char}/c$ is not small. And in the quantum mechanics section, we discussed what happens when the parameter $\kappa_{quant} = \hbar/S_{char}$ is not small.

A natural question is now what happens when *both* κ_{rel} and κ_{quant} are not small. Could the romance between c and $\hbar$ develop into a fruitful engagement? Would the offspring of the quantum father and relativistic mother (quantum mother and relativistic father?) be healthy and happy?

The answer to this question is the following. This romance is definitely not so merry and rosy as the love story of Papageno and Papagena. But it can well be compared to the story of Tamino and Pamina: after many adventures and passing the Trials, comes the happy ending, with the choir hailing the dawn of the new era of quantum relativistic theory.

But we are not there yet. Let us first try to merge the relativistic and quantum description in a way that seems to be the most natural and write down the relativistic Schrödinger equation. The standard Schrödinger equation (4.8) with the Hamiltonian (4.9) is invariant under the Galilei transformations (4.4), but not under the Lorentz transformations (4.3). It is not difficult, however, to modify it to ensure Lorentz invariance. Look at the equation

$$\left(\Box + \frac{m^2 c^2}{\hbar^2}\right) \Psi(x, t) = 0, \tag{4.18}$$

where

$$\Box = \frac{1}{c^2}\frac{\partial^2}{\partial t^2} - \Delta = \frac{1}{c^2}\frac{\partial^2}{\partial t^2} - \frac{\partial^2}{\partial x^2} - \frac{\partial^2}{\partial y^2} - \frac{\partial^2}{\partial z^2} \tag{4.19}$$

is the operator of d'Alembert (and Δ is, of course, the Laplacian).

One can then note that

1. The operator $\Box$ is invariant with respect to the transformations (4.3),

$$\frac{1}{c^2}\frac{\partial^2}{\partial t^2} - \Delta = \frac{1}{c^2}\frac{\partial^2}{\partial t'^2} - \Delta'. \tag{4.20}$$

   To understand that, it is sufficient to recall that the combination

$$ds^2 = c^2 dt^2 - (dx)^2 \tag{4.21}$$

   (the *interval* between two close events) is invariant with respect to Lorentz transformations. We will discuss all that again in more detail in Chap. 6.
2. Equation (4.18) can be represented in the Schrödinger form (with only first order time derivatives) as

$$i\hbar \frac{\partial \Psi}{\partial t} = c\sqrt{m^2 c^2 - \hbar^2 \Delta}\, \Psi \tag{4.22}$$

The square root that appears here is troublesome. First of all, it is unclear what sign of the square root should be chosen. In (4.22), we have chosen the plus sign.

But why? And what to do if the argument of the square root is close to zero? How to deal with its branch points?

Still, Eq. (4.22) can be attributed a quite definite meaning when the wave function is *smooth enough*, when its spatial derivatives are not too large, so that $|\triangle\Psi| \ll \frac{m^2c^2}{\hbar^2}|\Psi|$.

In that case, we can actually expand the right-hand side of (4.22) in

$$\kappa = \frac{\hbar^2\triangle}{m^2c^2}.$$

Keeping only the linear term of this expansion, we go back to the standard free nonrelativistic Schrödinger equation. Further terms give relativistic corrections.

In the presence of the potential we should just add it on the right-hand side of (4.22) to obtain

$$i\hbar\frac{\partial\Psi}{\partial t} = \left[c\sqrt{m^2c^2 - \hbar^2\triangle} + V(x)\right]\Psi.\tag{4.23}$$

If one wishes, one can get rid of the square root and write

$$\left[\frac{1}{c}\left(\frac{\partial}{\partial t} + \frac{i}{\hbar}V(x)\right)^2 - \triangle + \frac{m^2c^2}{\hbar^2}\right]\Psi(x, t) = 0.\tag{4.24}$$

We can now reassess the familiar textbook problems of quantum mechanics and find what happens when relativistic effects are taken into account. For example, for the hydrogen atom, the spectrum of the *relativistic* Hamiltonian displayed in (4.23) is not just the Bohr spectrum, but involves small shifts,[6]

$$E_n = -\frac{me^4}{2\hbar^2n^2} + \delta_n.\tag{4.25}$$

Not everybody knows that, while working on his seminal paper published in January 1926 in *Zeitschrift für Physik*, Erwin Schrödinger not only solved the nonrelativistic Schrödinger equation (4.8), but he also wrote down the relativistic equation (4.23) and solved it for hydrogen to determine the corrections δ_n. He was, however, much disappointed: the results for the latter did not coincide with experiment! The leading-order Bohr spectrum was obtained correctly in his approach, but the deviations were not.

By the standards of the third millennium, Schrödinger's behaviour in this situation was very bizarre. First, he hesitated to publish his paper. He probably thought that predicting just the Bohr spectrum was not a big deal, Bohr himself had already calculated it in the framework of the so-called *old quantum theory*. It is the relativistic corrections that Schrödinger was ambitious to find, and he failed. Finally, he still published the paper, but removed from it the relativistic discussion. That is why

---

[6]δ_n is of the order of $\alpha^2 E_{\text{Bohr}}$.

Eq. (4.18) [as well as its generalisation (4.24) for the problems with potential] does not carry the name of Schrödinger together with Eq. (4.8), but is usually called the Klein–Gordon equation after Oskar Klein and Walter Gordon, who rederived it on the basis of what they read in Schrödinger's paper.

But this is unjust to Vladimir Fock, who also rederived this equation in 1926, a couple of months after Klein, but before Gordon. Well, this story may serve as a gold-plated illustration of the well-known *Arnold principle*.[7] Not questioning the universality of the latter [as was mentioned, it was Schrödinger who discovered Eqs. (4.18) and (4.24) first], we will still call it the *Klein–Fock–Gordon* (KFG) equation in our book.

And we know now why Schrödinger's relativistic calculations failed to agree with experiment. The electron carries spin, and the relativistic wave equation that describes its motion is the *Dirac equation*, not the KFG equation. But in 1926, this was not known yet.

We will not write the Dirac equation here. It is technically more complicated and we will learn about it somewhat later (in Chap. 9) after we understand what a *relativistic spinor* is. But as far as physics is concerned, there is no big difference between the two equations. Both of them (the KFG equation for spinless particles and the Dirac equation for the particles carrying spin 1/2) can legitimately be applied when the relativistic effects are not too large.

The equation can be solved and the relativistic atomic spectrum can be calculated. But what does the wave function $\Psi(x, t)$ now mean? Unfortunately, $|\Psi(x, t)|^2$ cannot be interpreted any more as a probability density. If the time evolution of $\Psi(x, t)$ is described by (4.23), the integral (4.7) changes its value with time!

This setback and the as yet unclarified issue of the choice of sign in (4.23) are not accidental. They are related to the fact that the basic assumptions of quantum mechanics are not true any longer in the relativistic case.

In nonrelativistic quantum mechanics one cannot precisely measure the position and momentum *simultaneously*. But if one is interested only in the position, this is quite possible. This was explained in the previous section: strike a light on the electron and you will determine its position with the accuracy of the order of the light wavelength. If the wavelength is very small, the precision of the measurement is very high.

The problem is, however, that a very shortwave photon γ carries not only high momentum, but also high energy and, besides the simple-minded Compton scattering $\gamma + e^- \rightarrow \gamma + e^-$, processes creating extra electron-positron pairs, like

$$\gamma + e^- \rightarrow \gamma + e^- + e^+ + e^-,$$

are allowed. One can detect the scattered photon (if it is there, of course; processes when the photon is absorbed, like $\gamma + e^- \rightarrow e^- + e^+ + e^-$, are also possible), but the position of *which* electron are you going to determine in this way — the initial one

---

[7]The Arnold principle reads: *If a notion bears a personal name, then this name is not the name of the discoverer.* It goes without saying that this principle is self-referential.

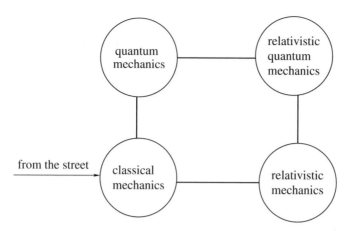

Fig. 4.1 The plan of the ground floor in the museum of theoretical physics (from Sophie's notepad).

or the additional electron created after scattering? In fact, one cannot determine the position of the electron with an accuracy better than the Compton wavelength (2.3).

This teaches us that the naïve relativistic quantum mechanics based on the KFG equation or on the Dirac equation can be applied only when processes with the creation of extra particles are not important.

"Not surprising," may think the reader, "Everybody knows that any physical theory has a realm of applicability". The reader is right, but not completely. Classical mechanics does not work at velocities of the order of c, but this is an *external* limitation. Classical mechanics by itself knows nothing about c. It is happy without it, being self-consistent on its own.

On the other hand, relativistic quantum mechanics is not self-consistent. It is a phenomenological approximation of something more complicated.

The four types of problems discussed so far can be represented as in Fig. 4.1.

A visitor coming from outside finds herself[8] in a large reception hall of classical mechanics, decorated with marble sculptures and tapestry. On one side she sees a system of large pulleys lifting an upside-down Roman trireme. She turns her head and sees on the opposite wall a large star chart. She also sees two doors.

On one of the doors she notices a gilded bas-relief representing a large letter $\hbar$. And the bas-relief with the letter C on another door. She opens the C door, follows the corridor and enters another large hall. Instead of the pulley with the trireme she sees a large model of a synchrophasotron. An equally large picture on the opposite wall represents two very similarly looking gentlemen, one of them wearing a corduroy dressing gown and another one — a spacesuit.

She comes back, follows the corridor behind the $\hbar$ door and enters the quantum hall. She feels very *uncertain* there; the walls are trembling, constantly changing

---

[8]I know that a politically correct formula would be him/herself. But it is somewhat boring to write he/she every time. Let it be a girl, let her name be Sophie and let her be clever (as her name says) and beautiful (Fig. 4.2)!

their colour. She notices there another door decorated with gilded C, opens it and penetrates the fourth hall. The feeling of uncertainty is increased there. Not only are the walls trembling, but also the floor, so that it is impossible to stay in place, to maintain a precise position. In the centre of the hall, there is, however, an exhibition area where no vibration is felt. Sophie goes there and finds a bronze statue of Paul Dirac. Then she watches a historical documentary: Schrödinger, Klein, Fock and Gordon are amicably discussing physics in the editorial office of *Zeitschrift für Physik*.

Sophie returns to the entrance hall of classical mechanics and looks up. She sees a staircase leading to the first floor...

4.5 Nonrelativistic Field Theory

The equations of motion (4.1) involve three scalar dynamical variables $x(t)$, $y(t)$, $z(t)$. For a system of several particles or of solids, the number of dynamical variables (the number of *degrees of freedom*) is larger. But as long as the number of planets or pulleys in our mechanical system is finite, the number of degrees of freedom also stays finite.

But what if we are dealing with a liquid? The number of degrees of freedom there becomes infinite and *continuous*. There are two ways to describe the flowing water — the Lagrange way and the Euler way. In the Lagrange description the dynamical variables are $x(x_0, t)$ — the positions of all infinitesimal liquid elements as a function of their initial positions $x_0 \equiv x(t = 0)$. We have three dynamical variables for each value of a continuously changing parameter x_0. The total number of variables is thus continuously infinite.[9] One can call the set $\{x(x_0)\}$ a *field* of the positions of the fluid particles.

In the Euler description (which might be less physical, but, for liquids, is technically more convenient), we are not interested in the history, in where an infinitesimal liquid particle came from, but only deal with the *velocity field* $v(x, t)$ — velocities of liquid particles at a given spatial point and a given time. One also introduces the pressure field $P(x, t)$.

The Navier–Stokes equation that describes the dynamics of an uncompressible liquid reads

$$\rho\left(\frac{\partial}{\partial t} + v \cdot \nabla\right) v = -\nabla P + \eta \triangle v, \qquad (4.26)$$

where ρ is the density of the liquid and η is its shear viscosity.

In contrast to (4.1), which was an ordinary differential equation for a vector quantity $x(t)$, (4.26) represents a vector equation in *partial derivatives*.

---

[9]The term "continuously infinite" must be intuitively clear. To please a mathematician who might read this text, we say that the *cardinality* of the infinite set $\{x(x_0)\}$ is the same as for the set of all real numbers and exceeds the cardinality of the (also infinite) set of all integer numbers.

Fig. 4.2 Curious to go in…

The Navier–Stokes equation is very nonlinear and very complicated. Thus, we will not discuss it further (we have many other sufficiently complicated issues to discuss in our book) and consider instead *elastic deformations* of a solid body, whose physics and mathematics are much simpler that for a flowing liquid.

For solids, the Lagrange approach is more appropriate. We introduce the *deformation vector* $u(x_0, t) = x(x_0, t) - x_0$. Let us assume now that $|u| \ll |x_0|$. By doing this, we have excluded: *(i)* A displacement of the solid as a whole; *(ii)* A rotation of the solid as a whole; *(iii)* Some special deformations, like bending of thin rods.

The partial derivative equation for the field $u(x, t)$ (we do not bother to endow the argument with the index "0" any more) is incomparably simpler than the Navier–Stokes equation. Especially so when one keeps only the terms linear in u. For homogeneous isotropic matter, we obtain:

$$\ddot{u} = c_t^2 \triangle u + (c_l^2 - c_t^2)\nabla(\nabla \cdot u). \tag{4.27}$$

Fig. 4.3 A spring mattress.

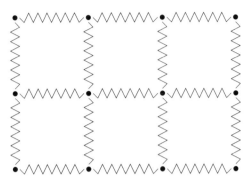

This is nothing but the wave equation describing the propagation of sound waves, with c_l being the longitudinal sound velocity for the deformations associated with compression and decompression and c_t the transverse sound velocity for shear deformations.

I have to confess now that the discussion above involved some cheating. I wrote: "a displacement of a liquid element", "a displacement of a solid element". But in real life, the "elements" of the material bodies are atoms and molecules. Thus, one cannot, strictly speaking, talk about a *continuous* deformation $u(x, t)$. For solids, one should talk instead about the displacements of the nodes of the crystal lattice. A good model for the latter is a spring mattress (Fig. 4.3).[10]

Thus, we are dealing in fact with a system involving a very large but finite number of degrees of freedom. The partial derivative equations of field theory, like (4.26) or (4.27), are obtained as the limit of a system of a large number of ordinary differential equations. However, if we are interested in the behavior of the medium at distances much larger than characteristic interatomic distances, the field theory description is appropriate and adequate!

Sophie goes upstairs and passes the nonrelativistic classical field theory hall full of aquariums and spring mattresses. She sees the familiar doors with the signs C and $\hbar$. She opens the C door ...

4.6   Relativistic Field Theory

Consider a complex[11] scalar field $\phi(x, t)$ (we postpone the discussion of its physical nature) which satisfies the equation

$$\left(\Box + \frac{m^2 c^2}{\hbar^2}\right) \phi(x, t) = 0. \tag{4.28}$$

---

[10]We took this from the book of Zee.

[11]It could be real, but we choose it to be complex to facilitate some technical moments in future.

Mathematically, it is the same KFG equation as (4.18), but the interpretation now is quite different. In (4.28), $\phi(\boldsymbol{x}, t)$ is not a quantum wave function. It is a classical field.

Anyway, this equation preserves its form after Lorentz transformations. The dynamical system described by (4.28) enjoys relativistic invariance. The solutions to the Eq. (4.28) are plane waves,

$$\phi(\boldsymbol{x}, t) \propto e^{i(\boldsymbol{k} \cdot \boldsymbol{x} - \omega t)}, \tag{4.29}$$

with $\boldsymbol{k}$ and ω satisfying the *dispersion relation*

$$\omega^2 - c^2 k^2 = \left(\frac{mc^2}{\hbar}\right)^2. \tag{4.30}$$

The phase velocity ω/k and the group velocity $d\omega/dk$ of such a wave are frequency-dependent.

One can generalize Eq. (4.28) by adding a cubic term $\propto \phi^2 \phi^*$ on the left-hand side. If ϕ is a *Lorentz scalar* (is not changed under Lorentz transformations), which we assume, a cubic term (or any other function of ϕ) does not affect the relativistic invariance of the equation. The solutions of the nonlinear equation thus obtained are complicated, representing interacting nonlinear waves.

A widely-known example of relativistic field theory is the Maxwell theory of the electromagnetic field. Maxwell's equations read (in Heaviside units)

$$\nabla \times \boldsymbol{E} = -\frac{1}{c} \frac{\partial \boldsymbol{B}}{\partial t},$$
$$\nabla \cdot \boldsymbol{B} = 0,$$
$$\nabla \cdot \boldsymbol{E} = \rho,$$
$$\nabla \times \boldsymbol{B} = \frac{\boldsymbol{j}}{c} + \frac{1}{c} \frac{\partial \boldsymbol{E}}{\partial t},$$

$$\tag{4.31}$$

Here ρ and $\boldsymbol{j}$ are the charge and current densities and $\boldsymbol{E}$ and $\boldsymbol{B}$ are the electric and magnetic fields. They are expressed via the scalar potential φ and the vector potential $\boldsymbol{A}$ as

$$\boldsymbol{E} = -\nabla \varphi - \frac{1}{c} \frac{\partial \boldsymbol{A}}{\partial t}$$
$$\boldsymbol{B} = \nabla \times \boldsymbol{A}. \tag{4.32}$$

The Eq. (4.31) look much more complicated than (4.28). The fact that they are invariant under Lorentz transformations (and they are!) does not seem to be obvious.

Well, neither did it seem obvious to Maxwell. To be more exact, Maxwell had absolutely no idea that the equations he derived enjoy beautiful relativistic invariance.

One cannot reproach him for that. The form (4.31) that we called complicated, the form in which Maxwell's equations are now written in university textbooks, is much simpler and more transparent than the form in which Maxwell first derived and presented them. A pioneer rarely cares about aesthetics, and the results of an original fresh research more often than not look ugly.

The relativistic invariance of Maxwell's equations was discovered by Hendrik Lorentz. And then Albert Einstein, Henri Poincaré and Hermann Minkowski *understood* where it comes from and what is its physical meaning. In Chap. 7, we will derive Maxwell's equations in two lines using the Lagrangian formalism. In this way of derivation relativistic invariance is seen immediately.

And here we only recall that the solutions to Maxwell's equations in the absence of external electric charges and currents represent electromagnetic waves. The electric and magnetic fields and also the scalar and vector potentials depend on x and t as in (4.29), but the dispersion relation is now

$$\omega^2 = k^2 c^2. \tag{4.33}$$

Light travels with the speed of light c!

In the previous section we noted that the nonrelativistic field theory Eqs. (4.26) and (4.27) in partial derivatives were obtained as the limit of a system of a large number of ordinary differential equations describing the motion of individual atoms. For Maxwell's theory the situation is the opposite. Electric and magnetic fields *are* continuous functions of x and t and the Maxwell equations are the *fundamental* equations. On the other hand, for practical purposes (for example, for solving them numerically with computer), one can quantize space and time and introduce a fictitious spring mattress there: approximate the partial differential equations by a system of a large number of ordinary differential equations.

4.7 Quantum Field Theory

We are now ready to start the discussion of the main subject of this book and to explore together with Sophie the two last halls on the first floor — the halls of relativistic and nonrelativistic quantum field theories.

What QFT *is* is clear from its name — its relationship to classical field theory is exactly the same as the relationship of ordinary quantum mechanics to ordinary classical mechanics. One can say that QFT is quantum mechanics of systems involving a continuously infinite number of degrees of freedom.

We give here some very basic clarifying remarks, taking as an example the theory of a free complex scalar field $\phi(x, t)$ satisfying (4.28). We will reexamine it in some more detail in Chap. 7. Fermion fields will be treated in Chap. 9.

The basic notion of quantum mechanics is not the classical trajectory $x(t)$, but the Schrödinger wave function $\Psi(x, t)$. By the same token, the basic notion of QFT is not the field $\phi(x, t)$, but a *wave functional*

$$\Psi \, [\{\phi(x)\}, t] \tag{4.34}$$

The wave functional (4.34) depends on the function $\phi(x)$, that is, on an infinite continuum of variables, the values of the field at all spatial points. It satisfies the dynamical Schrödinger equation (4.8) with a certain field Hamiltonian $\hat{H}$. The latter represents a differential operator involving *variational* derivatives $\delta/\delta\phi(x)$.

We will derive the exact expression for this Hamiltonian in Chap. 7. At this stage, we only mention that the most convenient way to solve the functional Schrödinger equation is to put our system in a finite spatial box of size L, and impose on the field $\phi(x)$ periodic boundary conditions [as in Eq. (4.16)]. We can then expand our field in the Fourier series,

$$\phi(x) \;=\; \sum_n c_n \, \exp\left\{\frac{2\pi i n \cdot x}{L}\right\}, \tag{4.35}$$

with the sum running over all integer-valued vectors n. Following the suggestion of Wolfgang Pauli and Victor Weisskopf,[12] let us choose the set of Fourier coefficients $\{c_n\}$ and their complex conjugates as the dynamical variables. This is convenient because, in the free case [with no extra cubic or higher-order terms in (4.28)], the field Hamiltonian can be expressed as an infinite sum,

$$\hat{H} \;=\; \sum_n \hat{H}_n, \tag{4.36}$$

with each term $\hat{H}_n$ depending only on the variables c_n, c_n^*. In the natural unit system it reads (see Chap. 7)

$$\hat{H}_n \;=\; -\frac{1}{V}\frac{\partial^2}{\partial c_n \partial c_n^*} + V\left[m^2 + \left(\frac{2\pi n}{L}\right)^2\right]c_n c_n^*, \tag{4.37}$$

where $V = L^3$ is the volume of our box.

The reader recognizes the Hamiltonian of the harmonic oscillator. The only distinction from the oscillator Hamiltonian found in the textbooks on quantum mechanics,

$$\hat{H}_{\text{osc}} \;=\; \frac{\hat{p}^2}{2m} + \frac{m\omega^2 x^2}{2}, \tag{4.38}$$

is that the oscillator (4.37) is *2-dimensional*, with a complex c_n involving two real degrees of freedom. One can write

---

[12]Their paper on the quantization of the free charged scalar field was published in 1934. There was an earlier paper by Enrico Fermi (1932) where this idea was put forward in application to the electromagnetic field.

$$c_n = \frac{X_n + iY_n}{\sqrt{2}},\tag{4.39}$$

in which case the Hamiltonian (4.37) represents two copies of (4.38), the volume V of the box playing the role of mass. The frequency of the oscillator is

$$\omega_n = \sqrt{m^2 + \left(\frac{2\pi n}{L}\right)^2}.\tag{4.40}$$

To avoid any possible confusion, we mention here that X_n, Y_n have nothing to do with the spatial coordinates. In contrast to the Hamiltonian (4.38) that describes the oscillations of a particle in ordinary space, the Hamiltonian (4.37) describes oscillations of a Fourier component of the field amplitude. These are oscillations in the functional Hilbert space.

The Hamiltonian (4.37) has the spectrum

$$E_{l_1^n, l_2^n} = (1 + l_1^n + l_2^n)\,\omega_n,\tag{4.41}$$

where $l_{1,2}^n = 0, 1, 2, \ldots$ are the quantum numbers describing the excitations of the X_n-oscillator and the Y_n-oscillator. The spectrum of the total Hamiltonian (4.36) is the infinite sum of the individual spectra (4.41).

The ground state of $\hat{H}$ corresponds to $l_{1,2}^n = 0$ for all n. Its energy is

$$E_0 = \sum_n \omega_n,\tag{4.42}$$

which is infinite!

Thus, our calculation seems to have shown that the *vacuum* (by definition, the physical vacuum is the ground state of the field Hamiltonian, the state with no excitations) has an infinite energy density. Does that make any sense? After all, we know from experience that the energy density of empty space is zero or very close to zero. Indeed, the energy, like mass, carries gravitational charge, while empty space does not: it does not attract anything and does not repel anything.

Or maybe it *does*? The last question is actually not so stupid. It is rather deep and deserves a detailed discussion.

Cosmological digression

Nonzero vacuum energy density is called the *cosmological constant*.[13] In contrast to ordinary masses, it can be either positive or negative. Einstein introduced a small positive cosmological constant in the hope of finding a stationary *cosmological* solution (i.e. a solution describing the whole Universe) to the equations for the

---

[13]To be more precise, the conventional definition for the cosmological constant Λ includes an extra numerical factor, $\Lambda = 8\pi \rho_{\text{vac}}$.

gravitational field that he derived. His idea was to introduce a universal gravitational repulsion in empty space[14] to balance the universal Newtonian attraction of massive bodies.

Without such a cosmological constant, only nonstationary solutions describing the expanding or expanding and then collapsing Universe exist. These solutions were found in 1922 and in 1924 by Alexander Friedmann, a mathematician from Petrograd.[15]

It is interesting that Einstein did not immediately accept the existence of nonstationary cosmological solutions to his equations. The story was dramatic. Einstein read the first Friedmann paper published in the *Zeitschrift für Physik* and, due to some error in the calculations, decided that the Friedmann solution was not consistent with energy conservation and was therefore wrong. He sent to the journal a brief comment including this statement. Friedmann was in Petrograd and Einstein was in Berlin. Skype sessions were not yet very common at that time, and it was difficult for Friedmann to convince Einstein that his paper was actually correct. However, Friedmann had a friend and colleague Yuri Krutkov, who was visiting in Berlin. By the request of Friedmann, he met Einstein, explained to him the essence of Friedmann's work and convinced him! Krutkov wrote in the letter to his sister on Monday, May 7, 1923: *I was reading, together with Einstein, Friedmann's article in the Zeitschrift für Physik*. And then on May 18: *I defeated Einstein in the argument about Friedmann. Petrograd's honor is saved!*

On May 31, 1923, Einstein sent another short note to the *Zeitschrift für Physik*. It read:

I have in an earlier paper (Einstein 1922) criticized the cited work (Friedmann 1922). My objection rested however — as Mr. Krutkov in person and a letter from Mr. Friedmann convinced me — on a calculational error. I am convinced that Mr. Friedmann's results are both correct and clarifying. They show that in addition to the static solutions to the field equations there are time-varying solutions with a spatially symmetric structure.

Personally, I find this story quite remarkable. *Errare humanum est* and even a great scientist can sometimes make a mistake. But not every scientist — great or not so great — admits his mistake publicly, as soon as s/he realizes that the mistake is there. Einstein did so without the slightest hesitation!

In 1922–1923, neither Einstein nor Friedmann could be sure which of the solutions — the stationary one with the cosmological term or a non-stationary solution,[16] — describes Nature. Well, already at that time one could *suspect* that the Universe was not stationary. Einstein's solution resembles in a sense a solution describing a ball

---

[14]In contrast to what we are used to for a gas cylinder, a positive vacuum energy density corresponds to a negative pressure. For pundits: this is because the vacuum energy-momentum tensor is proportional to the flat Minkowski metric $\eta_{\mu\nu}$, and $\eta_{11} = \eta_{22} = \eta_{33} = -\eta_{00} = -1$. See Chap. 15 for more details.

[15]That is how St. Petersburg was officially called in 1914–1924.

[16]We will see in Chap. 15 that there are two such solutions: the so-called *open cosmological model* where the Universe expands indefinitely and the *closed model* where the expansion is followed by the contraction and then by the Big Crunch.

placed on the top of a hill: it is *unstable*, small deviations in the initial conditions grow exponentially with time. But only experiment could give the ultimate answer.

This answer was obtained in 1929 when Edwin Hubble discovered the *red shift* — he observed that the spectral lines in the light emitted by distant galaxies are shifted to the red, and the more distant is a galaxy, the more significant is the shift. This showed that the Universe is not stationary; it expands and is described by the Friedmann solution rather than by the stationary Einstein solution. This also meant that the Universe had a finite age. As was mentioned earlier, it was created out of nothing about 13 billion years ago (the Big Bang).

Friedmann solved the equations both in the case of zero and nonzero cosmological constant. Its presence was an option for his scenario. For a long time, people thought that $\Lambda = 0$. It was just not required to describe the observations, and the principle of Occam's razor says that it is better not to introduce a notion when one does not really need it. However, recent precise measurements indicate that the Universe is not only expanding, but expanding with acceleration! To explain this, one *needs* a tiny positive cosmological constant (known also under the poetic name of *dark energy*). Its value is at the level of a dozen joules per cubic kilometer.

Having made this digression, we now go back to QFT and ask: how to reconcile a tiny observed value of the cosmological constant with the infinite *zero-point energy* (4.42)? A short answer is that *we do not know*. The following philosophical speculations may be noteworthy, however.

We considered so far only a bosonic field $\phi(x)$ satisfying the KFG equation. One can repeat the reasoning for a fermion field satisfying the Dirac equation. We will do so in Chap. 9, where we will see that the fermion zero-point energy is also infinite, but it is now *negative* infinity.

Among the elementary particles listed in Chap. 3 (and we will learn very soon that each such particle is associated with a certain elementary field), both bosons and fermions are present. Imagine now that the positive infinity due to bosons cancels exactly the negative infinity due to fermions.

This cancellation is exactly what happens in *supersymmetric* theories.[17] Unfortunately, the world that we see and know is not supersymmetric. True, many theorists believe that supersymmetry is there at the deep fundamental level, at ultrahigh energies probably inaccessible to experimental study. This might be so, but anyway, supersymmetry is broken at low energies. And it is absolutely unclear now why breaking of supersymmetry does not also bring about a huge (if not divergent) cosmological constant. This is one of the main unresolved mysteries in modern physics.

One can note, however, that the problem appears only when one incorporates gravitational interactions. In the world without gravity the absolute value of energy is not important and not observable. It is only *energy differences* that matter.

Let us look again at the spectrum (4.41) and consider the state where all quantum numbers vanish *but one*. If, say, $l_1^n = 1$ for some particular n, the energy of this state exceeds the vacuum energy by ω_n. Looking now at the expression (4.40) for

---

[17] Supersymmetry is the symmetry between bosons and fermions. We will discuss it in Chap. 14.

ω_n, we see that it can be interpreted as the energy of a free relativistic *particle* with mass m and momentum $p = 2\pi n/L$ or maybe $p = -2\pi n/L$ ($p = \pm 2\pi\hbar n/L$ in ordinary units). And the corresponding state in the Hamiltonian (4.36) can be called the *one-particle state*.

For a given n, there are actually *two* such states: a state where all quantum numbers but $l_1^n = 1$ are zero and the state where all quantum numbers but $l_2^n = 1$ are zero. The one-particle states with a given momentum represent their linear combinations. One usually defines them as follows[18]:

$$\Psi_{\text{particle with momentum } 2\pi n_0/L} \sim c_{n_0}^* \prod_n \exp\left\{-V\omega_n c_n c_n^*\right\}, \qquad (4.43)$$

$$\Psi_{\text{antiparticle with momentum } -2\pi n_0/L} \sim c_{n_0} \prod_n \exp\left\{-V\omega_n c_n c_n^*\right\}. \qquad (4.44)$$

The same product but without the factors $c_{n_0} \equiv c_0$ or $c_{n_0}^* \equiv c_0^*$ in front represents the vacuum wave function:

$$\Psi_{\text{vac}} = \prod_n \exp\left\{-V\omega_n c_n c_n^*\right\}. \qquad (4.45)$$

Incidentally, that is why we chose right at the beginning to deal with the complex scalar field and not with the real scalar field. In the latter case there would be only one quantum number for each momentum and there would be no antiparticles, or rather the particles would coincide with their antiparticles.

There are many other eigenstates (called the *Fock states*) of the Hamiltonian (4.36). One can have, for example, a state involving 4 particles with the momenta p_1, p_2, p_3, p_4 and 3 antiparticles with the momenta p_5, p_6, p_7. The wave function of such a state would involve the same product of exponentials as in (4.45) with an extra factor $c_1^* c_2^* c_3^* c_4^* c_{-5} c_{-6} c_{-7}$. The momenta can be different or they can coincide.[19]

We understand now (for free bosons, at least) how two inherent interpretational difficulties of the relativistic quantum mechanics mentioned in Sect. 4.4 — the impossibility of probabilistic interpretation for the relativistic wave function and the existence of eigenstates with negative energies [two possible signs in the square root in (4.23)] — are resolved:

- One should not consider the KFG equation as an equation for the wave function, but rather as an equation for the classical field $\phi(x, t)$. The quantity $|\phi(x, t)|^2$ does not then have a particular physical meaning.

---

[18] Yes, Sophie, it would be more natural to define it the other way round — the particle state with the factor $\sim c$ and the antiparticle state with the factor $\sim c^*$. But I am following here the standard convention arisen for historical reasons and used in all other books on QFT. Please wait until Chap. 7 (pp. 117–119) where we will accurately derive the Hamiltonian (4.36) and elucidate the structure of its spectrum, including the negative sign of momentum in (4.44).

[19] On the other hand, as Pauli taught us, identical fermions cannot have the same momentum. In Chap. 9, we will understand why.

- The true Hamiltonian of the system is not the Hamiltonian (4.22), but rather the Pauli–Weisskopf Hamiltonian (4.36). The latter has a ground state. When its energy is normalized to zero, all excited states have *positive* energies. This concerns also the one-particle and one-antiparticle states. In the absence of a potential, their energies coincide, being given by (4.40).

The trouble strikes back, however, when one considers motion in an external potential.[20]

For a strong attractive potential, with characteristic absolute value $|V(x)|$ of the order of the rest energy mc^2, the energy of the one-particle state can become negative. This is worrisome, but not a tragedy yet. A tragedy arises when the energy goes below $-mc^2$. Well, this is still not the global Armageddon-type killing-everything tragedy, but the one-particle approximation is definitely killed in this region of the parameters.

For an electron in the field of a positively charged nucleus, this happens if the nuclear charge exceeds[21]

$$Z_{\text{crit}} \approx 175 . \tag{4.46}$$

Or rather *would happen* — nuclei with such a large charge do not exist in nature: they would be immediately torn apart by electrostatic repulsion.

Suppose, however, that they existed. It would then be energetically favourable to *create* an extra electron-positron pair, with the electron placed at the strongly negative energy level, $E_{e^-} < -mc^2$, and the positron being endowed with a small momentum and removed to infinity, such that its energy was close to $E_{e^+} \approx mc^2$. If the nuclear charge exceeded essentially the critical charge, many such pairs would be created, until the nuclear charge was sufficiently screened. Then the newborn electrons, hovering below threshold near the nucleus, would eventually be absorbed by nuclear protons in the reaction $p + e^- \rightarrow n + \nu_e$. The nuclear charge would fall below the critical value (4.46). And things would calm down...

In the previous paragraphs we discussed the interaction of the *electron-positron* field with the classical Coulomb potential. But I have not yet explained how interactions arise in the QFT formalism — the Hamiltonian (4.36) describes only free fields. Let us now do so. Staying in the framework of our complex scalar field model, one can introduce field self-interaction by adding a quartic term in the Hamiltonian,[22]

$$H_{\text{int}} = \frac{\lambda}{4} \int (\phi^*)^2 \phi^2 \, dx . \tag{4.47}$$

This is equivalent to adding a nonlinear cubic term $-\frac{1}{2}\lambda\phi^2\phi^*$ in the right-hand side of the KFG equation (4.28).

---

[20]Or, better to say, when one takes into account the interaction of our quantum field $\phi(x, t)$ with a classical external potential. We prefer not to go into more details at this stage.

[21]A rough estimate is $Z_{\text{crit}} \approx 1/\alpha \approx 137$. The limit (4.46) was obtained in 1970 by Vladimir Popov, who solved the Dirac equation, taking into account a finite size of the nucleus.

[22]The factor $1/4$ is habitually introduced there to simplify the expression for the scattering amplitude. This will be explained in Chap. 10.

Substituting into (4.47) the Fourier expansion (4.35), we can express it as

$$H_{int} = \frac{\lambda V}{4} \sum_{n_1 n_2 n_3 n_4} c_3^* c_4^* c_1 c_2 \, \delta(n_1 + n_2 - n_3 - n_4). \tag{4.48}$$

If we add this to the free Hamiltonian (4.36), we obtain a complicated system of the infinite number of coupled oscillators. The Schrödinger equation for the wave functional can no longer be solved exactly. Still, if the interaction constant λ is sufficiently small, one can treat interactions perturbatively. To this end, one should evaluate the matrix elements of the interaction Hamiltonian (4.48) in the basis representing the eigenstates of the free Hamiltonian (4.36) — the particle and antiparticle states. For example, if we consider only the particle states (4.43), the matrix elements (see p. 119 for some more discussion) read[23]

$$\langle p_3, p_4 | H_{int} | p_2, p_1 \rangle \sim \lambda \delta(p_1 + p_2 - p_3 - p_4). \tag{4.50}$$

As the intelligent reader might have already guessed, the matrix elements (4.50) are related to the scattering amplitudes. Indeed, they seem to describe processes where two particles with momenta p_1, p_2 collide and scatter so that the final momenta of the particles are p_3 and p_4. Well, this is *almost*, but not *quite* so. The matrix element (4.50)[24] was calculated in the *Schrödinger picture*. It implements 3-momentum conservation, $p_1 + p_2 = p_3 + p_4$, but not energy conservation. To derive correctly the scattering amplitudes, implying $\epsilon_1 + \epsilon_2 = \epsilon_3 + \epsilon_4$ (especially, to evaluate the contributions of higher perturbative orders), it is better to go into the *Heisenberg picture* (more precisely, the *interaction picture*) with time-dependent field operators. This accurate derivation is beyond the scope of our book. But we will give later (in Chap. 10) a semi-heuristic derivation of the rules (Feynman rules) that allow one to write down analytic expressions for relativistic scattering amplitudes in any order.

For the time being, we only draw in Fig. 4.4 a simple picture associated with (4.50). This is the first Feynman diagram that the reader has encountered in our book. There will be more.

---

[23] In Eq. (4.48), the symbol $\delta(n_1 + n_2 - n_3 - n_4)$ meant that the summation was performed not over all n_j, but only over those that satisfy the condition $n_1 + n_2 = n_3 + n_4$. In Eq. (4.50), $\delta(p_1 + p_2 - p_3 - p_4)$ is the delta function introduced in mathematics by Dirac. The delta function $\delta(x)$ has the properties, $\delta(x \neq 0) = 0$, $\delta(0) = \infty$, $\int_{-\infty}^{\infty} \delta(x) dx = 1$. One can think of it as the limit

$$\delta(x) = \frac{1}{\pi} \lim_{\epsilon \to 0} \frac{\epsilon}{x^2 + \epsilon^2}. \tag{4.49}$$

And $\delta(p) = \delta(p_1)\delta(p_2)\delta(p_3)$.

[24] Note that the *final* momenta $p_{3,4}$ stand on the *left*, the site of arrival is mentioned *before* the site of departure. This convention, unnatural for an air traveller, follows from the definition,

$$\langle n | \hat{O} | m \rangle \sim \int \Psi_n^* \hat{O} \Psi_m \prod_k dc_k^* dc_k. \tag{4.51}$$

.

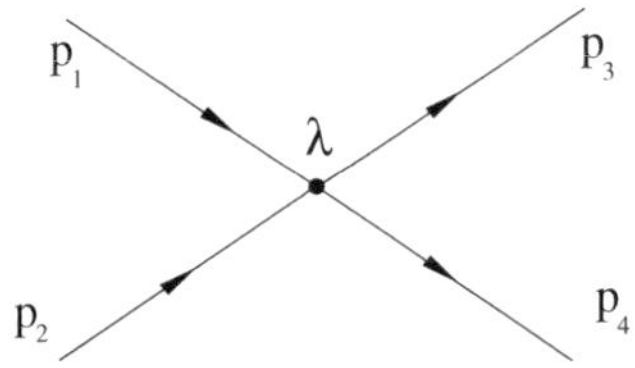

Fig. 4.4 The scalar $2 \to 2$ scattering amplitude to the lowest order.

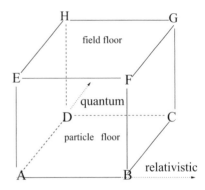

Fig. 4.5 The cubic scheme of physical theories drawn by Sophie after visiting the museum. $ABCD$ — particle floor. $EFGH$ — field floor. $ADEH$ — nonrelativistic wall, $BCFG$ — relativistic wall. $ABEF$ — classical wall, $DCGH$ — quantum wall.

So far, we have discussed only relativistic quantum field theory. And our discussion was unusually long by the standards of this chapter. The reader will forgive us. After all, we have just touched upon the main subject of this book, made the first exploratory reconnaissance to the fascinating world of relativistic QFT. But the architecture of the theoretical physics building (when Sophie came back, she made a sketch of what she saw, which we reproduce in Fig. 4.5) suggests the existence of a special hall devoted to nonrelativistic quantum field theory. What does this hall contain?

As was mentioned, nonrelativistic field theories provide an effective description of ordinary matter at distances substantially exceeding interatomic distances. The wave equation (4.27) describes solids, while the Navier–Stokes equation (4.26) describes liquids.

Both in liquids and in solids, there are quantum effects, and they can be taken into account using the formalism of QFT. For example (the first thing that comes to mind), quantization of the sound waves in a crystal produces *phonons*. The phonon wavelength cannot be less than the size of the crystal unit cell. Hence, the energy of a phonon cannot exceed some limit $\hbar\Omega_D$ (Ω_D standing for the so-called *Debye frequency*). And this limitation is important when calculating the specific heat of the crystal at low temperatures.

The theory of *superconductivity* is a nonrelativistic quantum field theory. First, superconductivity is due to some specific attraction of the electrons in a crystal due to their interaction with phonons, which at low temperatures may overcome their Coulomb repulsion. Second, the *Cooper pairs*, the bound states of two electrons that are formed as a result of this attraction, are described by an effective bosonic quantum field.

Quantum hydrodynamics is also an important branch of physics. It is extremely relevant for the description of the properties of liquid helium. Again, the phenomenon of superfluidity is best described in the framework of the formalism of QFT.

But this is all beyond the scope of our book...

4.8 Not in the Cube

Here follows a disclaimer: there are still at least two important branches of theoretical physics which do not fall within the proposed cubic scheme.

- First of all, there is gravity. Einstein's general relativity is a relativistic field theory. But it is a relativistic field theory of a very special kind. In the theories discussed above, space and time were a kind of stage where the action, field dynamics, unfolded. The dynamical field equations in a fixed reference system could be formulated as a Cauchy problem describing the evolution in a universal independent time.

  In general relativity, it is the fabric of spacetime, the metric $g_{\mu\nu}$ which is the dynamical variable.[25] As a result, one cannot introduce the notion of a universal time where evolution unfolds (one can do so for weak enough gravitational fields when the metric does not deviate much from the flat Minkowski metric, but not in general).

  The latter circumstance leads to paradoxes. For example, there exist rather strange solutions to Einstein's equations with *closed time loops*[26]: the flow of time is not linear, from past to future, but, following one such closed line you could find yourself in the past and kill your great grandfather. This deplorable act would not only be criminal, but also paradoxical. It would lead to an obvious *violation of causality*.

  Thus, the theory of gravity has logical problems even at the classical level. General relativity is not internally consistent and describes only a limited range of phenomena when gtavitational fields may be strong, but not *too* strong. And quantum gravity simply does not exist today! We will return to the discussion of this mystery at the end of the book.

- The second set of problems that we did not discuss is more mundane. We are referring to statistical physics. Indeed, the air in a room is a system with a really

---

[25]The metric tensor $g_{\mu\nu}$ determines the interval between nearby events in a curved spacetime. Instead of (4.21), we have

$$ds^2 = g_{\mu\nu}(x)\,dx^\mu dx^\nu \tag{4.52}$$

(the double summation over $\mu = 0, 1, 2, 3$ and $\nu = 0, 1, 2, 3$ is assumed).

[26]Such solutions were first found by Kurt Gödel, the same mathematician who proved the famous Gödel's theorem asserting the incompleteness of any sufficiently rich formal logical system — the existence of the statements that can neither be proved, nor disproved.

great number of degrees of freedom, but we do not treat them as field theorists, do not write partial differential equations to describe the gas behaviour. Well, we still can write the wave equation describing the propagation of the sound waves, but sound is only one of the aspects of gas dynamics. There are others, more basic. In particular, the Clapeyron–Mendeleev equation,

$$PV = Nk_B T ,\qquad\qquad (4.53)$$

does not follow from a solution of any ordinary or partial differential equation. It is derived as a result of statistic averaging.

There is classical statistics (Volume 5 of the Landau course in theoretical physics) and there is also quantum statistics (Volumes 9 and 11). In practice, bearing in mind that the nonrelativistic quantum field theories discussed at the end of the previous section are all effective theories obtained in the framework of a procedure that is akin to statistic averaging, it is difficult to really separate the domains of nonrelativistic QFT and the domain of quantum statistics. This explains why superconductivity and superfluidity are treated in Volume 9 of the Landau course. One can also note that a well-known manual for nonrelativistic QFT due to Abrikosov, Gor'kov and Dzyaloshinsky carries the name *Methods of Quantum Field Theory in Statistical Physics.*[27]

---

[27]This book is very good, but, being written in a very concise and dense manner, it is not an easy read. Russian students christened it the "Green monster" (all Russian editions of this book that I saw had a greenish cover).

Chapter 5
Genesis of the Standard Model

Our multi-course dinner unfolds and requires more and more refined and developed tastes to fully enjoy it. We assume in this long chapter that the reader has an idea about elementary relativistic kinematics and understands what quantum scattering amplitudes and differential cross sections are. These issues will be revised in the *Chef's Secrets* part — Chaps. 6 and 8. See also Chap. 10 for more details.

Alexis de Tocqueville once noticed: *History is a gallery of pictures where there are few originals and many copies.* He was right as far as political history is concerned. Man is known to be a political animal, and human passions have not essentially changed since the time of Aristotle or King Hammurabi.

But this remark of de Tocqueville does not seem to apply to the intellectual history of mankind, especially — to the history of sciences. Our knowledge *does* develop. King Hammurabi, Aristotle or de Tocqueville simply did not have the slightest idea about many things we know today. Including, of course, quantum field theory. This young branch of physics grew during the last century, between 1925–1926 (when modern quantum mechanics was born) and 1974–1975 (when the vast majority of experts realized that the microworld, the world of elementary particles, is described by what is now called the *Standard Model* — the quantum field theory of the strong, electromagnetic and weak interactions.

Its history is fascinating and sometimes dramatic. It is worth discussing.

5.1 Trees

Let us go back to 1926 and to the previous chapter, where the first attempts to build up relativistic quantum theory, associated with the names of Fock, Klein, and Gordon, were discussed. It was explained there that the quantum theory based on the KFG equation (4.18) replacing the Schrödinger equation works well in some situations, but not always. In fact, this theory is *paradoxical* (negative energies, the absence

© Springer International Publishing AG 2017
A. Smilga, *Digestible Quantum Field Theory*,
DOI 10.1007/978-3-319-59922-9_5

Fig. 5.1 An atomic
electromagnetic transition.

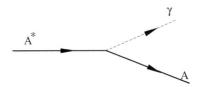

of the probabilistic interpretation for $|\Psi(x, t)|^2$ and so on). As also was explained
in the previous chapter, the remedy for these paradoxes was found by Fermi, Fock,
Pauli and Weisskopf, who introduced the notion of a field theory Hamiltonian [see
Eqs. (4.36), (4.37)] and found its spectrum.

But Eq. (4.36) is a free Hamiltonian, not involving interactions. The solution of
the quantum problem for interacting nonlinear Hamiltonians [like the Hamiltonian
(4.36) with the quartic term (4.47) added] is a complicated technical problem. One
can only do it perturbatively when the interaction is sufficiently weak.

The first problem of quantum field theory that was solved in this way was the
problem of atomic electromagnetic transitions. As is well known, excited atomic
states have a finite lifetime, they fall down to the ground state with the emission of
a photon or several photons. The process

$$A^* \rightarrow A + \gamma \qquad (5.1)$$

is associated with the creation of a particle (the photon) and is a genuine QFT
process.[1] It is described by the picture (Feynman diagram) drawn in Fig. 5.1:

But the atomic transition amplitudes can also be evaluated using the methods of
traditional nonrelativistic quantum mechanics and are treated in many traditional
quantum mechanics textbooks. It is sufficient to calculate the matrix element of the
perturbation brought about by the electromagnetic current operator,

$$j = \frac{i\hbar}{2m} \left(\Psi \nabla \Psi^* - \Psi^* \nabla \Psi \right) , \qquad (5.2)$$

between the initial and the final atomic eigenstates and to multiply it by the elec-
tron charge $e = -|e|$.[2] For hydrogen, where the analytic expressions for the wave
functions exist, this calculation was first performed by Gordon in 1929.

Generically, the interaction between the photons and electrons is described by the
operator

$$V_e = e j_e^\mu A_\mu , \qquad (5.3)$$

---

[1] That is why the authors of the Landau course prefered to treat it in Volume 4 (*Quantum Electrodynamics*) rather than in Volume 3 (*Quantum Mechanics*).

[2] To define j and its relativistic generalization, we are now using the convention usually adopted by
high-energy physicists. In the textbooks on *classical* electromagnetism, the definition of the current
usually includes the factor e. When writing down classical Maxwell equations (4.31), we also did
so.

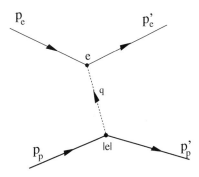

Fig. 5.2 Mott scattering.
p_e, p_p and p'_e, p'_p are the
initial and final electron and
proton 4-momenta;
$q = p'_e - p_e$ is the
4-momentum transfer. We
also marked the factors e and
$|e|$, associated with the
electron-photon and
proton-photon vertices, that
enter the amplitude.

where A_μ is the 4-vector describing the electromagnetic potential and j_e^μ is the relativistic operator of the electron electromagnetic current, which we prefer not to write here explicitly. Photons also interact with other charged particles, for example, with protons. The corresponding interaction is $V_p = |e| j_p^\mu A_\mu$. These interactions bring about nontrivial scattering processes.

Consider the scattering of electrons by heavy protons or *Mott scattering* (Neville Mott calculated its amplitude in 1929[3]). The corresponding Feynman diagram is drawn in Fig. 5.2.

It is now a proper time and place to disclose a secret. A Feynman diagram is not just a picture. In fact, each such diagram encodes an analytic expression for the scattering amplitude. We will learn more about how these analytic expressions are written in Chaps. 8 and 10. Right now, we only note that

- The thin solid lines in Fig. 5.2 describe the initial and the final electrons. The thick solid lines describe the initial and final protons.
- The electron vertex involves the factor e, and the proton vertex the factor $|e|$. The presence of two vertices indicates that the amplitude is obtained as a result of calculations in the second order of perturbation theory.
- The dashed line connecting two vertices is a new animal. It describes a *virtual photon*. Not going into detail now, we only say that another way to represent the amplitude is to say that the electron is scattered by the electromagnetic field emanating from the proton (or the other way round — as you wish!). Thus, the virtual photon line represents the electromagnetic field by means of which the charged particles interact.

The expression for the amplitude becomes especially simple in the nonrelativistic limit. When the proton is at rest and the electron velocity is small, we are dealing with scattering on the external Coulomb field. In this case, the nonrelativistic scattering

---

[3]To be more exact, Mott calculated the scattering amplitude of the electrons in the static Coulomb field. This applies to ep scattering when the energy of electrons does not exceed several hundred MeV, otherwise the effects due to internal proton structure come into play. But we now disregard these complications.

amplitude depends in the lowest order only on the momentum transfer $q = p'_e - p_e$. In the natural unit system, it is given by[4]

$$f(q) = \frac{2m\alpha}{q^2},\tag{5.4}$$

where m is the electron mass.

The square of this amplitude gives the *Rutherford formula* for the differential cross section:

$$\frac{d\sigma}{d\Omega} = \left(\frac{2m\alpha}{q^2}\right)^2.\tag{5.5}$$

Ernest Rutherford derived it to describe the scattering of α-particles on the nuclei in his famous experiment whereby the atomic nucleus was discovered.[5]

For sure, to derive (5.4) one needs neither the machinery of QFT, nor that of quantum mechanics. One can also derive it in *classical* electrodynamics, and that is what Rutherford did. In fact, the amplitude (5.4) is nothing but the Fourier transform of the Coulomb potential,

$$\int dx\, e^{-iq\cdot x}\left(-\frac{\alpha}{r}\right) = -\frac{4\pi\alpha}{q^2},\tag{5.6}$$

multiplied by a proper coefficient.

In the relativistic case, one has to evaluate the amplitude by the diagram in Fig. 5.2. When the mass of proton is much larger that the electron energy, the result represents a natural generalization of (5.4). For example, the square of the 3-vector q is replaced by the relativistic invariant $-q^2 = q^2 - q_0^2$. In addition, some factors taking into account the electron polarizations appear.

The calculation of the electron-electron scattering or electron-positron scattering in the kinematical region where the momenta of all particles are of the order of their mass or larger, is technically a somewhat more complicated problem, but it does not really require a lot of new insights or methods. This was done by Christian Møller (e^-e^- scattering) and Homi Bhabha (e^+e^- scattering) in the mid-thirties. The only complication that is worth mentioning is that, in these cases, the amplitude is given by the sum of two terms, each term being described by a distinct Feynman diagram. For the Møller scattering, these diagrams are schematically shown in Fig. 5.3. They differ by permutation of the final electrons.

Let us dwell on the process $e^+e^- \rightarrow \mu^+\mu^-$, the process of electron-positron *annihilation* into a muon pair. There is only one Feynman diagram, drawn in Fig. 5.4.

The interpretation is the following. First, the electron and positron merge together and form a virtual photon. Then the latter decays, creating two particles: the μ^+ and

---

[4]The amplitude has the dimension of inverse mass, i.e. of length. Its square $d\sigma/d\Omega = |f(q)|^2$, the differential cross section, has the dimension of area, as it should. See Chap. 8 for more detail.

[5]Of course, for a nucleus of charge $Z|e|$, the amplitude (5.4) should be multiplied by Z and the cross section (5.5) by Z^2.

Fig. 5.3 Two diagrams describing the $e^- e^-$ scattering.

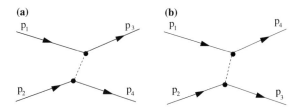

Fig. 5.4 The process $e^+ e^- \to \mu^+ \mu^-$.

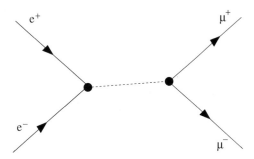

μ^-. When the energies of the initial and final particles are much larger than their masses, the differential cross section of this process is given by a simple expression

$$d\sigma_{e^+ e^- \to \mu^+ \mu^-} = \frac{\alpha^2}{16E^2}(1 + \cos^2 \theta)\, d\Omega, \tag{5.7}$$

where E is the energy of the electron and positron in the system of their centre of mass, and θ is the angle between the direction of the colliding e^+ and e^- and the direction of the produced μ^+ and μ^-. This cross section was evaluated by Vladimir Berestetsky and Isaak Pomeranchuk relatively late, in 1955. But the calculations are simpler than those for Møller or Bhabha scattering. The result (5.7) will serve us a reference point when we discuss at the end of Sect. 5.3 the physics of $e^+ e^-$ annihilation into strongly interacting particles.

To avoid confusion and anger of pundits, we have to make a comment about the arrows in Fig. 5.4. Their direction simply shows the direction of momenta meaning that the electron and positron with positive energies collide and the particles μ^+ and μ^- with positive energies are created. To actually *calculate* the amplitudes, it is convenient to draw the diagrams with an opposite direction of the arrows for the positron lines. Professionals do that. In addition, they usually reverse the overall direction of the arrows, drawing the outgoing particles on the left in accordance with the convention (4.50), (4.51). But we are amateurs, we will not calculate ourselves the amplitudes in this book and may direct our arrows sloppily.

5.2 Loops □ Feynman Technique □ Divergences and Renormalization

"Everything is quite clear", may say our reader, "but why was the previous section called *Trees*? It was all about the photons, electrons... No botanics or forestry..."

The answer is simple — the Feynman diagrams drawn in Figs. 5.1, 5.2, 5.3 and 5.4 all belong to the class of the so-called *tree diagrams*. They describe the scattering amplitudes in the lowest nontrivial order of perturbation theory.

But there are also higher-order corrections. An example of a higher-order contribution to the amplitude of *ep*-scattering is shown in Fig. 5.5. One can easily detect a closed loop in this graph and understand why the higher-order corrections in quantum field theory are called *loop corrections*.

Here the electron and the proton exchange two rather than one virtual photon. In contrast to the tree diagram in Fig. 5.2, where the 4-momentum of the virtual photon was fixed by the external momenta, $q = p'_e - p_e$, the momenta q_1 and q_2 in Fig. 5.5 are not fixed, only their sum is. The analytic expression for the amplitude involves an *integral*

$$\sim \int d^4 q_1 d^4 q_2 \, \delta^{(4)}(p'_e - p_e - q_1 - q_2) \, .$$

I should now confess that I simplified the history when I was discussing tree processes in the previous section. I illustrated them with Feynman diagrams, and the reader could conclude that Møller, Bhabha and others obtained their results for the amplitudes and cross sections by evaluating Feynman diagrams, as is done in the modern textbooks.

But Feynman diagrams were not yet invented in the thirties, and people performed calculations in a different, much more cumbersome way. They used what is now called the *old diagram technique*. Basically, that was perturbation theory in ordinary non-relativistic quantum mechanics adapted to describe particle scattering processes. As I said, this method was cumbersome. For example, the amplitude in Fig. 5.5 was described not by one, but by many diagrams distinguished by the time order in which the four perturbations (5.3) occur. One had to evaluate separately

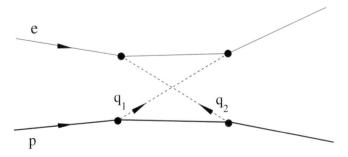

Fig. 5.5 A loop correction to the *ep* scattering amplitude.

Fig. 5.6 A diagram with creation of an extra e^+e^- pair in the old diagram technique. The vertex associated with the matrix element $\langle e^+e^- | \hat{V} | \gamma \rangle$ is made "thicker". The *left dotted vertical line* marks the intermediate state $|e^-\gamma p\rangle$ and the *right line* marks the state $|e^-e^+e^-p\rangle$.

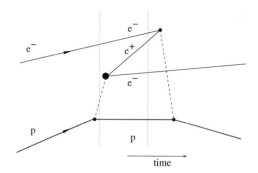

the contribution depicted in Fig. 5.6 involving among other factors the matrix element $\langle e^+e^- | \hat{V} | \gamma \rangle$ with creating an extra e^+e^- pair in the final state. [Here $\hat{V}$ is the interaction Hamiltonian (5.3).]

In Feynman's relativistically invariant technique, there is only one diagram for this process. Only topology matters and, say, for the diagram in Fig. 5.6, it is obviously the same as for that in Fig. 5.5.

A distinguishing feature of the Feynman technique compared to the nonrelativistic technique is the conservation of all components of 4-momentum in the vertices. On the other hand, internal lines describe *virtual* particles, i.e. the particles for which the square of their 4-momentum is different from m^2. For example, in the graph in Fig. 5.4 for the annihilation $e^+e^- \rightarrow \mu^+\mu^-$ considered in the centre-of-mass system, the 3-momentum of the virtual photon is zero. But its energy $q_0 = E_{e^-} + E_{e^+} = 2E$ is not. As a result, $q^2 = 4E^2 > 0$.

And in the old diagram technique only 3-momentum is conserved in the vertices, but not the energy. On the other hand, there are no virtual particles, one calculates there the matrix elements between the eigenstates of the free Hamiltonian (Fock states) involving only the real particles with $p^2 = m^2$.

The technical simplifications brought about by the Feynman technique are enormous. Let me quote here a paragraph from Feynman's Nobel Lecture. Or rather half a paragraph. In the first half that I omit, Feynman tells how he attended a seminar where the speaker, Murray Slotnick, reported his calculations of some amplitude in two different theoretical models. Feynman decided that it was a nice opportunity to check his own methods. So he went home and calculated it *his* way.

... The next day at the meeting I saw Slotnick and said, "Slotnick, I worked it out last night, I wanted to see if I got the same answers as you do. I got a different answer for each coupling — but, I would like to check in detail with you because I want to be sure of my methods". And, he said, "What do you mean you worked it out last night, it took me six months!" And, when we compared the answers he looked at mine and he asked, "What is that Q in there, that variable Q?" (I had expressions like (arctan Q)/Q, etc.) I said, "that's the momentum transferred by the electron, the electrons deflected by different angles. "Oh", he said, "no, I only have the limiting value as Q approaches zero; the forward scattering." Well, it was easy enough to

Fig. 5.7 Renormalization of
Coulomb potential.

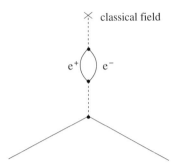

substitute Q equals zero in my form and I then got the same answers as he did. But it took him six months to do the case of zero momentum transfer, whereas, during one evening I had done the finite and arbitrary momentum transfer. That was a thrilling moment for me, like receiving the Nobel Prize, because that convinced me, at last, I did have some kind of method and technique and understood how to do something that other people did not know how to do. That was my moment of triumph in which I realized I really had succeeded in working out something worthwhile.

Going back to the loop corrections to *ep* scattering (or, better to say, loop corrections to the amplitude of electron scattering in an external static field), it was evaluated by several people in 1948–1950 using the new methods. The calculation was not difficult.

The inconvenience of the nonrelativistic perturbation theory was a technical problem. But there was also (and, to some extent, *there is*) a much more serious problem of a fundamental nature. It turned out that the integrals associated with *some* loop diagrams diverge!

In particular, this concerns the diagram in Fig. 5.7 describing the one-loop correction to the electron scattering in the static external field, but the correction of another type, compared to that in Fig. 5.5. The amplitude involves an integral over momenta of the virtual electron and positron. And this integral behaves at large momenta as

$$\sim \int_0^\infty \frac{dp^2}{p^2 + m^2} \tag{5.8}$$

displaying a logarithmic ultraviolet (i.e. large momenta) divergence.[6]

Consider the Coulomb external field created by the static electric charge e. The sum of the tree graph and the divergent loop graph in the electron scattering amplitude can be interpreted as a modification of the Coulomb potential,[7]

---

[6]If one does not calculate sufficiently carefully, even a *power* UV divergence pops up. It took some time for people to understand why it does not in fact appear.

[7]The charge is measured in the Heaviside units.

$$\frac{e_0^2}{4\pi r} \rightarrow \frac{e_0^2}{4\pi r}\left(1 - \frac{e_0^2}{6\pi^2}\ln\frac{\Lambda}{m}\right) + \text{finite terms}, \qquad (5.9)$$

where Λ is a new kind of strange monster called an *ultraviolet cutoff*. We just integrate in (5.8) not over all momenta, but with the restriction $p^2 < \Lambda^2$.

But why should we do that? And what is the physical meaning of Λ? I must tell you right away that today there is *no* clear and exhaustive answer to this question. Moreover, the necessity to introduce an unphysical ultraviolet cutoff[8] makes quantum electrodynamics an internally inconsistent theory!

On the other hand, one can also observe that, if we keep Λ very large, but finite, the loop correction to the potential is still small. For the correction to be equal in absolute value to the leading term, one would need

$$\Lambda \sim m e^{3\pi/(2\alpha)} \approx 10^{250}\,g \qquad (5.10)$$

— larger than anything one can imagine (the mass of the Universe is just $\sim 10^{57}\,g$). Thus, the problem acquires a somewhat academic flavour: in any case, we do not expect for quantum electrodynamics or anything else to make any sense at such large energies.

You probably noticed that we endowed the electric charge in (5.9) with the index 0. This emphasizes that e_0 is a "bare" unphysical and unobserved charge. The observed physical charge

$$e_{\text{phys}}^2 \approx e_0^2\left(1 - \frac{e_0^2}{6\pi^2}\ln\frac{\Lambda}{m} + \cdots\right) \qquad (5.11)$$

is less than the bare one.

Another name for (5.11) is *renormalized* charge. The procedure of *renormalization* consists in expressing everything in terms of the renormalized physical charge and the renormalized physical mass. For quantum electrodynamics and many other physical theories, one can prove (even though the rigorous proof is not so simple) that all ultraviolet divergences are hidden in these two quantities. Once renormalization is done, the divergences disappear.

We should now warn the reader. In old popular and not so popular books, one can meet a statement that renormalization consists in *subtracting* certain UV-divergent constants. Indeed, this is the way the pioneers, Richard Feynman, Julian Schwinger, Sin-Itiro Tomonaga[9] and Freeman Dyson formulated it. But we understand now that renormalization is *not* an arbitrary subtraction. I tried to explain what it *is* in the previous paragraph.

---

[8] Aggravated by the negative sign of the correction in (5.9); we will see in the next section that, in theories like QCD, where this correction has a positive sign, it *is* eventually possible to get rid of Λ — see Eqs. (5.17)–(5.19).

[9] In 1965, these three people got a Nobel Prize "for their fundamental work in quantum electrodynamics, with deep-ploughing consequences for the physics of elementary particles". I recommend to a reader who wants to learn more about the history of the creation of quantum electrodynamics the book by S. Schweber, *QED and the Men Who Made It*. It is fascinating reading!

Fig. 5.8 Vacuum
polarization. The blob in the
centre is a positively charged
source. It is surrounded by
the cloud of polarized virtual
e^+e^- pairs.

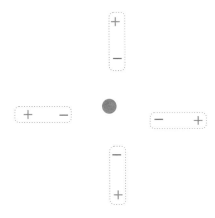

I also want to write here another formula. It turns out that one can evaluate not only the first nontrivial one-loop correction to the electric charge, but, in a certain approximation, also higher-order corrections (2-loop, 3-loop, 4-loop, ...). The result is a geometric series. Its sum gives

$$e_{\text{phys}}^2 \approx \frac{e_0^2}{1 + \frac{e_0^2}{6\pi^2} \ln \frac{\Lambda}{m}} . \tag{5.12}$$

In contrast to (5.11), it never vanishes. As Λ grows, while e_0^2 is kept fixed, it just gets smaller and smaller. In the (academic) limit $\Lambda \to \infty$, the physical charge vanishes. This phenomenon was first observed by Lev Landau, Alexei Abrikosov and Isaak Khalatnikov in 1954. It is sometimes called the *Moscow zero* and sometimes the *Landau pole*. The latter notion is a mirror reflection of the former. Equation (5.12) can also be represented as

$$e_0^2 \approx \frac{e_{\text{phys}}^2}{1 - \frac{e_{\text{phys}}^2}{6\pi^2} \ln \frac{\Lambda}{m}} . \tag{5.13}$$

If e_{phys}^2 is kept fixed and Λ grows, e_0^2 grows. At some point it becomes infinite — it runs into a pole.

One can interpret (5.12) by saying that the bare electric charge is *screened* and illustrate it with the picture in Fig. 5.8. It is the same picture that describes the screening of the external charge in a dielectric, say, in water. The positive external charge affects the orientation of water molecules, attracting the oxygen ions and repelling the hydrogen ions. In our case, space is empty, there are no molecules and their role is taken by the *virtual e^+e^- pairs* present in the vacuum.[10]

---

[10]In quantum field theory, the vacuum is defined as the ground state of the field Hamiltonian [cf. the comment after Eq. (4.44)]. Its wave function depends on *all* field variables. Thus, the vacuum contains *everything* in some sense.

The charge of the source surrounded by the cloud of the virtual pairs is smaller than the bare charge. The latter is screened. If one gets inside this cloud, s/he will see some increase of the charge compared with (5.12) observed from a large distance. Basically, if the distance to the source r is much smaller that the Compton wavelength m^{-1}, one should substitute $\ln(\Lambda r)$ for $\ln(\Lambda/m)$ in the expression (5.12).

5.3 From Nuclei to Gluon Jets

Even before people learned how to calculate loops in quantum electrodynamics, they tried to apply the QFT methods to the other known interactions — strong and weak. The strong interaction is stronger, and we consider it first.

The first phenomenological theory of the strong interaction was constructed by Hideki Yukawa in 1935. By analogy with QED, where the interaction between charged particles is mediated by the electromagnetic field whose quanta are photons, Yukawa suggested that the interaction of the nucleons is mediated by some new field whose quanta (the *mesons*) represent a new kind of particle. Yukawa's theory differed from QED in three points:

- We are dealing with the *strong* interaction, and that means that the meson–nucleon coupling constant g (the analogue of electric charge) is large.
- The electromagnetic field is vectorial in nature. That leads to the repulsion of same–sign charges, and only charges of opposite signs attract. However, the nucleus contains only nucleons, no antinucleons, and their interaction is attractive. That can only be realized if the quanta of the mediating field have even spin.[11] The simplest possibility is zero spin.
- Photons are massless and the Coulomb interaction is long-range. But the nuclear force has a short range, of order 1 fm. That means that our meson should be massive, with mass μ of order $(1 \text{ fm})^{-1} \sim 200 \text{ MeV}$.

Let us discuss the last very important point in more detail. Our discussion will be heuristic, involving logical gaps. A reader who stays with us until Chap. 8 will see these gaps filled out.

We have seen that the amplitude of the electron-proton scattering is proportional to the Fourier transform of the Coulomb potential (5.6). (No wonder, it is just the Born formula (8.20) for the scattering amplitude.) As was also mentioned, in the relativistic case the factor q^2 in the denominator of (5.4) goes over into $-q^2 = \mathbf{q}^2 - q_0^2$. Note now that multiplying any function of momentum by q^2 is equivalent to acting on its Fourier transform by the operator of d'Alembert:

$$(q_0^2 - \mathbf{q}^2) f(q) \quad \leftrightarrow \quad -\left(\frac{\partial^2}{\partial t^2} - \Delta\right) \tilde{f}(x) = -\Box \tilde{f}(x). \tag{5.14}$$

---

[11] Please believe me now at this point.

This makes us think about Maxwell's equations for the electromagnetic 4-vector potential.

Now, a field describing a massive particle (let its mass be μ) satisfies the KFG equation (4.28). In momentum representation, the operator $\Box + \mu^2$ corresponds to multiplying by $-q^2 + \mu^2$. Pursuing the analogy with QED, where the relativistic scattering amplitude is $\propto e^2/q^2$, it is natural to suggest that the amplitude corresponding to the exchange by a massive particle is proportional to $g^2/(q^2 - \mu^2)$, which gives

$$-\frac{g^2}{\mu^2 + \boldsymbol{q}^2} \, . \tag{5.15}$$

in the nonrelativistic limit. This should coincide with the Fourier transform of the potential. The latter is obtained from (5.15) by the inverse transformation. We obtain the *Yukawa potential*,

$$V_{\text{Yukawa}}(r) \;=\; -g^2 \int \frac{d\boldsymbol{q}}{(2\pi)^3} \, \frac{e^{i\boldsymbol{q}\boldsymbol{x}}}{\boldsymbol{q}^2 + \mu^2} \;=\; -\frac{g^2}{4\pi r} \, e^{-\mu r} \, , \tag{5.16}$$

which decays exponentially fast at large distances.

A diligent reader can check this mathematical fact. Not so diligent ones can still be convinced that a Fourier transform of (5.15) gives a short-range potential by noticing that *(i)* The large r behaviour of any function of coordinates is determined by the small q behaviour of its Fourier transform; *(ii)* For small $\boldsymbol{q}$, the amplitude (5.15) is just a constant. *(iii)* The Fourier transform of a constant is $\propto \delta(\boldsymbol{r})$, which vanishes everywhere except at the origin.

The theory of Yukawa seemed to be brilliantly confirmed in 1947 when a meson with the required properties was discovered in cosmic rays. That was the π meson with a mass around 150 MeV.[12] However, it very soon became clear that, even though the Yukawa theory correctly describes some features of nuclear interactions, it cannot be the final truth.

First of all, after the discovery of the π mesons many more mesons and baryons were discovered, first in cosmic rays and then in accelerator experiments. Yukawa's theory had no place for them.

Second, the large phenomenological value of the meson-nucleon coupling constant ($g^2/4\pi \approx 14$, which is ~ 1000 times larger than $e^2/4\pi \equiv \alpha$) brought about serious practical and conceptual problems. The practical problem was the impossibility of calculating the amplitudes in the perturbative framework. And the conceptual problem was the presence of the Landau pole and the impossibility of getting rid of the ultraviolet cutoff and sending Λ to infinity. For QED, that was only of academic importance, as the trouble occurred there only far away, at the absolutely unphysical

---

[12]Actually, there are three particles $\pi^\pm$, π^0 with the same strong interaction, but distinguished by their electric charges and having slightly different masses — the same story as with the proton and neutron.

values (5.10). But, for the meson strong-coupling theory, the trouble was right on the doorstep.

Theorists just did not know what to do. Many of them thought at that time (we are now in the fifties) that quantum field theory was *inherently* sick, and that the correct theory describing the properties of elementary particles should be founded on other principles. For example, people seriously considered the idea of the *bootstrap* (that there are no fundamental particles and no fundamental fields whatsoever, and everything depends on itself and everything else in some self-consistent way).

For about 20 years, from the mid-fifties to the mid-seventies, physicists, leaving aside the ambition to reveal a fundamental field theory, occupied themselves with a more pragmatic task — to handle the rapidly growing zoo of new "elementary" particles. If not in a refined theoretical fashion, then at least in some phenomenological way. It was basically the same job as had been done by Dmitry Mendeleev for the zoo of chemical elements a century earlier.

The "Mendeleev Table" of hadrons had been constructed by the early sixties due to the efforts of many people, of whom Gell-Mann probably deserves the most credit. It was found that all known *hadrons* (particles participating in the strong interaction) can be grouped into the *multiplets of SU(3)*, mainly octets and decuplets (there are also some singlets).

We will explain later (in Chap. 6) what $SU(3)$ is and what its multiplets are. But in more mundane language, this meant that the observed properties of hadrons could be rather well understood by assuming that they are not fundamental entities, but represent bound states of truly fundamental objects, the quarks. At first, only three quarks, u, d, s (see Table 3.1) were known, and that is where the number "3" in the notation $SU(3)$ comes from.

It is now a proper time to make a linguistic and terminological digression. By the same token that *lepton* is derived from Greek λεπτός (light), the word *hadron* comes from ἁδρός (bulky), *baryon* — from βαρὺς (heavy) and meson from μέσος (intermediate).

Mesons are typically (not always) heavier than leptons, but lighter than baryons. When they decay, they produce leptons and photons (examples: $\pi^+ \to e^+\nu_e$, $\pi^0 \to \gamma\gamma$). An unstable baryon has in addition a proton among the products of its decay. For example, the so-called Σ^- hyperon decays due to the weak interaction into a neutron, electron and antineutrino. In $\sim$17 min, the neutron decays further into a proton, another electron and another antineutrino [See Eq. (3.11)].

In the *constituent quark model* representing a natural interpretation of Gell-Mann's classification, baryons are made of 3 quarks, and mesons of a quark and an antiquark. For example, the π^0 meson represents a superposition of the quantum states $|\bar{u}u\rangle$ (a u quark accompanied by its anticomrade) and $|\bar{d}d\rangle$. The π^0 decays due to annihilation $\bar{u}u \to \gamma\gamma$, $\bar{d}d \to \gamma\gamma$. Its charged partner, the π^+, is made of quark u and antiquark $\bar{d}$. The underlying process leading to the decay $\pi^+ \to e^+\nu_e$ is $u \to de^+\nu_e$ with the subsequent $\bar{d}d$ - annihilation leaving in this case no further traces. Another possibility is that the pair $|\bar{d}d\rangle$ does not annihilate, but becomes a π^0. This gives the decay mode $\pi^+ \to \pi^0 e^+\nu_e$.

There are also hadrons involving strange quarks in their composition. For example, $\Sigma^- = |dds\rangle$. Here the underlying process responsible for its decay is $s \to ue^-\bar{\nu}_e$. As was mentioned, we are dealing with the weak interaction here, so the lifetime of Σ^- and other particles containing the strange quark or antiquark is comparatively large ($\sim 10^{-10}$ s), much larger than, for example, the lifetime of the particle $\Delta^{++} = |uuu\rangle$, which decays almost immediately (in $\sim 10^{-23}$ s) into the proton, $p = |uud\rangle$, and $\pi^+ = |u\bar{d}\rangle$.

This was strange for the experimentalists who observed Σ^- and other such long-living hadrons in the fifties. So they dubbed them *strange*. According to the modern understanding and definition, the strange particles are the particles containing strange quarks.

Quarks carry fractional electric charges. Indeed, if one admits the quark decompositions $p = |uud\rangle$ and $n = |udd\rangle$ for the nucleons, it immediately follows that the electric charges of u and d quarks are $q_u = +2|e|/3$ and $q_d = -|e|/3$. The s quark has the same charge as the d quark. This follows, for example, from charge conservation in the process $s \to ue^-\bar{\nu}_e$.

When Mendeleev constructed his Table, not all chemical elements were yet known. Thus, he left empty squares in the Table and made predictions: new elements with such and such chemical properties must exist. These squares were soon filled out — they were gallium, scandium and germanium. The constituent quark model has also had its gallium (its Neptune, if you will). That was the baryon Ω^- made of three strange quarks and discovered experimentally 2 years after its existence was predicted by Murray Gell-Mann and Yuval Ne'eman.

But the most direct and obvious prediction of the model was the existence of quarks. And *this* prediction had a hard time to be confirmed. A *lot* of experimental efforts were devoted to the search of fractionally charged particles in accelerators, in cosmic rays, in meteorites and in the ice of Greenland: to no avail...

The model also had a technical difficulty. Take, for example, the baryon Δ^{++} which, according to the constituent quark model, is made of three u quarks with zero angular momenta and the same spin orientation. But the Pauli principle (quarks are fermions) forbids it. By the same token, it forbids the existence of all other baryons, including nucleons.

The remedy for the second trouble was found rather soon. It was suggested that the quark of each type or *flavour*[13] (u, d, and s) is found in nature in three *colour* forms. Call the colours red, green and blue. Then Ω^- is made of three nonidentical particles: a red s quark, a green s quark and a blue s quark. Pauli's principle is no longer violated.

It is interesting that, looking from afar, the red, green and blue colours mix up, as in the solar light, and we observe a *white* object. Some time in the last millennium, I saw at CERN a toy called a "hadron". It was a disc with three coloured spots endowed with a pivot. When you turned this disc rapidly, the colours mixed up for your eyes, and you saw a white circular band (Fig. 5.9).

---

[13]It is interesting that the French equivalent of this physical term is *saveur* (taste) rather than *aromate*.

Fig. 5.9 A dynamical model of baryon.

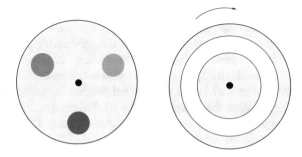

It was postulated that for some miraculous reason only such colourless states (one can attribute a mathematical meaning to this notion) are allowed to exist in nature in unbound form. And the coloured quarks are not allowed — they are doomed to stay confined within the hadrons. The word *confinement* was coined, but no satisfactory explanation for this phenomenon was suggested. In the mid-sixties (and even much later) many theorists found this picture so crazy that they did not believe in the existence of quarks. Their position was pretty much the same as the position of Copernicus who said that the astronomical data could be explained more easily if one adopted a mathematical model that the Earth and other planets go around the Sun. But he never said that it *was* the case. (Incidentally, that is why Copernicus, in contrast to Galilei, was never bothered by the Church.)

This viewpoint actually prevailed until 1969, when experiments on the scattering of energetic electrons on nucleons were performed at Stanford Linear Accelerator Centre (SLAC). It was shown that the *inclusive*[14] cross section of this process is the same as the cross section of the scattering on pointlike charged particles. This property (known as *Bjorken scaling*) meant that there *are* some pointlike constituents within the nucleons. They were at first called *partons*, but it soon became clear that everything works wonderfully well if one assumes that these partons have the quantum numbers of quarks. Thus, quarks surpassed the status of an amusing mathematical model and displayed themselves as dynamical entities which interact and, hence, are really there (*interago ergo sum*).

But what about field theory? Was it indeed true that theorists, being frustrated by the Moscow zero, abandoned all field theory studies? Of course not. The latter were no longer mainstream, but the nature of science is such that one can always find some people standing aside from the crowd, doing unpopular things. Sometimes, this happens to be the right thing to do.

The field theory on whose basis the Standard Model was eventually built had actually been invented by Chen Ning Yang and Robert Mills long before that, in 1954. The time has not yet come for us to describe it in detail, but basically it was a generalization of quantum electrodynamics involving several types of photons that interact between themselves. They are thus not neutral like the ordinary photon, but

---

[14]"Inclusive" means that we are only interested in what happens with the scattered electron and pay no attention to other created particles.

charged in a certain sense. Yang and Mills tried to associate these "charged photons" with the ρ mesons.[15] This attempt did not work. As a result, Yang–Mills theory was not perceived as a realistic physical theory during the following 10–15 years.

But the model as such was so interesting that many theorists continued to play with it. It turned out that the quantization of such a theory is far from being easy. Proceeding in a naïve way led to nonsensical results, including violation of unitarity. At the beginning of the sixties Feynman tried to resolve this problem (he was interested in Yang–Mills theory as a model for quantum gravity; the nonlinear interaction of gauge bosons in the former carry some features of the nonlinear interaction of gravitons in the latter). Feynman did not completely succeed, but he had some important insights which helped Ludwig Faddeev and Viktor Popov to develop their *ghost* method in 1967. Including the ghosts made perturbation theory self-consistent, and it became possible to calculate scattering amplitudes and other quantities at, in principle, any order of the coupling constant.

The idea that Yang–Mills theory describes the physics of the strong interactions was suggested in 1973 in three independent papers by Jogesh Pati and Abdus Salam; Harald Fritzsch, Murray Gell-Mann and Heinrich Leutwyler; and Steven Weinberg. These people suggested that the coloured quarks interact by the exchange of coloured *gluons* that are nothing but the "charged photons" of Yang and Mills. The only difference was that Yang and Mills had three such "photons", while there are eight colour varieties of gluons. Using the terminology of group theory (and group theory is indispensable to really understand what Yang–Mills theory is about — see Chaps. 6, 11 for detailed derivations and explanations), gluons belong to the adjoint representation of the $SU(3)$ group associated with the three quark colours. And quarks belong to the fundamental representation of the same group.

The real breakthrough occurred when the charge renormalization in this theory was calculated by David Gross, Frank Wilczek and David Politzer in 1974 and the phenomenon of *asymptotic freedom* was discovered. It turned out that, in Yang–Mills theory, the charge is not screened at large distances, but is rather *antiscreened* — it grows with the distance! This means that the effective charge decreases when the characteristic distance diminishes and the characteristic energy grows. For sufficiently large energies, the coupling constant is small. To the leading order, one can just neglect it — when they are hit hard, quarks behave as almost free particles, and this explains the Bjorken scaling effect. In the limit $E \to \infty$ the coupling constant vanishes: the quarks become completely free for asymptotically high energies.

History is rather dramatic at this point. The first person who correctly calculated the renormalization of the coupling constant in pure Yang–Mills theory was Iosif Khriplovich as early as in 1969. However, he did not understand the meaning of his result, and did not think of applying Yang–Mills theory to strong interaction physics. There was an even earlier paper in 1965 by Mikhail Terent'ev and Vladimir Vanyashin, who found indications that charge may decrease as energy grows, but they did the calculation in a not quite consistent theory (without ghosts, which were

---

[15]There are 3 ρ mesons, $\rho^\pm$ and ρ^0, which are vector particles (carry spin 1), like photons. Unlike photons, they are, however, massive, made of quarks and participate in the strong interactions.

not yet known at that time) and ascribed the strange behaviour of the charge to this. Finally, Gerald't Hooft also calculated it about a year before Gross, Wilczek, and Politzer, but he did not at first understand the overwhelming significance of his result and did not publish it. The interested reader having access to the internet can find more details in a historical review lecture by't Hooft in hep-th/9808154.[16]

At any rate, by the end of 1974 the dust settled, and the picture became clear. Asymptotic freedom of the theory:

- Explained the results of the SLAC experiment.
- Explained why the strong interaction was strong. If the coupling falls off at small distances, it grows at large distances, becoming roughly of order 1 at the distances of about 1 fm relevant for nuclear physics. This does not explain yet the phenomenon of confinement (we cannot *prove* it even now), but made it probable. If the interaction is strong at large distances, who knows how it behaves there? It is quite conceivable that the spectrum of asymptotic physical states does not resemble the set of the fundamental fields in terms of which the theory is formulated, and, in particular, does not include coloured states.
- Made the theory self-consistent. We have seen that quantum electrodynamics is not self-consistent due to the Landau pole. Inconsistent at unphysically large energies, but still. But *quantum chromodynamics*[17] does not have any conceptual problems. Indeed, the counterpart of (5.13) now reads

$$g_0^2 \approx \frac{g^2(\mu^2)}{1 + b \frac{g^2(\mu^2)}{8\pi^2} \ln \frac{\Lambda}{\mu}}, \tag{5.17}$$

where the numerical constant b depends on the number of quark flavours, $b = 11 - (2/3)N_f$, and $g^2(\mu)$ is an effective charge for processes with characteristic energy scale μ. One can take μ^2 sufficiently large so that the effective charge at this scale is already small enough [see Eq. (5.20) below] and perturbation theory applies. We see that, as the ultraviolet cutoff Λ grows, the "bare charge" g_0^2 decreases. There is no pole and no problem. Well, both of them *could* emerge if the number of different quark species were too large, $N_f > 15$, but, according to Table 3.1, there are six quark flavours, the coefficient b is positive and asymptotic freedom holds.

Equation (5.17) can be rewritten as

$$g_0^2 = \frac{8\pi^2}{b \ln \frac{\Lambda}{\Lambda_{\text{QCD}}}}, \tag{5.18}$$

where

---

[16]"hep-th/9808154" is the number of't Hooft's preprint in the SLAC data base, but never mind. Just feed it to Google!

[17]This is how the modern theory of the strong interactions (with the abbreviation *QCD*) is called. "Chromodynamics" means the dynamics of colour.

$$\Lambda_{\text{QCD}} = \Lambda \exp\left\{-\frac{8\pi^2}{bg_0^2}\right\}. \tag{5.19}$$

The parameter Λ_{QCD} should not be confused with the unphysical ultraviolet cutoff Λ. It is very physical, defining the scale μ where the effective charge becomes large. This scale is estimated at the level of $\Lambda_{\text{QCD}} \approx 200$ MeV.[18]

A brief inspection of (5.19) tells us that one can easily get rid of the ultraviolet cutoff Λ by sending it to infinity, while simultaneously sending the bare charge g_0 to zero, such that the value of Λ_{QCD} stays fixed. This remarkable phenomenon is called *dimensional transmutation*. In the original formulation, the theory did not involve any dimensionful parameter. (This is so in the limit when the quark masses vanish; in reality they do not, but, for the present discussion, this fact is irrelevant.) We had instead a dimensionless coupling constant, an analog of the electric charge. But we have seen that, after loops have been calculated, this effective charge acquires an energy dependence. And Λ_{QCD} determines the scale where the effective charge becomes large. It *is* the true fundamental constant of QCD!

The reader is invited at this point to recall the discussion in Chap. 3. We stated there that the fundamental strong interaction constant was the proton mass. Now we understand things a little better. It is more reasonable to say that the fundamental constant is Λ_{QCD} rather than m_p. But these two quantities are of the same order and, for the crude order-of-magnitude estimates that we made in Chap. 3, this difference is irrelevant.

Once the ultraviolet cutoff Λ and the bare coupling g_0^2 are excluded, the only physically relevant question is how the effective strength of the interaction depends on energy. In the one-loop approximation, it is easy to derive that

$$\alpha_s(\mu) = \frac{g^2(\mu)}{4\pi} = \frac{2\pi}{b \ln\frac{\mu}{\Lambda_{QCD}}}. \tag{5.20}$$

As was announced, $\alpha_s(\mu)$ (the strong counterpart of the fine structure constant α) decreases with the increase of μ, which allows one to perform perturbative calculations at high energies and compare them with experiment.

Since 1974 many other experimental tests of QCD have been carried out, and there is not the slightest doubt any more that QCD is correct. Probably the most spectacular confirmation comes from jet physics. Consider the process of e^+e^- annihilation with the creation of a lot of hadrons in the final state. If the energy of colliding electron and positron is large, the process goes, according to the theory, in two stages. In the first stage the fundamental particles, the quarks and gluons are produced. This fundamental process can be $e^+e^- \to q\bar{q}$, $e^+e^- \to q\bar{q}g$, $e^+e^- \to q\bar{q}q\bar{q}$ and so on. The cross section of these processes can be calculated within the framework of QCD perturbation theory.

---

[18] One should not forget, however, that formulas like (5.17), (5.19) are approximate. Performing the accurate 2–loop and 3–loop computations, one can write improved formulas. These improvements affect, of course, the definition of Λ_{QCD} and its numerical estimate.

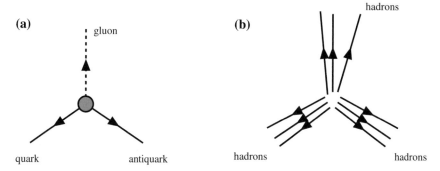

Fig. 5.10   Production of 3 jets in e^+e^- annihilation. (**a**) quark and gluon stage, (**b**) hadron stage.

But since quarks and gluons do not exist in free form, we observe only hadrons in the final state. Thus, in the second stage each energetic quark or gluon creates a *jet* of hadrons going roughly in the same direction as the parent coloured particles.

Such jet processes are indeed seen. The most common is the process when only 2 jets are created.[19] The underlying fundamental process is $e^+e^- \to q\bar{q}$. The 3-jet process, depicted in Fig. 5.10 and corresponding to the QCD process $e^+e^- \to q\bar{q}g$ is less probable because the cross section involves an extra power of the strong coupling constant (5.20), which is small at large energies. An event where four jets are created (corresponding to the QCD processes $e^+e^- \to q\bar{q}gg$ and $e^+e^- \to q\bar{q}q\bar{q}$) is still less probable and so on.

Experimentalists can measure the jet cross sections — the probabilities of producing a given number of jets with given energies (the energy of a jet is the sum of the energies of the jet particles) and forming given angles. There is quantitative agreement between such measurements and the theoretical QCD results for the differential cross sections of the corresponding underlying processes. For example, the theoretical differential cross section of the process $e^+e^- \to q\bar{q}$ is given by the same expression (5.7) as for the muon pair production, but involves the extra factor $N_c Z_q^2$ ($Z_q = 2/3$ or $Z_q = -1/3$ being the quark charge and $N_c = 3$ being the number of colours). To obtain the theoretical prediction for the process $e^+e^- \to 2$ jets, one has to sum over all quark flavours sufficiently light for creation of quark pairs to be kinematically possible. And it coincides with what is measured in experiment, including a nontrivial angular distribution $\sim 1 + \cos^2 \theta$ in the angle between the jet axis and the direction of the colliding electrons and positrons.

---

[19]This was first observed in 1975 at the Stanford e^+e^- collider when its total centre-of-mass energy achieved the value $E \sim 3$ GeV (for smaller energies, there are no jets and the angular distribution of the outgoing hadrons is roughly isotropic).

Fig. 5.11 Fermi 4-fermion
interaction.

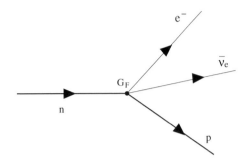

5.4 From β Decay to the Higgs Boson

We now go back a hundred years to 1914. At that time, the energy of electrons in
different β decays was measured. James Chadwick expected to observe a particular
energy value for each decay — a spectral line. And he was very surprised not to find
it. The spectrum (energy distribution) was continuous — the electron could have
any energy in some interval. This was very bizarre, it meant that the energy was
not conserved. Niels Bohr even tried (in 1924) to explain this nonconservation by
evoking quantum mechanical spirits — he suggested that the conservation holds only
in a statistical sense.

But Pauli had a different opinion. In 1930,[20] Pauli suggested that the energy was
actually conserved and that the observed nonconservation was due to the creation of
an extra light neutral particle. Pauli called it the "neutron", but later Fermi rechris-
tened it to *neutrino* ("small neutron") to avoid confusion with the particle discovered
by Chadwick in 1932 that is now called the neutron. Pauli also suggested that the
neutrino interacts rather weakly with matter and that it escaped detection for that
reason.

The quantitative theory of β decay was built by Fermi in 1934. In modern language, Fermi's calculation amounted to evaluating the diagram in Fig. 5.11. The analytical expression for the corresponding amplitude will be written later (in Chap. 12).
Describing it in words, Fermi constructed out of the electron and (anti)neutrino fields
a vector combination — an expression that carried a vector index and had the same
symmetry properties as the electromagnetic potential $A_\mu(x)$. And he suggested that
e^- and $\bar{\nu}_e$ interact with the nucleons not separately, but via this combination, which
replaces $A_\mu(x)$ in the expression (5.3).

Fermi calculated the spectrum of the emitted electrons and found an excellent
agreement with experimental data, already reasonably precise at that time. The emission probability is proportional to the square of the amplitude and hence the square
G_F^2 of the interaction constant. The latter carries dimension in this case, the same

---

[20]Well, by that time quantum mechanics was already firmly established and understood. It became
clear that the energy could be not conserved only for a brief moment, $\tau \sim \hbar/(\Delta E)$, and this effect
is irrelevant for measurements of the energies of asymptotic scattering states.

dimension $\sim m^{-2}$ as Newton's gravitational constant (2.6). Fermi's constant is, however, much larger:

$$G_F \approx 1.2 \cdot 10^{-5}\,\text{GeV}^{-2}. \tag{5.21}$$

The same diagram describes the crossed channels, the scattering processes $\nu_e n \rightarrow e^- p$ and $\bar{\nu}_e p \rightarrow e^+ n$ [cf. Eq. (3.14)]. For the former process, it is sufficient to trade the outgoing antineutrino line in Fig. 5.11 for the ingoing neutrino line. For the scattering $\bar{\nu}_e p \rightarrow e^+ n$, one should trade the outgoing electron for the ingoing positron and then exchange the initial and the final states. For neutrino energies smaller than the nucleon mass, the total cross section is estimated as

$$\sigma_{\nu N}^{\text{tot}} \sim G_F^2 E_\nu^2. \tag{5.22}$$

Pauli was right, it is very small. When $E_\nu \sim 1\,\text{MeV}$ (a characteristic energy of the neutrino produced in β decays, in nuclear reactors and in the solar core), the mean free path of neutrinos in tungsten is of order of 100 light years. No wonder they were not detected in first experiments.

Neutrino scattering was first observed directly only in 1956 by Clyde Cowan and Frederick Reines. No, the size of their installation was a little smaller than 100 light years. But they used a nuclear reactor as the neutrino source so that the neutrino flux they dealt with was very large $\sim 5 \cdot 10^{13}$ neutrinos per second. And the count that Cowan and Reines observed was at the level of 3 neutrino interactions per hour.

By 1956 physicists already knew that the Fermi theory was not quite exact because it did not include the phenomenon of *parity nonconservation* discovered in the mid-fifties. In physics, parity does not mean the same as in number theory. It is a certain characteristic (quantum number) of a quantum state indicating whether the wave function of this state stays the same after a reflection in a mirror (then the parity is said to be positive) or changes sign (then the parity is negative). In the world where the laws of physics are invariant under reflection, a symmetric state stays symmetric forever; parity is conserved. Parity nonconservation means that the laws are not invariant.

It is easy to give an example of an *object* that is not invariant under reflection. Sophie, the heroine of our fourth chapter, would probably think at this point of her thin velvet gloves, but somehow the first thing which comes to my own mind is a corkscrew. Its profile is helical. Thus, one can call (and calls!) the property of being not symmetric under mirror reflection *helicity*. For a microparticle, the helicity is defined as the projection of the spin of the particle, its intrinsic angular momentum, along the direction of its motion:

$$h = \frac{\boldsymbol{s} \cdot \boldsymbol{p}}{|\boldsymbol{p}|}. \tag{5.23}$$

It was shown in different experiments that neutrinos are almost perfect left-handed helices, the direction of their spin being almost exactly opposite to the direction of their momentum, $h_\nu \approx -1/2$. The electrons created in β decays also carry negative helicity, though not the maximal possible one. On the other hand, antineutrinos are almost perfect right-handed helices.

The definite helicity of corkscrews is related to the helicity of humans who use them. Indeed, most of us are right-handers, and it is more convenient for us to turn a corkscrew clockwise, rather than counterclockwise. We are helical also at deeper molecular level: the sugar molecules, proteins and DNA can all exist in two helical mirror-symmetric forms, but only one of these forms is present in living organisms. Such asymmetry is probably a matter of chance. Our sugars may inherit the helicity of the first ever sugar molecule formed in the primordial ocean. One can well imagine a world where the sugars and/or the people are left-handed. The latter also may be or may be not right-hearted.

But the neutrinos are another story. It is the neutrinos and not antineutrinos that are produced in the interior of the Sun and other stars, and their helicity is negative, not positive. That means that our world is helical, left-right asymmetric at the intrinsic, fundamental level!

A natural question arises. Does the reflection of a left-handed neutrino give exactly a right-handed antineutrino? Is it true that our world is still invariant under the combination of two operations: *(i) parity symmetry* or symmetry under reflection in a mirror and *(ii) charge symmetry* or symmetry under replacing all particles by antiparticles and vice versa? This is the hypothesis of *CP-symmetry* put forward by Landau. It is *almost* correct, but not quite, due to a rather subtle effect related to the existence of three quark and lepton generations. We will not discuss it here.

I would like to make a digression now and try to pose a rather provocative question. The strong and electromagnetic interactions are left-right symmetric, only the weak interaction is not. Thus, asymmetry is small and seems to be not relevant in our everyday life. Why then are people interested in it and study it?

There is, of course, a standard (and quite correct) answer. Not everything that people do has a rational explanation. Why do people play music? Why do they pray to God? Why did they try to prove Fermat's theorem and Poincaré's conjecture and are now trying to unravel the conceptual foundations of string theory? The aspirations of the human spirit are versatile and have no bounds. One should be proud of that and not seek immediate practical gains. Let us agree on that.

Still, one can note that these general wisdoms do not completely apply to the particular problem we are discussing now. The effect of parity nonconservation turned out to be relevant for quite mundane technological purposes. I am talking about the μSR (*muon spin rotation*) method. Consider a proton accelerator. Not necessarily a large accelerator, not the Large Hadron Collider at CERN, but an ordinary accelerator with the energy of the proton beam at the level of ~ 1 GeV. Proton-nuclear collisions at such energies are inelastic and a lot of extra particles are created, in the first case — π mesons. Charged π mesons decay into muons and neutrinos (this is the dominant decay channel) by the weak interaction.

The muons produced in these decays are polarized: the μ^- have negative helicity and the μ^+ have positive helicity. The muon lifetime $\tau \approx 2.2 \cdot 10^{-6}$ s is very large by nuclear standards. If a muon with an energy of several hundred MeV is injected into the medium, it has plenty of time to give away its momentum and to stop. Up to that point it still keeps its polarization. Thereafter the fate of the negative and positive muons is different.

Negative muons, being heavy relatives of electrons, are captured by the nuclei; after emitting several photons, they occupy the ground state of the hydrogen-like ($\mu^- A$) atom, where they form a close bond with the nucleus and are unaware of and uninterested in whatever may happen around them.

The life of positive muons is more interesting. They behave roughly like protons: they may just hang between atoms, may form with them loose chemical bonds, or they may capture an electron, forming *muonium*, a hydrogen-like atom where the μ^+ plays the role of the proton. Positive muons are affected by the electric and magnetic fields acting at the microscopic atomic scale. The magnetic field rotates the muon polarization. Finally, the μ^+ decays into a positron and a pair of neutrinos, as in (3.13). Its nonzero polarization together with parity nonconservation leads to an angular asymmetry — most of the positrons go along the spin direction of the decaying muon.

By studying the angular distributions of positrons produced in the decays of positive muons injected in matter, one can therefore extract information about their polarization and hence about the microscopic magnetic fields that acted on the μ^+ between the moment it stopped and the moment it decayed. This phenomenon gives us a tool to study the microscopic structure of matter. It allows one to extract nontrivial information on the structure of ferromagnets, superconductors and so on. My father Voldemar Smilga was among the pioneers who ~50 years ago understood the importance of this method and performed the first theoretical calculations in that framework.

We now go back to the "aspirations of human spirit" in fundamental physics — the main subject of our book. More precisely, to the modifications of the Fermi theory brought about by parity nonconservation. As was mentioned before, in its original variant the Fermi theory involved a certain vector combination of the electron and neutrino fields. Physicists call this combination a *charged vector lepton current*. But one can also build up a combination which is not an ordinary polar vector, but an axial vector.[21] It was found in 1957 (in two independent papers — by Robert Marshak and George Sudarshan and by Feynman and Gell-Mann) that the pure vector current V_μ should be replaced by the combination $V_\mu - A_\mu$, the difference of the vector and axial currents. And this concerns not only the lepton current, but also the hadron current made of the proton and neutron fields (or rather, as we know now, of quark fields).

In contrast to the original Fermi theory, the modified V–A theory explained the polarizations and the angular asymmetries discussed above, whereas its predictions for the spectra of the emitted electrons did not essentially differ from the original Fermi predictions.

We have to say now that both the original Fermi theory and the modified Fermi theory suffer from a serious illness — they are not renormalizable. This means that the higher-loop corrections are plagued there by incurable *power* ultraviolet divergences.

---

[21] Axial vectors gain an additional flip of direction under reflection, compared to polar vectors. For example, the angular momentum is an axial vector (see Fig. 5.12).

Fig. 5.12 Corkscrews,
neutrinos and the
looking-glass.

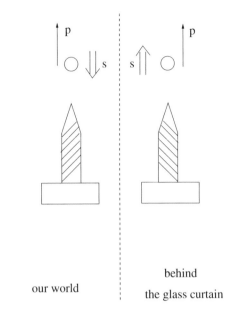

our world

behind

the glass curtain

Fig. 5.13 A divergent
1-loop graph in Fermi theory.

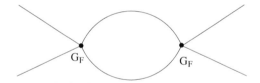

For example, the one-loop diagram drawn in Fig. 5.13 involves an extra dimensionless
factor $\sim G_F \Lambda^2$, compared to the diagram in Fig. 5.11. Two-loop graphs involve an
extra power of G_F multiplied by the extra factor Λ^2. This gives a quartic power
ultraviolet divergence. And so on. With power ultraviolet divergences, one cannot
carry out the renormalization programme described earlier for QED and QCD. This
means that higher-order corrections are incalculable.[22]

A related problem is associated with the power growth of cross sections, as in
(5.22). It is fine while the energy is small. But the power growth cannot go on
indefinitely, much beyond the energy

$$E \sim G_F^{-1/2}. \tag{5.24}$$

Such growth would contradict *unitarity*, a fundamental property of any sensible
theory that the total probability of all possible outcomes in a scattering process is

---

[22] See Sect. 16.1.2 for a detailed discussion of this point.

Fig. 5.14 νe scattering:
(a) in the Fermi theory. (b)
with W exchange.

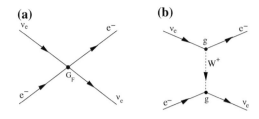

equal to 1. One can say, of course, that the tree-level formula does not work at the scale (5.24), that the loop corrections become important there. But as the corrections are incalculable, we are in a deadlock anyway.

People suggested the following remedy to this illness. Imagine that the Fermi theory is not fundamental, but *effective*. And that the fundamental theory of the weak interaction involves the interaction of the lepton and quark currents with some charged vector boson field W_μ. It is the same kind of interaction as in QED and QCD, but the difference is that the particle W_μ (it was dubbed the "intermediate boson") is massive.

Consider the process of elastic scattering $\nu_e e^- \to \nu_e e^-$ and compare the two graphs depicted in Fig. 5.14. The left-hand graph, the Fermi theory graph, gives the relativistic scattering amplitude $M \sim G_F E^2$, where E is the centre-of-mass energy of the scattered particles. In the right-hand graph, one vertex is split up into two. Each of the new vertices involves a dimensionless factor g of the same nature as the electron charge. There is also a line of the virtual W-boson. It is called a *propagator* — this notion will be explained in Chaps. 8 and 10. For the time being, we just notice the resemblance of the graph in Fig. 5.14b to the graph in Fig. 5.2, with the massive W now playing the role of the massless photon. As was already mentioned [see Eq. (5.15) and discussion thereabout], such a graph with massive particle exchange involves the factor

$$\frac{g^2}{m_W^2 - q^2}$$

(q_μ being the 4-momentum transfer). We see that, for small q^2, this factor is a constant. This constant is naturally associated with G_F,

$$\frac{g^2}{m_W^2} \to G_F. \tag{5.25}$$

In other words, for small energies the amplitude described by the graph in Fig. 5.14b is indistinguishable from the Fermi theory graph in Fig. 5.14a. Fermi theory with its dimensionful constant can hence well be (and we know now that it *is*) an effective low-energy approximation to a completely different theory involving a dimensionless coupling g and extra degrees of freedom associated with the W field.

We have already made brief remarks in this book about effective theories of different kinds. For example, when together with Sophie we visited the theoretical physics animation and went to the hall dedicated to nonrelativistic field theories, we

Fig. 5.15 Photon-photon
scattering. (**a**) QED diagram;
(**b**) effective vertex.

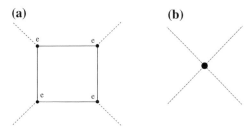

mentioned that these are effective theories describing ordinary matter at distances
much larger than interatomic distances.

It is proper to discuss now the notion of an effective theory with more attention.
Actually, this is one of the basic, most important notions in physics. Before solving
any physical problem, we should construct a theoretical model for it. To this end,
we have to determine which variables and parameters are essential and which can
be safely neglected and forgotten. For example, in celestial mechanics, we neglect
the finite size of the Sun and planets and consider them as pointlike objects. When
determining the trajectory of a stone thrown by hand, we also consider it pointlike
and usually neglect air resistance as well. And so forth.

In quantum mechanics, the first example of a low-energy effective Hamiltonian
was worked out by Max Born and Julius Oppenheimer. They considered the hydrogen
molecule. This is a system made of four particles, and the exact quantum problem is
complicated. Born and Oppenheimer noted that, if we are interested only in the low-
energy dynamics of the system, the problem can be very essentially simplified. To
begin with, let us fix the positions of the heavy protons (mentally *nail* them to some
particular spatial points). Then we solve the Schrödinger equation for two electrons
moving in the static field generated by the protons.[23] The ground state energy will
depend on the distance R between the protons. The function $E(R)$ thus obtained
may be renamed $V(R)$ and serve as an *effective potential*. Adding the proton kinetic
energies, we obtain the effective Born–Oppenheimer Hamiltonian

$$H_{\text{eff}} = \frac{P^2}{m_p} + V(R), \tag{5.26}$$

where P is the momentum of one of the protons in the system of their centre of mass.

There is no trace of electrons any more. Using the theorists' slang, the electrons
were *integrated out*.

Let me give here one more example, this time a field theory one. One of the
predictions of quantum electrodynamics is a nonzero amplitude for *photon-photon
scattering*. It is described by the "box" graph in Fig. 5.15a. It turns out that, in the
case when the frequency of the photons is much less than the electron mass (which

---

[23]This problem is not exactly solvable, but it is much simpler than the original one and can be
solved using some approximative methods.

means that their wavelength is much larger than the electron Compton wavelength), one can integrate out the electron degrees of freedom and describe the scattering by an effective 4-photon vertex, as in Fig. 5.15b. The structure of this effective vertex was theoretically determined by Werner Heisenberg and Hans Euler in 1936. Experimentally, such scattering has not been observed yet, since the amplitude is very small.

By the same token that the effective $\gamma\gamma$ scattering vertex in Fig. 5.15b has no trace of electrons, the effective fermion-fermion scattering vertex in Fermi theory has no trace of W bosons. They are integrated out!

As the coupling constant in the new theory including W bosons is dimensionless, one can hope that the theory turns out to be renormalizable. Another indication that the new theory is better than the old one is the energy dependence of cross sections. The cross section of $\nu_e e$ scattering calculated by the diagram in Fig. 5.14b first grows with energy as in (5.22), but at the scale $E \sim m_W$ [which, for small g that we assume, and bearing in mind (5.25), is still much smaller than the unitary limit (5.24)], it levels off and does not grow further.

Unfortunately, life is sometimes harder than it seems. Theories with massless vector mediating fields, like QED or QCD, are, indeed, renormalizable and benign. But there is no simple way to make these vector fields massive, while keeping the theory renormalizable. The mathematical underlying reason for this is that giving mass to vector fields in an immediate simple-minded way destroys a remarkable symmetry that a theory with massless fields enjoys — the *gauge invariance*.[24]

I have said that there is no *simple* way to make W boson massive. Still, there *is* a way. The main idea is that the mass should be *generated* due to a specially chosen interaction that respects gauge invariance. The effect of dynamical mass generation is not something very bizarre, inconceivable and new. Very similar things are known to occur in mundane low-energy physics.

The photon is massless, and the dispersive law of the electromagnetic waves in vacuum is $\omega^2 - c^2k^2 = 0$. The same electromagnetic field in water or in glass has a different velocity, and the corresponding dispersive equation is $n^2\omega^2 - c^2k^2 = 0$, n being the refractive index. In a *plasma* the dispersive equation for the oscillations is more involved. It reads

$$\omega^2 - c^2k^2 = \Pi(\omega, k), \tag{5.27}$$

where $\Pi(\omega, k)$ is a rather complicated function. For very large ω, k, the right-hand side of (5.27) can be neglected, and we obtain ordinary electromagnetic waves. But when $\omega \sim ck \sim \Pi^{1/2}$, the dispersive law becomes complicated. The remarkable fact is that $\Pi(\omega, 0) \neq 0$. This means that a *stationary* oscillatory solution exists. The frequency of these oscillations, the solution to the equation $\omega^2 = \Pi(\omega, 0)$ is called the *plasma frequency*. For an ordinary plasma,

---

[24]The reader who studied Maxwell's theory at university is probably already familiar with the notion of *Abelian* gauge invariance. We have chosen not to discuss it in detail here, but we will recall what it is in Chap. 7 and generalize it to the non-Abelian case in Chap. 11.

$$\omega_{\text{pl}}^2 = \frac{ne^2}{m}, \tag{5.28}$$

where n is the density of electrons in plasma, m is the electron mass and the electron charge e is measured in the Heaviside units. In quantum theory, ω_{pl} can be interpreted as the dynamically generated photon mass.

Another physical situation where photons acquire mass is more relevant for our weak interaction studies. I am talking about the *Meissner effect* in superconductors. It is known that magnetic fields cannot penetrate superconductors of a certain type;[25] they are screened by induced circular currents. There is, however, a finite length (*London length*) at which the magnetic field can still penetrate. This finite screening length can be associated with the inverse nonzero photon mass!

In more quantitative terms, superconductors are described by the *Ginzburg–Landau effective theory*. A relativistic variant of this theory is called the Abelian *Higgs model*, and dynamical generation of the photon mass in this model — the Abelian *Higgs effect*.

In this *Hors d'Oeuvres* part of the book, we stay at a semi-philological level and will not now write down the Lagrangian of this model, which would allow the reader to understand exactly what the Higgs effect means and how it works. But our reader is not yet supposed to read Chap. 7 and to learn what is the Lagrangian in field theory. We will enjoy digesting a detailed description of Abelian and non-Abelian Higgs models only in Chap. 12, one of our *Entrées*. We will still try now to describe some salient features of these models in words.

- Of all the salient features of the non-Abelian Higgs models mentioned in the previous paragraph, the most salient is, definitely, the non-Abelian Higgs mechanism that gives masses to the non-Abelian gauge fields ("charged photons" of Yang and Mills). The theory of the weak and electromagnetic interactions, constructed by Weinberg in 1967 and by Salam in 1968 is based on this.
  "But why 'weak and electromagnetic' interactions? What does electromagnetism have to do with the weak interaction we are discussing now?" the irritated reader may ask, "The theory of the electromagnetic interaction was constructed back in the forties. We have already learned about it in Sect. 5.2. Why do we need to go back to it?"
  The answer is the following. W^+ and W^- are charged particles, so they interact with photons. Thus, the photons should be treated right from the beginning roughly on an equal footing with $W^\pm$ — otherwise one would not be able to keep gauge invariance and renormalizability. And that is why the theory we are talking about now is not just the theory of the weak interaction, but the theory of the *unified weak and electromagnetic interaction*. In 1979 Glashow, Salam and Weinberg got the Nobel Prize for their contributions to this theory.[26]

---

[25]Type I if you want me to be more precise.

[26]Sheldon Glashow published in 1961 a paper that involved remarkable insights about the symmetry structure of the future theory. In particular, he predicted the existence of two neutral vector bosons, the massless photon and the massive Z.

- It turns out that, besides the massive $W^{\pm}$ and the massless photon, the electroweak theory involves another neutral massive mediating boson Z^0. The masses of $W^{\pm}$ and Z^0 are of the same order. The theory predicts the values

$$m_W = \frac{1}{\sin\theta_W}\sqrt{\frac{\pi\alpha}{G_F\sqrt{2}}} \approx \frac{37.3}{\sin\theta_W}\, \text{GeV},$$

$$m_Z = \frac{m_W}{\cos\theta_W} \approx \frac{74.6}{\sin 2\theta_W}\, \text{GeV}, \qquad (5.29)$$

where θ_W is a free theoretical parameter called the *Weinberg angle*. We will derive these exact relations in Chap. 12.

An order-of-magnitude estimate for the W, Z masses follows from the approximate relation (5.25) [the exact one will be written in (12.48)] and the fact that we are dealing with the unified weak and electromagnetic theory so that the constant g is of the same order as the electron charge e.

The W and Z bosons were discovered at CERN in 1983. The experimental values for the masses,

$$m_W = 80.385 \pm 0.015\,\text{GeV}, \qquad m_Z = 91.187 \pm 0.002\,\text{GeV}, \qquad (5.30)$$

are in perfect agreement with (5.29) and allow one to determine the Weinberg angle precisely:

$$\sin^2\theta_W = 0.23120 \pm 0.00015. \qquad (5.31)$$

The presence of Z^0 brings about so-called *neutral current* processes. One of them, the process of inelastic neutrino-proton scattering, is depicted in Fig. 5.16. The Fermi theory, which deals only with the interaction of charged currents has no place for this process. The latter was experimentally observed at CERN in 1973, which was a brilliant experimental confirmation of the *new* electroweak theory.[27] Z interacts also with the electrons (and with other leptons and quarks). Processes with Z exchange are "admixed" to the electromagnetic processes, and this brings about parity nonconservation effects not only in high-energy, but also in low-energy atomic physics. In almost all cases, these effects are too tiny to be observed, but there are exceptions. Parity nonconservation in atomic transitions was first observed in Novosibirsk by Lev Barkov and Mark Zolotarev in 1978.

- The Higgs mechanism of mass generation implies the presence of fundamental scalar fields. In a superconductor, this scalar field is called the *order parameter* and is associated with the condensate of Cooper pairs. The electroweak theory predicts the existence of a heavy scalar particle called the *Higgs boson* (after Peter Higgs,

---

[27]The experimental discovery of W and Z bosons was also, of course, a brilliant confirmation, but it came later, in 1983, when the direct production of these bosons was observed at the new proton-antiproton collider at CERN. One can say that most experts believed in the validity of the Weinberg and Salam scheme since 1973, while the discovery of W and Z in 1983 eliminated the last doubts about it.

Fig. 5.16 A neutral current
process: inelastic νp
scattering.

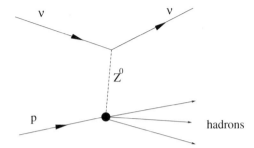

who among others constructed the relativistic model with a dynamical generation
of the photon mass back in 1960).

Note that the masses of the fermions — quarks and leptons — are also not introduced in the theory by hand (they cannot, that would break gauge invariance and
renormalizability), but are generated due to the interaction with the scalar fields.
As the reader has probably already heard, the Higgs boson with mass ≈ 125 GeV
was discovered at CERN five years ago. This was another predictable triumph of
the Standard Model. Well, probably the words "another triumph" are not quite
proper here. The Higgs boson was a kind of keystone in the central vault in the
Standard Model building. Its discovery marked, if you will, the end of an epoch.
Most of what was predicted has now been found and we do not expect to have a
lot of other predictable discoveries in future.[28]

We all hope to live to see unpredictable and unpredicted discoveries. But they will
be the subject of some other book...

---

[28]I have written "most" and not "all" because not all parameters of the Standard Model (they will be
discussed in more detail in Chap. 12) are known precisely yet. This applies in particular to neutrino
masses.

Part III
Chef's Secrets

Chapter 6
Groups and Algebras

In the previous part, we discussed different field theories, simple and complicated, but we never actually *defined* what we were talking about. Any mechanical or field theory system is characterized by its Lagrangian. Once the Lagrangian is written, everything else follows. To write such Lagrangians, we need to recall the methods of analytical mechanics; this will be done in Chap. 7. But we also need to understand the guiding principles allowing one to write different candidate Lagrangians and choose between them — which Lagrangian describes Nature and which does not. And the most basic of these principles is the *symmetry* principle.

The appropriate mathematical language to deal with symmetries is provided by group theory. In this chapter, we introduce and explain it. This chapter also includes a small section devoted to *Grassmann algebra* — a necessary mathematical tool for dealing with the fermion fields.

6.1 Scalars □ Vectors □ Tensors

But before addressing new mathematical ideas and methods, we recall the basic notions of vector and tensor algebra, with which our reader is probably already familiar.

6.1.1 Euclidean Space

Consider n-dimensional Euclidean space. Different Cartesian coordinate systems can be chosen there. They are all related by orthogonal transformations

© Springer International Publishing AG 2017
A. Smilga, *Digestible Quantum Field Theory*,
DOI 10.1007/978-3-319-59922-9_6

$$x_i' = \sum_{j=1}^{n} O_{ij} x_j . \tag{6.1}$$

An orthogonal matrix O satisfies the condition

$$O O^T = \mathbb{1} \tag{6.2}$$

(the superscript T means transposition and $\mathbb{1}$ is the unit matrix). It follows from (6.2) that det $O = \pm 1$. If det $O = 1$, we are dealing with a proper rotation. If det $O = -1$, it is an improper rotation — a proper rotation supplemented by a mirror reflection.

There are many interesting things that do not depend on the coordinate choice: geometrical quantities like distances and angles and physical quantities like mass and electric charge. Such quantities are called *scalars*. There are also *pseudoscalars* [like the helicity (5.23)]. The latter are invariant under proper rotations, but change sign after reflection.

There are many vector physical quantities. A vector is not just a set of its components. When going to a different coordinate system, the components of a vector transform in the same way as the coordinates:

$$V_i' = \sum_{j=1}^{n} O_{ij} V_j . \tag{6.3}$$

More exactly, the law (6.3) holds for ordinary polar vectors (like velocity). For axial vectors (like angular momentum or the magnetic field), one should rather write

$$A_i' = (\det O) \sum_{j=1}^{n} O_{ij} A_j ; \tag{6.4}$$

their transformation law involves an extra sign change for improper rotations.

There are also tensors, objects carrying several indices. In physics, the best known example is the tensor of moment of inertia — a symmetric tensor of rank 2, $I_{ij} = I_{ji}$. When the coordinate system is changed, it transforms as

$$I_{ij}' = \sum_{k,l=1}^{n} O_{ik} O_{jl} I_{kl} . \tag{6.5}$$

Another example is the stiffness tensor of rank 4 used in elasticity theory.

The law of transformation of a proper tensor of rank r is

$$I_{i_1 \cdots i_r}' = \sum_{k_1, \cdots, k_r = 1}^{n} O_{i_1 k_1} \cdots O_{i_r k_r} I_{k_1 \cdots k_r} . \tag{6.6}$$

The laws of transformation written above involve the summation sign. But this is the last time we use it to denote the summation over indices. To indulge our sloth and to economise on paper, we use the *Einstein convention* throughout the book: whenever the index name is repeated twice in a product, the summation over this index is tacitly assumed. Equation (6.3) then acquires the form $V_i' = O_{ij}V_j$.

One can add tensors of the same rank and multiply tensors. In the latter case, the product can keep all the indices of the factors or lose some of them due to contraction. For example, taking a tensor A_{ijk} of rank 3 and a tensor B_{mnlp} of rank 4, one can define

$$C_{imn} = A_{ijk}B_{mnjk}. \tag{6.7}$$

It is a simple exercise to show that, if A and B are tensors (i.e. they transform as in (6.6) under rotations), C is also a tensor. While doing that, one should use the orthogonality condition (6.2).

There are two distinguished tensors:

1. The *Kronecker delta* tensor, whose only nonzero components are $\delta_{11} = \cdots = \delta_{nn} = 1$.
2. The antisymmetric tensor $\epsilon_{i_1 \cdots i_n}$ defined by the requirement of antisymmetry (it changes sign under permutation of any pair of indices) and the normalization $\epsilon_{12 \cdots n} = 1$.

It is again a simple exercise to show that these tensors are *invariant*, do not change their form under rotations. Note the useful 3-dimensional relations

$$\epsilon_{ijk}\epsilon_{mnp} = \begin{vmatrix} \delta_{im} & \delta_{in} & \delta_{ip} \\ \delta_{jm} & \delta_{jn} & \delta_{jp} \\ \delta_{km} & \delta_{kn} & \delta_{kp} \end{vmatrix},$$

$$\epsilon_{ijk}\epsilon_{mnk} = \delta_{im}\delta_{jn} - \delta_{in}\delta_{jm},$$

$$\epsilon_{ijk}\epsilon_{mjk} = 2\delta_{im},$$

$$\epsilon_{ijk}\epsilon_{ijk} = 6. \tag{6.8}$$

In this notation, the familiar scalar and vector products are expressed as $A \cdot B = A_i B_i$, $[A \times B]_i = \epsilon_{ijk}A_j B_k$. As an exercise, prove the relation

$$A \times (B \times C) = B(A \cdot C) - C(A \cdot B).$$

6.1.2 Minkowski Space

We are now in Minkowski (3+1)-dimensional space. Different inertial reference frames are related by Lorentz transformations,

$$x'^{\mu} = O^{\mu}_{\ \nu}x^{\nu}, \tag{6.9}$$

where $x^\mu = (t, \mathbf{x})$ and $O^\mu_{\ \nu}$ is the matrix of a Lorentz transformation satisfying the condition

$$\eta_{\mu\rho} O^\mu_{\ \nu} O^\rho_{\ \sigma} = \eta_{\nu\sigma}, \tag{6.10}$$

where $\eta_{\mu\nu} = \mathrm{diag}(1, -1, -1, -1)$ is the Minkowski space metric. You may check that a particular Lorentz transformation (4.3) describing the boost along the x axis with velocity v can be represented as

$$O = \begin{pmatrix} \cosh\psi & -\sinh\psi & 0 & 0 \\ -\sinh\psi & \cosh\psi & 0 & 0 \\ 0 & 0 & 1 & 0 \\ 0 & 0 & 0 & 1 \end{pmatrix} \tag{6.11}$$

with some ψ and indeed satisfies the condition (6.10).

The role of the length of the vector, $l^2 = x_i x_i$, which was invariant under ordinary Euclidean rotations, is now played by the interval

$$s^2 = \eta_{\mu\nu} x^\mu x^\nu = t^2 - \mathbf{x}^2, \tag{6.12}$$

which is invariant under (6.9).

You noticed that the indices are now written at two levels — upstairs and downstairs. This is done to distinguish *contravariant* and *covariant* indices. A contravariant vector [like 4-momentum $p^\mu = (E, \mathbf{p})$] carries an upper index and transforms under Lorentz transformations in the same way as x^μ, $p'^\mu = O^\mu_{\ \nu} p^\nu$. And a covariant vector (like the vector potential A_μ or the gradient operator ∂_μ) transforms as $A'_\mu = O^\nu_{\ \mu} A_\nu$. A generic Minkowski tensor has r contravariant and q covariant indices and transforms as

$$\left(T^{\mu_1 \cdots \mu_r}_{\nu_1 \cdots \nu_q} \right)' = O^{\mu_1}_{\ \alpha_1} \cdots O^{\mu_r}_{\ \alpha_r} O^{\beta_1}_{\ \nu_1} \cdots O^{\beta_q}_{\ \nu_q} T^{\alpha_1 \cdots \alpha_r}_{\beta_1 \cdots \beta_q}. \tag{6.13}$$

With any contravariant vector is associated a scalar invariant ("Minkowski length") $\eta_{\mu\nu} V^\mu V^\nu$.[1] With any covariant vector, one can associate the invariant $\eta^{\mu\nu} V_\mu V_\nu$, where $\eta^{\mu\nu}$ is the inverse Minkowski metric tensor, $\eta^{\mu\nu} \eta_{\nu\alpha} = \delta^\mu_\alpha$. $\eta^{\mu\nu}$ has the same components as $\eta_{\mu\nu}$.

One can observe that, for any contravariant V^μ, the vector $V_\mu = \eta_{\mu\nu} V^\nu$ is covariant. And the other way round, one can raise the index by multiplying a covariant vector by the inverse metric tensor, $V^\mu = \eta^{\mu\nu} V_\nu$. In Euclidean space with $\eta_{\mu\nu} \to \delta_{ik}$, V^i and V_i obviously coincide.

The distinction between the co- and contravariant indices is indispensable for curved space-time in general relativity (see Chap. 15). In flat Minkowski space it is not so important. Still, paying no attention to this distinction may lead to confusions, which we will try to avoid. Equation (5.3) of the previous chapter was written bearing this in mind. And, for example, the KFG equation (4.28) can be expressed in natural units as $(\partial_\mu \partial^\mu + m^2)\phi = 0$.

---

[1] For example, for the 4-momentum p^μ, it is just the mass of the particle, $\eta_{\mu\nu} p^\mu p^\nu = m^2$.

Table 6.1 Multiplication table for the group Z_2.

	e	a
e	e	a
a	a	e

6.2 Finite Groups

A formal definition of a group is the following:

A group is a set of elements G where a binary operation (call it multiplication) is defined. A "binary operation" is in fact a mapping of an ordered pair (a, b) of the elements of G to an element $c \in G$. Call c the product of a and b and denote it

$$c = ab. \tag{6.14}$$

The following defining properties hold:

1. There is a distinguished element $e \in G$, called the *unit* or *identity* element, such that, for any $a \in G$,

$$ae = ea = a. \tag{6.15}$$

2. For any element a, there is an *inverse element* a^{-1}, such that

$$aa^{-1} = a^{-1}a = e \tag{6.16}$$

(if $a = e$, then a and a^{-1} coincide).
3. The multiplication is associative:

$$(ab)c = a(bc). \tag{6.17}$$

The simplest nontrivial group (called Z_2) involves only two elements: the unit element e and the element $a = a^{-1}$. The following multiplication table holds (Table 6.1).

The table is symmetric, which means that the group is commutative.[2] An arithmetic realization of this abstractly defined group consists of the numbers $e = 1$, $a = -1$. But many other interpretations are possible. For example, one can associate the elements of Z_2 with transformations of asymmetric geometric objects like corkscrews and gloves: e is the identity transformation (a right glove stays right and the left glove stays left) and a is the reflection in a mirror (the right and left gloves are interchanged).

---

[2]Commutative groups are also called *Abelian*.

Not all groups are commutative. Consider the group S_3 of permutations of three different objects.[3] This group consists of 6 elements: the identity element e that does not change the order of the objects, which one can denote by (123), and then there are 5 nontrivial elements (132), (213), (321), (231), (312) in an obvious notation — a permutation is characterized by the result of its action on the three numbers set in the standard order (123).

Then $(132)(213) = (231)$, which does not coincide with $(213)(132) = (312)$.

A very important notion is the *subgroup*. A subgroup of G is a subset F of G such that the operation of multiplication defined in G can also be "projected" onto F. In other words: if two elements (a, b) belong to F, their product also belongs to F; the identity element of G also belongs to F; if $a \in F$, it is also true that $a^{-1} \in F$.

For example, the group S_3 has a commutative subgroup consisting only of cyclic permutations: $e = (123)$, $a = (231)$, and $a^{-1} = a^2 = (312)$. This group has the name Z_3.

6.3 Lie Groups

Simple finite groups are simple, but complicated finite groups are very complicated. There are groups way more involved than, for example, the finite group describing the transformations of Rubik's cube (Fig. 6.1).

Fortunately, the mathematical theory of complicated finite groups turned out to be irrelevant for physics.

But there are also groups with an infinite continuum of elements. These groups (called *Lie groups* after Sophus Lie, a Norwegian mathematician who created their theory at the end of the 19th century) *are* relevant. In particular, some of these groups represent groups of symmetry for QCD and for the electroweak theory.

6.3.1 Orthogonal Groups

The simplest example is the group of rotations on the plane. It is called $SO(2)$ and may be represented by the group of orthogonal 2×2 matrices with unit determinant,

$$g_\phi = \begin{pmatrix} \cos\phi & \sin\phi \\ -\sin\phi & \cos\phi \end{pmatrix} . \tag{6.18}$$

The composition of two rotations by angles ϕ and ψ is represented by the product of the matrices

$$g_\phi g_\psi = g_\psi g_\phi = g_{\phi+\psi} . \tag{6.19}$$

---

[3] The objects, as is common in mathematics, may have different natures. One can think, for example, of wolf, sheep and cabbage. Or just of three numbers 1,2,3.

Fig. 6.1 Irrelevant but beautiful.

If we add mirror reflections, we obtain the group called $O(2)$. Both $SO(2)$ and $O(2)$ are commutative (Abelian).

An element of $SO(2)$ is characterized by one parameter $\phi \in [0, 2\pi)$. Topologically, $SO(2)$ represents a circle. And $O(2)$ — two disconnected circles.

The simplest noncommutative Lie group is $SO(3)$, the group of 3-dimensional rotations. Any element of the group can be represented as a superposition of three "elementary" rotations: around the first, the second and the third spatial axes,

$$g = g_1 g_2 g_3 =$$
$$\begin{pmatrix} 1 & 0 & 0 \\ 0 & \cos\phi_1 & \sin\phi_1 \\ 0 & -\sin\phi_1 & \cos\phi_1 \end{pmatrix} \begin{pmatrix} \cos\phi_2 & 0 & -\sin\phi_2 \\ 0 & 1 & 0 \\ \sin\phi_2 & 0 & \cos\phi_2 \end{pmatrix} \begin{pmatrix} \cos\phi_3 & \sin\phi_3 & 0 \\ -\sin\phi_3 & \cos\phi_3 & 0 \\ 0 & 0 & 1 \end{pmatrix}.$$
$$(6.20)$$

An element of the group is thus characterized by three continuous parameters[4] and the group represents a 3-dimensional *manifold*.[5]

---

[4]In physics, a more common choice for such parameters is the Euler angles. And aircraft engineers prefer to describe the dynamics of aircraft flight in terms of *roll, pitch and yaw*.

[5]Mathematicians call by this strange word a smooth enough multidimensional surface. The group $SO(3)$ represents a particular manifold that is topologically equivalent to a 3-dimensional sphere S^3 with opposite points identified. This manifold is compact (its volume is finite). Not all groups are compact. The Lorentz group is not.

Noncommutativity can be seen explicitly. Let us show it for *small* rotations, $\phi^a \ll 1$.[6] A group element (6.20) can then be represented as[7]

$$g \;=\; \mathbb{1} + i\phi^a t^a + o(\phi^a)\,,\tag{6.21}$$

where $\mathbb{1}$ is the unit matrix, and the three Hermitian matrices

$$t^1 = \begin{pmatrix} 0 & 0 & 0 \\ 0 & 0 & -i \\ 0 & i & 0 \end{pmatrix},\quad t^2 = \begin{pmatrix} 0 & 0 & i \\ 0 & 0 & 0 \\ -i & 0 & 0 \end{pmatrix},\quad t^3 = \begin{pmatrix} 0 & -i & 0 \\ i & 0 & 0 \\ 0 & 0 & 0 \end{pmatrix},\tag{6.22}$$

are called the *generators* of the group. The generators are traceless and satisfy the property

$$\mathrm{Tr}\{t^a t^b\} \;=\; 2\delta^{ab}\,.\tag{6.23}$$

Consider the *group commutator*

$$g(\phi)g(\psi)g^{-1}(\phi)g^{-1}(\phi) \;\approx\; \mathbb{1} - \phi^a \psi^b [t^a, t^b]\,.\tag{6.24}$$

The generators do not commute,

$$[t^a, t^b] = i\epsilon^{abc} t^c\,,\tag{6.25}$$

hence the group commutator (6.24) is different from unity, and hence $g(\phi)g(\psi) \neq g(\psi)g(\phi)$. The group is non-Abelian.

Note that the infinitesimal rotations of the coordinates $\boldsymbol{x}$, associated with the action of the generators (6.22), can also be regarded as the result of the action of the operators of angular momentum

$$\hat{j}^a = \epsilon^{abc} x^b \hat{p}^c \;=\; -i\epsilon^{abc} x^b \frac{\partial}{\partial x^c}\,,\tag{6.26}$$

on the functions $f(\boldsymbol{x})$. In fact, the *Lie algebra* (6.25) is nothing but the algebra of the angular momentum operators known from a course on quantum mechanics.

The group of rotations in Euclidean space of arbitrary dimension n is characterized by $n(n-1)/2$ parameters. An "elementary" rotation occurs in the plane $(\mu\nu)$. The generator of such a rotation is the Hermitian matrix with the elements

$$\left(t_{\mu\nu}\right)_{\alpha\beta} \;=\; i(\delta_{\mu\beta}\delta_{\nu\alpha} - \delta_{\mu\alpha}\delta_{\nu\beta})\,.\tag{6.27}$$

The generators (6.27) satisfy the commutation relations

---

[6]A remarkable observation of Lie was that one can learn almost everything about a continuous group [any group, not only $SO(3)$] by looking at what happens in a small neighbourhood of its unit element.

[7]Here the upper position of the indices carries a purely aesthetic and not a mathematical meaning.

$$[t_{\mu\nu}, t_{\alpha\beta}] = i \left(\delta_{\nu\beta} t_{\mu\alpha} + \delta_{\mu\alpha} t_{\nu\beta} - \delta_{\mu\beta} t_{\nu\alpha} - \delta_{\nu\alpha} t_{\mu\beta} \right). \tag{6.28}$$

Generically, for a D-dimensional compact Lie group, there are D generators represented by Hermitian matrices $t^{a=1,\dots,D}$. They satisfy the algebra

$$[t^a, t^b] = i f^{abc} t^c, \tag{6.29}$$

with real totally antisymmetric f^{abc}. The latter objects are called the *structure constants*.

Knowing the generators of the group not only allows one to express a group element in the neighbourhood of unity as in (6.21). One can show that an *arbitrary* group element can be represented as

$$g = \exp\{i\phi^a t^a\}. \tag{6.30}$$

The reader might be perplexed. S/he probably understands how to add or multiply matrices, but might have never seen before a formula with a matrix in the exponent. But there is no mystery in this formula. Any smooth function of matrix argument can be understood as the corresponding Taylor series. In particular,

$$\exp\{i\phi^a t^a\} = \mathbb{1} + i\phi^a t^a + \frac{1}{2}(i\phi^a t^a)^2 + \frac{1}{6}(i\phi^a t^a)^3 + \dots \tag{6.31}$$

When $\phi^a \ll 1$, we can keep only the first two terms in this expansion, obtaining (6.21).

·*Exercise*: Show that the 3×3 matrix (6.30) with the generators (6.22) is orthogonal. (*Hint*: it is sufficient to observe that $g^T = \exp\{-i\phi^a t^a\}$.)

6.3.2 Lorentz Group

Being so important for physics, this group deserves a special subsection.

As was mentioned, the group of Lorentz transformations is isomorphic to the group of real 4×4 matrices satisfying the condition (6.10). This group is akin to the group of 4-dimensional rotations and, bearing in mind that the invariant (6.12) is a quadratic form with one positive and three negative eigenvalues, it is denoted $O(3, 1) \equiv O(1, 3)$. Imposing the condition $\det O = 1$ gives the *restricted* Lorentz group $SO(3, 1)$.

An element of $SO(3, 1)$ in the neighbourhood of unity has the form

$$O = \mathbb{1} + i\boldsymbol{\theta} \cdot \boldsymbol{J} + i\boldsymbol{v} \cdot \boldsymbol{K} + o(\boldsymbol{\theta}, \boldsymbol{v}), \tag{6.32}$$

where $\boldsymbol{J}$ are the generators of the ordinary spatial rotations [they are given by (6.22), where one should add an extra row above and an extra column on the left consisting of zeros], and K are the generators of "hyperbolic" rotations, or Lorentz boosts,

$$K^1 = i \begin{pmatrix} 0 & 1 & 0 & 0 \\ 1 & 0 & 0 & 0 \\ 0 & 0 & 0 & 0 \\ 0 & 0 & 0 & 0 \end{pmatrix}, \quad K^2 = i \begin{pmatrix} 0 & 0 & 1 & 0 \\ 0 & 0 & 0 & 0 \\ 1 & 0 & 0 & 0 \\ 0 & 0 & 0 & 0 \end{pmatrix}, \quad K^3 = i \begin{pmatrix} 0 & 0 & 0 & 1 \\ 0 & 0 & 0 & 0 \\ 0 & 0 & 0 & 0 \\ 1 & 0 & 0 & 0 \end{pmatrix}.$$

(6.33)

We see that K are anti-Hermitian. The presence of both Hermitian and anti-Hermitian generators is related to the alternating signs in the invariant (6.12). Another corollary of the alternating signs is the fact that the manifold describing the Lorentz group is not compact — its volume is infinite.

The six generators J^a, K^a satisfy the algebra

$$[J^a, J^b] = i\epsilon^{abc} J^c, \quad [K^a, K^b] = -i\epsilon^{abc} J^c, \quad [J^a, K^b] = i\epsilon^{abc} K^c. \tag{6.34}$$

We now define the Hermitian matrices

$$M^a = \frac{J^a + iK^a}{2}, \quad N^a = \frac{J^a - iK^a}{2} \tag{6.35}$$

and observe that the algebra (6.34) is rewritten in terms of M and N as

$$[M^a, M^b] = i\epsilon^{abc} M^c, \quad [N^a, N^b] = i\epsilon^{abc} N^c, \quad [M^a, N^b] = 0. \tag{6.36}$$

In other words, the algebra of the Lorentz group is equivalent to two copies of independent $so(3)$ algebras.[8] In mathematical language, the former represents a *direct sum* of the latter.

When expressed in terms of M and N, the group element (6.32) acquires the form

$$O = \mathbb{1} + M^a (i\theta^a + v^a) + N^a (i\theta^a - v^a)] + \ldots \tag{6.37}$$

6.3.3 Unitary Groups

An orthogonal group describes transformations of coordinates in a real vector space leaving the lengths and scalar products intact. Likewise, a unitary group describes the transformations of coordinates z_j in a complex vector space leaving the norm $z^{*j} z_j$ [we defined $z^{*j} = (z_j)^*$] and the scalar products intact. Unitary matrices satisfy the constraint [cf. (6.2)]

$$U U^\dagger = U^\dagger U = \mathbb{1}, \tag{6.38}$$

$U^\dagger$ being the Hermitian conjugated matrix.

---

[8]It is customary to denote the Lie algebras in the same way as the corresponding groups, but with lower-case letters.

For complex spaces of dimension n, the constraint (6.38) represents n^2 real conditions for n^2 complex or $2n^2$ real parameters in a generic complex matrix of order n. Thus, a unitary matrix depends on n^2 independent real parameters. The unitary matrices form a group [called $U(n)$]. Indeed, it is easy to check that the product of two unitary matrices is unitary, etc. The group $U(1)$ is Abelian. The groups $U(n > 1)$ are non-Abelian.

It follows from (6.38) that $\det U = e^{i\theta}$ with some real θ. If we impose a further constraint $\det U = 1$, we obtain *special* unitary matrices characterized by $n^2 - 1$ real parameters. They belong to a subgroup of $U(n)$ denoted $SU(n)$. Besides the norm $z^{*j} z_j$, special unitary transformations leave invariant the structure

$$\epsilon^{j_1 \ldots j_n} z_{j_1} \cdots z_{j_n} . \tag{6.39}$$

Consider the simplest nontrivial special unitary group $SU(2)$. It has 3 parameters and 3 generators. In the neighbourhood of unity an element of $SU(2)$ can be represented as in (6.21), where the generators t^a are now Hermitian 2×2 matrices. They can be chosen as $t^a = \frac{1}{2}\sigma^a$, σ^a being the Pauli matrices. Explicitly,

$$t^1 = \frac{1}{2}\begin{pmatrix} 0 & 1 \\ 1 & 0 \end{pmatrix}, \quad t^2 = \frac{1}{2}\begin{pmatrix} 0 & -i \\ i & 0 \end{pmatrix}, \quad t^3 = \frac{1}{2}\begin{pmatrix} 1 & 0 \\ 0 & -1 \end{pmatrix}. \tag{6.40}$$

The generators t^a satisfy the same orthogonality condition as (6.23), but with a different coefficient:

$$\text{Tr}\{t^a t^b\} = \frac{1}{2}\delta^{ab} . \tag{6.41}$$

One can check that the generators (6.40) do not commute and that their commutators satisfy exactly *the same* relation (6.25) as the generators (6.22) of the group $SO(3)$.[9] A mathematician would say that the Lie algebras of the matrices (6.22) and of the matrices (6.40) are equivalent. This coincidence cannot be accidental, and it is not.[10] One can show that, for any $U \in SU(2)$, the matrix

$$O^{ab} = 2\text{Tr}\{U t^a U^\dagger t^b\} \tag{6.42}$$

is orthogonal. It seems that Lie was right: as the algebra (6.25) (describing the behaviour of a group in the neighbourhood of unity) is the same, we are dealing here in fact with the same group, $SU(2)$ is equivalent to $SO(3)$, isn't it?

Well, almost, but not quite. True, the relation (6.42) represents a mapping $SU(2) \to SO(3)$. But this mapping is not bijective; two different matrices, U and

---

[9]We can now understand the origin of the factor $1/2$ in (6.40). Were it not there, the coefficients in the orthogonality condition (6.41) would be the same as in (6.23), but the commutation relations would acquire an extra factor of 2 on the right-hand side. People prefer to avoid it.

[10]If something or someone buzzes, it is not without a reason, as Winnie the Pooh once wisely mentioned.

$-U$ give one and the same orthogonal matrix. One can say that $SU(2)$ is two times larger than $SO(3)$!

We will discuss very important implications of this fact for physics very soon, but let us first make acquaintance with another interesting unitary group, which is relevant for physics, the group $SU(3)$. $SU(3)$ has $3^2 - 1 = 8$ parameters and 8 generators. The latter can be chosen as

$$t^1 = \frac{1}{2} \begin{pmatrix} 0 & 1 & 0 \\ 1 & 0 & 0 \\ 0 & 0 & 0 \end{pmatrix}, \quad t^2 = \frac{1}{2} \begin{pmatrix} 0 & -i & 0 \\ i & 0 & 0 \\ 0 & 0 & 0 \end{pmatrix}, \quad t^3 = \frac{1}{2} \begin{pmatrix} 1 & 0 & 0 \\ 0 & -1 & 0 \\ 0 & 0 & 0 \end{pmatrix},$$

$$t^4 = \frac{1}{2} \begin{pmatrix} 0 & 0 & 1 \\ 0 & 0 & 0 \\ 1 & 0 & 0 \end{pmatrix}, \quad t^5 = \frac{1}{2} \begin{pmatrix} 0 & 0 & -i \\ 0 & 0 & 0 \\ i & 0 & 0 \end{pmatrix}, \quad t^6 = \frac{1}{2} \begin{pmatrix} 0 & 0 & 0 \\ 0 & 0 & 1 \\ 0 & 1 & 0 \end{pmatrix},$$

$$t^7 = \frac{1}{2} \begin{pmatrix} 0 & 0 & 0 \\ 0 & 0 & -i \\ 0 & i & 0 \end{pmatrix}, \quad t^8 = \frac{1}{\sqrt{12}} \begin{pmatrix} 1 & 0 & 0 \\ 0 & 1 & 0 \\ 0 & 0 & -2 \end{pmatrix}. \tag{6.43}$$

These matrices are all Hermitian, traceless and satisfy the orthogonality condition (6.41). Their commutators are given by (6.29) with

$$f^{123} = 1,$$

$$f^{147} = -f^{156} = f^{246} = f^{257} = f^{345} = -f^{367} = \frac{1}{2},$$

$$f^{458} = f^{678} = \frac{\sqrt{3}}{2}, \tag{6.44}$$

the other components being obtained by antisymmetry.

6.3.4 Representations

Consider first the group $SO(3)$. Its elements are 3×3 matrices that rotate real 3-dimensional vectors. But we have seen that this rotation can also be represented by a 2×2 unitary matrix U (or $-U$) rotating complex 2-component vectors. What is the physical meaning of this complex vector space?

The reader may already know the answer. The elements of this space, two-component complex objects, are nothing but *spinors*. A spinor is a 2-component wave function of a spin 1/2 particle. It transforms nontrivially under rotations — namely, it is multiplied on the left by a unitary matrix, related to the orthogonal matrix acting on the spatial coordinates as in (6.42).[11] Thus, when studying the $SU(2)$ group,

---

[11] Note that, when the rotation angle is changed continuously from 0 to 2π, the corresponding unitary matrix is changed from $\mathbb{1}$ to $-\mathbb{1}$ (it is like going onto another sheet of a Riemann surface).

we simultaneously constructed a *spinor representation* of the 3-dimensional rotation group.

As for the 3×3 orthogonal matrices, they realize its vector representation. Note that these matrices act not only on the ordinary classical vectors, but also on the wave functions of a quantum particle of spin 1, which have vector nature. Indeed, such a particle can have 3 different spin projections onto any given spatial axis (say, the z axis), $S_z = 1, 0, -1$. The wave function has 3 components which can be presented as[12]

$$\Psi_{S=1} = \begin{pmatrix} \Psi_+ \\ \Psi_0 \\ \Psi_{-1} \end{pmatrix} = \begin{pmatrix} -(V_x + i V_y)/\sqrt{2} \\ V_z \\ (V_x - i V_y)/\sqrt{2} \end{pmatrix}. \tag{6.45}$$

One can then be convinced that the object (V_x, V_y, V_z) transforms as a vector under rotations.

But there are also particles of spin 3/2 (an example is the Ω^- hyperon, the "Neptune" of the quark model that we talked about in Sect. 5.3) with 4 possible spin projections, $S_z = \pm 3/2, \pm 1/2$. The wave function of such a particle involves 4 components and is rotated by a 4×4 matrix. There are particles of spin 2 with 5 spin projections (their wave functions can be represented as traceless symmetric tensors of rank 2) and so on. We thus obtain a sequence of representations characterized by spin S, which can be integer or half-integer.

As a particle of spin S can represent a bound state of $2S$ $S = 1/2$ particles, a representation of spin S can be made out of $2S$ "elementary" spinor representations. Take as an example the Ω^- hyperon. Its wave function represents an antisymmetrized product of the wave functions of its fermion quark constituents — this is required by the Pauli principle. The quark internal wave functions (or the quark fields — whichever you prefer) carry spin and colour indices. The wave function of Ω^- should be antisymmetrized in colour (in that case, Ω^- has zero colour charge, is white as in the picture in Fig. 5.9). Hence, it should be symmetrized in spin indices:

$$\left(\Psi^{3/2} \right)_{ijk} = \psi_{\{i} \chi_j \phi_{k\}}. \tag{6.46}$$

$\Psi^{3/2}$ has 4 independent components, as it should: $\Psi_{111}, \Psi_{112}, \Psi_{122}, \Psi_{222}$.

Note also the existence of the trivial spin-0 or scalar representation.

It is now proper to make a remark on terminology. Physicists and mathematicians speak related but different languages. A translation is often required. In many cases the same mathematical objects are called by different names. On the other hand, sometimes the same word (a *false friend*) is used by the members of two different communities in a different way.

(Footnote 11 continued)
In other words, the spinor wave function changes sign after the 2π rotation. This is a known quantum mechanical effect. It was confirmed in experiment.

[12]The negative sign in the upper component on the right is not a mistake. It appears under the standard convention $\Psi_+ = \hat{J}^+ \Psi_0 / \sqrt{2} = (\hat{J}^+)^2 \Psi_- / 2$.

When a mathematician says: "representation", he thinks of a group element represented by a matrix of a particular size. The column-vectors by which these matrices can be multiplied belong for him to "representation space". And when a physicist says: "representation", he thinks in the first place about the *objects* on which the symmetry transformations act. We say: "...fields in a spinor representation, in a vector representation" and so on, omitting the word "space".

We now go over to the $SU(n)$ groups. The unitary $n \times n$ matrices act on complex n-dimensional column-vectors:

$$\psi'_j = U_j{}^k \psi_k . \tag{6.47}$$

These vectors belong to the *fundamental representation* of $SU(n)$. In addition, there is also the *anti-fundamental* representation, the row-vectors ψ^{*j}, which transform as

$$(\psi^{*k})' = \psi^{*j} (U^\dagger)_j{}^k . \tag{6.48}$$

In short, one can write, $\psi \to U\psi$, $\psi^* \to \psi^* U^\dagger$, so that the norm $\psi^* \psi$ is invariant due to (6.38).

The anti-fundamental representation is related to the fundamental one by complex conjugation. These two representations are not equivalent when $n \geq 3$. But for $SU(2)$, they are. Their equivalence can be established by the linear relation, $\psi^{*j} = \epsilon^{jk} \chi_k$, where ϵ^{jk} is the invariant $SU(2)$ tensor [see Eq. (6.39)]. We have already discussed the (anti)fundamental representation of $SU(2)$ calling it the spinor representation.

$SU(n)$ has many other representations. An important role is played by the *adjoint representation*. It can be constructed from the fundamental and anti-fundamental ones:

$$A_j{}^k = \psi_j \chi^{*k} - \frac{1}{3} \delta_j^k \psi_l \chi^{*l} , \tag{6.49}$$

such that the property that $A_j{}^j = 0$ holds. The matrices (6.49) belong to $su(n)$, the Lie algebra of $SU(n)$,[13] and can be presented as

$$A_j{}^k = A^a (t^a)_j{}^k , \tag{6.50}$$

where $(t^a)_j{}^k$ are the $n^2 - 1$ generators of $SU(n)$. The matrices (6.50) transform under $SU(n)$ as

$$\hat{A} \to U \hat{A} U^\dagger . \tag{6.51}$$

---

[13]Probably, that is why this representation was called adjoint. In French this word has a meaning of "attached" or "united", and one can imagine that a Lie group attaches as a representation its own Lie algebra to make them both happy. Evidently, the adjoint representation exists for any Lie group, not only for $SU(n)$.

For any $n \geq 2$, one can think of the adjoint representation as of a set of $n^2 - 1$ real numbers A^a. The action (6.51) of the group can then be realized by a real matrix of size $n^2 - 1$ given by the familiar expression (6.42). Thus, for $n = 2$, the adjoint representation coincides with the vector representation.

The fundamental and adjoint representations of $SU(n)$ are almost everything that we need. But in Chap. 11, when discussing the phenomenology of the hadron spectrum, we will also use the decuplet representation of $SU(3)$. It is described by the *symmetric* 3rd rank $SU(3)$ tensor Ψ_{ijk}. The latter has 10 independent components:

$$\Psi_{111}, \ \Psi_{222}, \ \Psi_{333}, \ \Psi_{112}, \ \Psi_{113}, \ \Psi_{221}, \ \Psi_{223}, \ \Psi_{331}, \ \Psi_{332}, \ \Psi_{123} \ . \tag{6.52}$$

Finally, we discuss the representations of the Lorentz group. We have seen [see Eq. (6.36)] that the algebra $so(3, 1)$ represents the direct sum of two $so(3) \equiv su(2)$ algebras. A mathematical corollary of this is the fact that the group $SO(3, 1)$ represents *in some sense* a product of two $SU(2)$ groups.

To understand in what particular sense it is a product (and what the product of two groups means) and to have fewer doubts in the validity of the proposition above, note that a generic element of $SO(3, 1)$ can be written as the product of two commuting factors:

$$O \ = \ \exp\left\{i M^a u^a\right\} \exp\left\{i N^a u^{*a}\right\} \ , \tag{6.53}$$

where u^a are *complex* parameters. This follows from: *(i)* the expression (6.37) of an infinitesimal element; *(ii)* the general formula (6.30); *(iii)* the fact that M^a and N^b commute.

Given that both M^a and N^a satisfy the standard $su(2)$ algebra, each factor in (6.53) can be interpreted as an element of a *complexified $SU(2)$* group (the group with complex transformation parameters!). It is important that these parameters for the group generated by M^a and for the group generated by N^a are related by complex conjugation.[14]

The formula (6.53) will allow us to construct the representations of the Lorentz group — the objects which the group may transform. As a generic element of $SO(3, 1)$ is a product of two independent commuting factors [call them the left $SU(2)$ and the right $SU(2)$], a generic representation of $SO(3, 1)$ is a tensor product of two independent $SU(2)$ representations. For example, take an object $\xi_{\alpha=1,2}$ that transforms as a spinor under the action of the left $SU(2)$ group and does not transform at all under the action of the right one. By the same token, we may consider an object $\xi_{\dot{\alpha}=1,2}$ that transforms as a spinor under the action of the right $SU(2)$ group and does not transform at all under the action of the left one.

We will see in Chap. 9 that the "undotted" and "dotted" fermion spinor fields describe the states with definite helicity, like the left-handed neutrino and right-handed anti-neutrino. Thus, the names "left" and "right" for the two $SU(2)$ groups

---

[14]The presence of complex transformation parameters is a complication associated with the fact that the Lorentz group is not compact. In a similar construction for the compact $SO(4)$ group, all the parameters are real.

in (6.53) are associated not only with the typographic fact that one of these factors stays on the left and another on the right in this formula, but have a more deep physical justification.

But in this mathematical chapter, we only note that a general representation of the Lorentz group is a tensor product of some representations of the left and the right groups. It is characterized by two numbers (j_L, j_R). Each of them can be either integer or half-integer. The spinor representation ξ_α is thus called the representation $(\frac{1}{2}, 0)$ and the representation $\eta_{\dot\alpha}$ — the representation $(0, \frac{1}{2})$. Bearing in mind that the parameters of the two $SU(2)$ groups in (6.53) are related by complex conjugation, the representations $(\frac{1}{2}, 0)$ and $(0, \frac{1}{2})$ are also related by complex conjugation: the spinor $(\xi_\alpha)^*$ transforms in the same way as $\eta_{\dot\alpha}$.

Note that one can raise and lower the undotted and dotted $SU(2)$ indices by the invariant $SU(2)$ tensors, $\epsilon_{\alpha\beta} = -\epsilon^{\alpha\beta}$ and $\epsilon_{\dot\alpha\dot\beta} = -\epsilon^{\dot\alpha\dot\beta}$. Then

$$\xi_\alpha = \epsilon_{\alpha\beta}\xi^\beta, \quad \xi^\alpha = \epsilon^{\alpha\beta}\xi_\beta, \quad \eta_{\dot\alpha} = \epsilon_{\dot\alpha\dot\beta}\xi^{\dot\beta}, \quad \eta^{\dot\alpha} = \epsilon^{\dot\alpha\dot\beta}\xi_{\dot\beta}. \tag{6.54}$$

A legitimate question that the reader can ask at this point is the following. The matrix O in Eq. (6.53) and both factors there are 4-dimensional, and it is not evident how exactly the spinors ξ_α and $\eta_{\dot\alpha}$ transform.

To determine the action of ordinary rotations on the ordinary spinors, one had to replace the orthogonal 3×3 matrix O by a unitary 2×2 matrix U, related to O according to (6.42). Likewise, to determine the action of Lorentz transformations on the spinors ξ_α and $\eta_{\dot\alpha}$, one has to replace O in (6.53) by a 2×2 complex matrix

$$g = \begin{pmatrix} \alpha & \beta \\ \gamma & \delta \end{pmatrix} \tag{6.55}$$

with unit determinant,

$$\alpha\delta - \beta\gamma = 1. \tag{6.56}$$

Without imposing the latter condition, the group of matrices (6.55) would have $4 \cdot 2 = 8$ real parameters. One complex or two real constraints (6.56) restrict their number to 6, as we know should be the case for the Lorentz group. An analogue of the formula (6.42) relating g and O can be written, though we prefer not to do so here. After all, our book is semi-popular, and we are not obliged to (or rather are obliged not to) go into all the technical details.

In these terms, the action of the Lorentz group on the spinors is expressed in a very simple way

$$\xi \to g\xi, \quad \eta \to \eta g^\dagger. \tag{6.57}$$

Consider now the representation $(\frac{1}{2}, \frac{1}{2})$. A field belonging to this representation carries two indices, dotted and undotted. We denote it $V_{\alpha\dot\alpha}$. It transforms as

$$V \to gVg^\dagger \tag{6.58}$$

under Lorentz transformations. A general complex matrix V has 8 parameters, but this is, as mathematicians say, a *reducible* representation. "Reducible" means that the number of independent parameters characterizing the representation can be reduced by imposing certain constraints. In our case, we may impose the Hermiticity constraint — a Hermitian matrix stays Hermitian under the transformation (6.58). A Hermitian matrix V has 4 independent parameters. It is nothing but a 4-vector! Explicitly,

$$V_{\alpha\dot\alpha} = \begin{pmatrix} V_0 - V_3 & -V_1 + i V_2 \\ -V_1 - i V_2 & V_0 + V_3 \end{pmatrix} = V_0 - V_j \sigma_j = \sigma^\mu V_\mu, \tag{6.59}$$

where we have introduced the matrix 4-vector

$$\sigma^\mu = (\mathbb{1}, \boldsymbol{\sigma}) \tag{6.60}$$

[see (2.11)].

6.4 Grassmann Algebra

And now for something completely different. We could explain what are the Grassmann numbers and Grassmann algebra in Chap. 9, where we need them, but it is probably more logical to make one heap of all the mathematical preliminaries and to amass them all in one chapter.

The Grassmann numbers are *anticommuting* numbers. This mathematical notion should not surprise the reader too much, as s/he already knows that multiplication is not always commutative in mathematics (matrices do not commute, for example). The Grassmann numbers represent a particular class of such objects.

The main definitions are the following:

- Let $\{a_i\}$ be a set of n basic anticommuting variables: $a_i a_j + a_j a_i = 0$. The elements of the Grassmann algebra can be written as the functions $f(a_i) = c_0 + c_i a_i + c_{ij} a_i a_j + \ldots$ where the coefficents $c_0, c_i, \ldots$ are ordinary (real or complex) numbers. The series terminates at the n-th term: the $(n + 1)$ - th term of this series would involve a product of two identical anticommuting variables (say, a_1^2), which is zero. Note that even though a Grassmann number like $a_1 a_2$ commutes with any other number, it still cannot be treated as a normal number but represents instead an *even element* of the Grassmann algebra. There are, of course, also odd anticommuting elements. The variables $\{a_i\}$ are called the *generators* of the algebra.
- One can add Grassmann numbers, $f(a_i) + g(a_i) = c_0 + d_0 + (c_i + d_i)a_i + \ldots$, as well as multiply them. For example, $(1 + a_1 + a_2)(1 + a_1 - a_2) = 1 + 2a_1 - 2a_1 a_2$ (the anticommutation property of a_i was used).

- One can also differentiate the functions $f(a_i)$ with respect to Grassmann variables: $\partial/\partial a_i \, (1) \overset{\text{def}}{=} 0, \partial/\partial a_i \, (a_j) \overset{\text{def}}{=} \delta_{ij}$, the derivative of a sum is the sum of derivatives, and the derivative of a product of two functions satisfies the Leibnitz rule, except that the operator $\partial/\partial a_i$ should be thought of as a Grassmann variable, which sometimes leads to a sign change when $\partial/\partial a_i$ is pulled through to annihilate a_i in the product. For example, $\partial/\partial a_1 (a_2 a_3 a_1) = a_2 a_3$, but $\partial/\partial a_1 (a_2 a_1 a_3) = -a_2 a_3$.
- One can also integrate over Grassmann variables. In contrast to the usual bosonic case, the integral cannot be obtained here as a limit of integral sums, cannot be calculated numerically through "finite limits" (this makes no sense for Grassmann numbers) with the Simpson method and so on. What one can do, however, is to integrate over a Grassmann variable in the "whole range ("from $-\infty$ to ∞" if you wish, though this is, again, meaningless). The definition (due to Felix Berezin) is $\int da_j f(a) \overset{\text{def}}{=} \partial/\partial a_j f(a)$.
- If the Grassmann algebra involves an even number of generators $2n$, one can divide them in two equal parts, $\{a_{j=1,\dots,2n}\} \rightarrow \{a_{j=1,\dots,n}, a^\dagger_{j=1,\dots,n}\}$ and introduce an operation of involution, $a_j \leftrightarrow a^\dagger_j$, which we will associate with complex conjugation. We will assume in particular that, simultaneously with the involution of the generators, the ordinary numbers c_0, etc. are replaced by their complex conjugates:

$$f(a) = c_0 + c_i a_i + d_i a^\dagger_i + \dots \longrightarrow f^\dagger(a) = c_0^* + c_i^* a^\dagger_i + d_i^* a_i + \dots$$

It is also convenient to assume that, for any two elements f, g of the Grassmann algebra, $(fg)^\dagger = g^\dagger f^\dagger$, as for Hermitian conjugation (hence the notation $\dagger$).

Chapter 7
Lagrangians and Hamiltonians

The reader probably followed a course on analytical mechanics during his/her university studies. If not, there are a great number of textbooks where its principles are explained in a detailed pedagogical way. I recommend two of them. For a lucid physical introduction, see Chap. 19 of volume 2 of Feynman's lectures on physics. And for a serious (and at the same time succinct) exposition, consult *Mechanics* by Lev Landau and Evgeny Lifshitz — the first volume of the Landau course of theoretical physics.

Or you may not look for other sources, but stay with us. In order to make our book self-contained, we give a brief review of this subject in the first half of this chapter. In the second half, we apply this technique to relativistic mechanical systems and generalize it to field theories.

7.1 Nonrelativistic Mechanics

In the previous chapters we already explored to some extent the cubic building of physics depicted in Fig. 4.5. We walked through its halls and corridors in different directions. But now we return to the first entrance hall. Paying no more attention to funny pictures and installations, we go to the bookshelf, pick up Feynman's and Landau's books, sit down and start to read...

We start with the simplest nontrivial example: the motion of a stone, thrown vertically up to the skies. Consider all possible trajectories $z(t)$, such that, at the initial moment t_0, the elevation of the stone is fixed to be z_0, while at a later moment $t_1 > t_0$ the elevation is fixed to be z_1.

Calculate the *action* integral

$$S = \int_{t_0}^{t_1} dt \, (T - U) = \int_{t_0}^{t_1} dt \left[\frac{m\dot{z}^2}{2} - mgz \right], \qquad (7.1)$$

© Springer International Publishing AG 2017
A. Smilga, *Digestible Quantum Field Theory*,
DOI 10.1007/978-3-319-59922-9_7

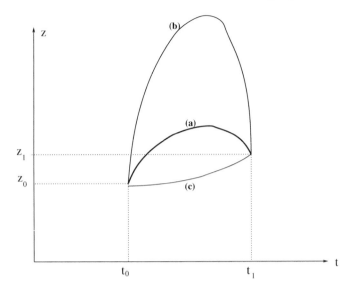

Fig. 7.1 Different trajectories of the stone: (**a**) the trajectory with minimal action; (**b**) too *high* kinetic energy; (**c**) too *low* potential energy.

T being the kinetic and U the potential energy.

The statement (the *least action principle*) is that, for the actual physical trajectory $z^{\text{phys}}(t)$, this integral takes the minimal value. This is illustrated in Fig. 7.1. To minimize the action (7.1), the stone would like to increase its potential energy and to climb higher. But, bearing in mind that it has an appointment at z_1 at fixed time t_1, going too high would entail too high a velocity and a large positive contribution to the integral (7.1) coming from the first term. The real trajectory is a negotiated compromise between these two effects.

Consider now a general mechanical system with n dynamical variables q_i. Choose the action functional[1] $S[q_i(t)]$ in the form

$$S = \int_{t_0}^{t_1} dt\, L(\dot{q}_i, q_i)\,. \tag{7.2}$$

It is not the most generic possible form of the action and implies two assumptions:

- We assumed the *Lagrangian L* not to involve an explicit time dependence. If it did, the system would not be *conservative*: the energy would change with time. And we are not interested in nonconservative systems in our book.
- We assumed that the Lagrangian depends only on q_i (*generalized coordinates*) and $\dot{q}_i$ (*generalized velocities*), but not on $\ddot{q}_i$, $\dddot{q}_i$ and so on. Systems involving higher time derivatives have interesting dynamics — your author played with them for

---

[1] A functional is a function whose argument is itself a function.

some time and we will discuss them in Chap. 16.[2] But our main task in this book is to present and explain the Lagrangian of the Standard Model, which involves only the fields and their first derivatives.

The problem is to find a trajectory $q_i(t)$ satisfying the boundary conditions

$$q_i(t_0) = q_i^{(0)}, \qquad q_i(t_1) = q_i^{(1)}, \tag{7.3}$$

such that the integral (7.2) acquires the minimal value.

For a usual smooth function f, an extremum is achieved only when all its partial derivatives and hence the differential df vanish. For a functional like the action functional, its *variation* should vanish. We have[3]

$$\delta S = \int_{t_0}^{t_1} dt \left[\frac{\partial L}{\partial q_i} \delta q_i(t) + \frac{\partial L}{\partial \dot{q}_i} \delta \dot{q}_i(t) \right]. \tag{7.4}$$

It is convenient to integrate the second term by parts. The boundary term vanishes due to the fixed boundary conditions (7.3) — we perform the variation only over the trajectories that satisfy them — so that

$$\delta q_i(t_0) = \delta q_i(t_1) = 0. \tag{7.5}$$

We obtain

$$\delta S = \int_{t_0}^{t_1} dt \left[\frac{\partial L}{\partial q_i} - \frac{d}{dt} \frac{\partial L}{\partial \dot{q}_i} \right] \delta q_i(t). \tag{7.6}$$

This should be zero for *any* $\delta q_i(t)$ satisfying the constraint (7.5). It follows that

$$\frac{d}{dt} \frac{\partial L}{\partial \dot{q}_i} - \frac{\partial L}{\partial q_i} = 0, \qquad i = 1, \ldots, n. \tag{7.7}$$

We have derived the *equations of Lagrange*. The Eq. (7.7) represent a system of n ordinary second-order differential equations. For the simple system (7.1), we obtain second Newton's law, $m\ddot{z} = -mg$.

A Lagrangian not exhibiting an explicit time dependence is invariant under time translation, $t \to t + \Delta t$. By Noether's theorem,[4] any invariance (alias, symmetry) of a Lagrangian entails a conservation law. The invariance with respect to time shifts

---

[2] There are some reasons to believe that the Theory of Everything — a Holy Grail unified theory of all interactions including gravity — involves higher derivatives.

[3] We keep using the Einstein summation convention even though the set q_i is by no means a vector.

[4] Emmy Noether was a remarkable mathematician. Arguably, the most important woman in the history of mathematics. She worked in Germany in the beginning of the 20th century. In 1919 she got a professorship position at the University of Göttingen. But that was not easy for her. At that time, there were no women at the universities, and her colleagues hesitated to create a precedent. One of the faculty members protested, "What will our soldiers think when they return to the university and find that they are required to learn at the feet of a woman?" The controversy was settled by

brings about the conservation of energy. Indeed, you may check that the energy, defined as

$$E = \dot{q}_i \frac{\partial L}{\partial \dot{q}_i} - L, \tag{7.8}$$

represents an *integral of motion* — its time derivative vanishes on the trajectory $q_i(t)$ satisfying the equations of motion (7.7).

Introduce now the *canonical momenta*

$$p_i = \frac{\partial L}{\partial \dot{q}_i}, \tag{7.9}$$

express the generalized velocities $\dot{q}_i$ in terms of the p_i and substitute them into (7.8). We obtain the *Hamiltonian* — the energy expressed in terms of the canonical coordinates and momenta. They parametrize together what is called *phase space*. The equations of motion in phase space (*Hamilton's equations*) are

$$\dot{q}_i = \frac{\partial H}{\partial p_i}$$

$$\dot{p}_i = -\frac{\partial H}{\partial q_i}. \tag{7.10}$$

Equation (7.10) represents a system of $2n$ first-order differential equations. They are equivalent to the Lagrange equations (7.7). To see that, evaluate the differential of the Hamiltonian. Bearing in mind (7.7) and the definition (7.9), we can write

$$dH = d(p_i \dot{q}_i - L) = \dot{q}_i dp_i + p_i d\dot{q}_i - \frac{\partial L}{\partial q_i} dq_i - \frac{\partial L}{\partial \dot{q}_i} d\dot{q}_i$$

$$= \dot{q}_i dp_i - \dot{p}_i dq_i \tag{7.11}$$

from which (7.10) follows.

Let $f(q_i, p_i)$ be a function on phase space. Taking into account the evolution of the system $q_i(t)$, $p_i(t)$ according to the equations of motion (7.10), its total time derivative is

$$\frac{df}{dt} = \frac{\partial f}{\partial q_i} \dot{q}_i + \frac{\partial f}{\partial p_i} \dot{p}_i = \frac{\partial f}{\partial q_i} \frac{\partial H}{\partial p_i} - \frac{\partial f}{\partial p_i} \frac{\partial H}{\partial q_i}. \tag{7.12}$$

The sum on the right-hand side is called the *Poisson bracket* $\{H, f\} = -\{f, H\}$. Thus, for the function f to be an integral of motion, it is necessary and sufficient that the Poisson bracket of this function and the Hamiltonian vanishes.

This is a "classical" chapter, but it is still difficult to conceal from the sagacious reader that, in quantum mechanics, phase-space functions become operators. The

(Footnote 4 continued)
the intervention of David Hilbert, the leading, most respected German mathematician at that time. "Aber, meine Herren", said he, "the university is still not a bath house!".

explicit expressions for the latter are obtained from the expressions for the former by substituting $p_i \rightarrow -i\hbar\partial/\partial q_i$. In the simple cases, like for the Hamiltonian

$$H = \frac{p^2}{2m} + V(x), \tag{7.13}$$

describing the motion of a particle in an external potential field, this correspondence rule can be applied straightforwardly to obtain (4.9). In more fancy cases, like for an exotic Hamiltonian $H = p^2 q^2$, one should resolve *ordering ambiguities* and decide whether the derivatives should be placed on the left of q^2, on the right, or in the middle. This complication is beyond the scope of our discussion.

We also note that the quantum counterpart of the classical Poisson bracket is the commutator. A quantum operator describes a conserved quantity if it commutes with the Hamiltonian.

7.2 Lorentz Force

To show how the Lagrange and Hamilton methods work, we derive here the equation of motion for a relativistic charged particle in external electric and magnetic fields. For an electron carrying the electric charge $e = -|e|$, it reads

$$\frac{d\boldsymbol{p}}{dt} = \boldsymbol{F} = e\left(\boldsymbol{E} + \frac{1}{c}[\boldsymbol{v} \times \boldsymbol{B}]\right). \tag{7.14}$$

You know that from your electromagnetism course, but probably the expression (7.14) for the force (the *Lorentz force*) was just postulated, but not derived there.[5] And we will derive it, using the methods of the preceding section and recalling at the same time some basic notions of special relativity.

Consider first the case when the external fields are absent and we are dealing with the free relativistic particle. We start by writing the action. It should be a Lorentz scalar (invariant under Lorentz transformations). Well, we know one such scalar — it is just the interval (6.12) between the initial and final points of the trajectory. For a free particle, no other invariant is in sight and the action should be proportional to (6.12), which we can represent as

$$S = Cmc \int_0^1 ds, \tag{7.15}$$

with ds written in Eq. (4.21) and the limits 0,1 do not refer to the initial and final *values* of s (that would make no sense), but just indicate that the integral is done along a trajectory connecting the initial and final points 0 and 1 in Minkowski space.

---

[5]Even Feynman did not do so in his Lectures. Unfortunately, the courses of analytical mechanics and electromagnetism in the standard university curriculum do not usually "commute".

The expression (7.15) has the right dimension. The numerical constant C is a pure convention (it does not affect the dynamics), but the most convenient choice is $C = -1$. We will shortly see that, in the nonrelativistic limit, the expression

$$S = -mc \int_0^1 ds \qquad (7.16)$$

goes over to the nonrelativistic expression for the free action plus (or rather minus) an irrelevant constant.

To write the Lagrangian, we should trade the integral over the invariant interval in (7.16) for the time integral. We know that

$$ds = \sqrt{c^2 dt^2 - dx^2} = cdt\sqrt{1 - v^2/c^2}, \qquad (7.17)$$

which gives

$$L = -mc^2\sqrt{1 - v^2/c^2}. \qquad (7.18)$$

For small velocities, it reduces to $L = -mc^2 + mv^2/2 + \cdots$, as promised.

The canonical momentum for the relativistic Lagrangian is

$$p = \frac{\delta L}{\delta v} = \frac{mv}{\sqrt{1 - v^2/c^2}}. \qquad (7.19)$$

This is the known expression for the relativistic 3-momentum. The equation of motion is

$$\frac{d}{dt}\left(\frac{\partial L}{\partial \dot{x}}\right) = \frac{dp}{dt} = 0. \qquad (7.20)$$

The relativistic energy is

$$E = p \cdot v - L = \frac{mc^2}{\sqrt{1 - v^2/c^2}}. \qquad (7.21)$$

It is expressed via the canonical momentum (7.19) as

$$E = \sqrt{m^2c^4 + p^2c^2}. \qquad (7.22)$$

This is the Hamiltonian of the free relativistic particle.

We now switch on the electromagnetic field. It is characterized by the 4-vector potential $A^\mu = (\varphi, A)$ [and $A_\mu = \eta_{\mu\nu} A^\nu = (\varphi, -A)$] and by the antisymmetric field strength tensor

$$F_{\mu\nu} = \partial_\mu A_\nu - \partial_\nu A_\mu = \begin{pmatrix} 0 & E_x & E_y & E_z \\ -E_x & 0 & -B_z & B_y \\ -E_y & B_z & 0 & -B_x \\ -E_z & -B_y & B_x & 0 \end{pmatrix},$$

$$F^{\mu\nu} = \begin{pmatrix} 0 & -E_x & -E_y & -E_z \\ E_x & 0 & -B_z & B_y \\ E_y & B_z & 0 & -B_x \\ E_z & -B_y & B_x & 0 \end{pmatrix}, \tag{7.23}$$

where E and B are the electric and magnetic fields. In 3-dimensional notation,

$$E = -\nabla\phi - \frac{1}{c}\frac{\partial A}{\partial t}, \qquad B = \nabla \times A. \tag{7.24}$$

In the presence of the field, we can write besides (7.16) *another* Lorentz invariant $\propto \int A_\mu dx^\mu$. Adding this in the action affects the trajectory of the particle. It is clear that the coefficient with which it enters is proportional to the electric charge. The standard convention is

$$S = -mc \int_0^1 ds - \frac{e}{c} \int_0^1 A_\mu dx^\mu, \tag{7.25}$$

Rewriting it as the time integral, we derive

$$L = -mc^2\sqrt{1 - v^2/c^2} - e\phi + \frac{e}{c}v \cdot A. \tag{7.26}$$

The interactive part of this Lagrangian can also be represented as [cf. Eq. (5.3)!]

$$L^{\text{int}} = -e \int A_\mu j^\mu d^3x, \tag{7.27}$$

where

$$j^\mu(x) = \left(j^0, \frac{j}{c}\right) = \left(1, \frac{v}{c}\right)\delta[x - x_0(t)] \tag{7.28}$$

is the current density associated with the pointlike particle moving along the trajectory $x_0(t)$. Note the property[6]

$$\partial_\mu j^\mu = \frac{1}{c}\left(\frac{\partial}{\partial t}j^0 + \nabla \cdot j\right) = 0 \tag{7.29}$$

This is an important law of *current conservation* (it amounts here to the conservation of the number of particles and hence to the conservation of the electric charge).

---

[6]One easily derives it using $v = \dot{x}_0$.

The canonical momentum following from the Lagrangian (7.26) is

$$P = \frac{\partial L}{\partial v} = \frac{mv}{\sqrt{1 - v^2/c^2}} + \frac{e}{c}A. \tag{7.30}$$

Note that, in the presence of the vector potential implying the presence of the magnetic field, the canonical momentum P does not coincide with the *kinetic* momentum p given by (7.19).

The Lagrange equations of motion are

$$\frac{d}{dt}\left[p_i + \frac{e}{c}A_i\right] = -e\frac{\partial \phi}{\partial x_i} + \frac{e}{c}\frac{\partial A_j}{\partial x_i}v_j. \tag{7.31}$$

Note now that the total time derivative of the vector potential is a sum of two terms:

$$\frac{dA_i(x, t)}{dt} = \frac{\partial A_i}{\partial t} + \frac{\partial A_i}{\partial x_j}\frac{\partial x_j}{\partial t} = \frac{\partial A_i}{\partial t} + \frac{\partial A_i}{\partial x_j}v_j. \tag{7.32}$$

The first term of this expression, the partial derivative $\sim \partial_0 A_i$, describes how the vector potential is changed at a fixed spatial point. And the whole expression describes how it is changed at the point moving with the particle.

We finally obtain

$$\frac{dp_i}{dt} = e\left(-\frac{1}{c}\frac{\partial A_i}{\partial t} - \frac{\partial \varphi}{\partial x_i}\right) + \frac{e}{c}\left(\frac{\partial A_j}{\partial x_i} - \frac{\partial A_i}{\partial x_j}\right)v_j, \tag{7.33}$$

which coincides with (7.14).

We also write down the expression for the canonical Hamiltonian of this system:

$$H = \frac{1}{2m}\left(P - \frac{e}{c}A\right)^2 + e\varphi. \tag{7.34}$$

7.3 Field Theories

This book is devoted to quantum field theories. To learn this subject, one has to understand well what *classical* field theory is. The reader already knows something about it from his/her previous studies and from Chap. 4, where the equations of motion for different systems of fields were presented and briefly discussed. However, the equations of motion were just postulated and not derived there. In this section we will derive the KFG equation and also Maxwell's equations from the fundamental least-action principle. In addition, we will give the expressions for the corresponding classical Hamiltonians.

We start with the theory of a complex scalar field, which is simpler than Maxwell's theory. The action functional now represents not just a time integral, but an integral over time and spatial coordinates:

$$S = \int dt \int d\mathbf{x}\, \mathcal{L}. \tag{7.35}$$

The integrand is the spatial Lagrangian density. The Lagrangian of the system is given by the integral

$$L = \int d\mathbf{x}\, \mathcal{L}. \tag{7.36}$$

However, lazy field theorists often use a sloppy language and call $\mathcal{L}$ rather than L the Lagrangian. In subsequent chapters (but mostly not in this chapter), we will also follow their example.

We want the action to be invariant under Lorentz transformations. To ensure that, $\mathcal{L}$ should also be a Lorentz scalar.

Note the following. In classical mechanics, we put the limits t_0 and t_1 for the time integral and imposed fixed boundary conditions at the limits. The integral (7.2) then had the meaning of the action on a trajectory describing the transition from $q^{(0)}$ at t_0 to $q^{(1)}$ at t_1. We could do the same for field theories.[7] But imposing finite limits for the time integral and not putting any limits for the spatial integral is not a Lorentz-invariant procedure, which is not "aesthetic". We prefer to integrate over the whole space-time, $-\infty < t, \mathbf{x} < \infty$. We require, however, that the field $\phi(t, \mathbf{x})$ vanish at spatial infinity, in the distant past and in the distant future.

Based on our experience with the mechanical systems, we suppose that the Lagrangian density $\mathcal{L}$ depends on the fields ϕ, ϕ^* and their first derivatives $\partial_\mu \phi$, $\partial_\mu \phi^*$. Consider the variation of the action. We can write

$$\delta S = \int d^4 x \left[\frac{\delta \mathcal{L}}{\delta \phi} \delta\phi(x) + \frac{\delta \mathcal{L}}{\delta(\partial_\mu \phi)} \partial_\mu(\delta\phi(x)) \right] + \text{complex conjugate} \tag{7.37}$$

[we used $\delta(\partial_\mu \phi) = \partial_\mu(\delta\phi)$]. In this expression, $\delta \mathcal{L}/\delta\phi$ and $\delta \mathcal{L}/\delta(\partial_\mu \phi)$ are the partial derivatives of the function $\mathcal{L}(\phi, \phi^*; \partial_\mu \phi, \partial_\mu \phi^*)$ with respect to its arguments. Here we are using the symbol δ rather than ∂ so as not to confuse these partial derivatives with the ordinary space-time derivatives $\partial_\mu \phi = \partial\phi/\partial x^\mu$.

In the same way as we did for mechanical systems, we now integrate the second term in (7.37) by parts to obtain

$$\delta S = \int d^4 x\, \delta\phi(x) \left[\frac{\delta \mathcal{L}}{\delta \phi} - \partial_\mu \left(\frac{\delta \mathcal{L}}{\delta(\partial_\mu \phi)} \right) \right] + \text{c.c.} \tag{7.38}$$

---

[7]People do so to define and derive quite rigorously scattering amplitudes in quantum field theory. Such a derivation is beyond the scope of our book.

The boundary term vanishes here due to the vanishing of all the fields at infinity — the boundary condition that we imposed.

The integral (7.38) should vanish for *any* variation $\delta\phi(t, \boldsymbol{x})$. This implies

$$\partial_\mu \left(\frac{\delta\mathcal{L}}{\delta(\partial_\mu\phi)} \right) - \frac{\delta\mathcal{L}}{\delta\phi} = 0 \,. \tag{7.39}$$

This partial derivative equation is the Lagrange equation of motion for our system. It represents a rather transparent generalization of the ordinary Lagrange equations (7.7).[8]

Let us choose $\mathcal{L}$ in the following simple form (we revert to the natural units)

$$\mathcal{L} = (\partial_\mu\phi^*)(\partial^\mu\phi) - m^2\phi^*\phi \,. \tag{7.40}$$

The equation (7.39) coincides in this case with the KFG equation (4.28)![9] Thus, the Lagrangian density (7.40) describes the free complex scalar field.

It is also not difficult to write $\mathcal{L}$ for the free real scalar field. One should only remove the stars in (7.40) and introduce for convenience a common factor $1/2$.

In natural units, the action (7.35) is dimensionless and the density $\mathcal{L}$ has the dimension m^4. Bearing in mind (7.40), the field ϕ has the dimension of mass, the same as for the electromagnetic potential A_μ.

Consider now the theory[10]

$$\mathcal{L} = (\partial_\mu\phi^*)(\partial^\mu\phi) - m^2\phi^*\phi - \frac{\lambda}{4}(\phi^*\phi)^2 \,. \tag{7.41}$$

with positive λ (see below). The equation of motion

$$(\Box + m^2)\phi + \frac{\lambda}{2}\phi^2\phi^* = 0 \,, \tag{7.42}$$

is now nonlinear and complicated. We are dealing with a nontrivial theory of an interacting scalar field.

---

[8]Anticipating applications to high-energy physics, we have written this equation in relativistic notation using the 4-vector ∂_μ. But one can also write the equations of motion for a nonrelativistic field theory by replacing

$$\partial_\mu \left(\frac{\delta\mathcal{L}}{\delta(\partial_\mu\phi)} \right) \rightarrow \frac{\partial}{\partial t} \left(\frac{\delta\mathcal{L}}{\delta\dot\Phi_i} \right) + \frac{\partial}{\partial \boldsymbol{x}} \left(\frac{\delta\mathcal{L}}{\delta(\nabla\Phi_i)} \right) \,,$$

where Φ_i is a relevant set of nonrelativistic fields. For example, the Eq. (4.27) could be derived in this way.

[9]Well, the complex conjugate of that equation, but that does not matter.

[10]We already met it in Chap. 4 — see Eq. (4.47) and the following discussion.

We now construct the canonical Hamiltonian. Note that, in contrast to the Lagrangian density, the Hamiltonian density is not a Lorentz scalar.[11] The Hamiltonian itself represents the 0th component of a 4-vector. Therefore, it is no surprise that the canonical procedure to construct the field theory Hamiltonian is not manifestly Lorentz-covariant.

At the first step, we define the canonical momenta

$$\Pi = \frac{\delta \mathcal{L}}{\delta(\partial_0 \phi)} = \partial_0 \phi^*, \qquad \Pi^* = \frac{\delta \mathcal{L}}{\delta(\partial_0 \phi^*)} = \partial_0 \phi. \qquad (7.43)$$

The Hamiltonian density is

$$\mathcal{H} = \Pi^* \partial_0 \phi^* + \Pi \partial_0 \phi - \mathcal{L} = \Pi^* \Pi + (\partial_i \phi^*)(\partial_i \phi) + m^2 \phi^* \phi + \frac{\lambda}{4}(\phi^* \phi)^2. \quad (7.44)$$

It is positive definite. This explains our sign choice for m^2 in (7.40) and, especially, for λ in (7.41). With negative λ, the Hamiltonian (7.44) would have no lower bound. What is *worse*,[12] the classical field equations (7.42) would have no solution — the field ϕ would run to infinity at finite time. Such theories are bad classically and do not become better when one tries to quantize them (Fig. 7.2c).

On the other hand, there is absolutely nothing wrong with the theory (7.41) when $\lambda > 0$ and $m^2 < 0$ (Fig. 7.2b). The potential part of the Hamiltonian is

$$V(\phi^*, \phi) = \frac{\lambda}{4}(\phi^* \phi)^2 - |m^2| \phi^* \phi. \qquad (7.45)$$

True, the minimum of the potential energy is now not at zero, as was the case for $\lambda, m^2 > 0$, but at

$$\phi^* \phi = \frac{2|m^2|}{\lambda}. \qquad (7.46)$$

This is unusual, but does not lead to contradictions or inconsistencies. On the contrary, a theory with the potential (7.45) displays a very interesting and nontrivial *Goldstone effect*, related to the Higgs effect mentioned in Chap. 5 — the delicious *foie gras* in our set of *Hors d'Oeuvres*. We will discuss all that in detail in Chaps. 11 and 12.

The Hamiltonian $H = \int d^3x \, \mathcal{H}(\mathbf{x})$ is classical. But it is easy to write down its quantum counterpart — one only has to substitute the canonical momenta $\Pi(\mathbf{x})$ and

---

[11] It represents the component T_{00} of a certain tensor $T_{\mu\nu}$ called the *energy-momentum tensor*. We will meet it again in Chap. 15.

[12] The Hamiltonian in a free theory with $\lambda = 0$ and the potential $V = -m^2 \phi^* \phi$ also has no bottom. In this exotic *tachyonic* case, the field ϕ does not oscillate, but grows exponentially with time. However, in *free* tachyonic theory (the presence of interactions change that), it never runs into a singularity; there is no collapse and there are no internal inconsistencies...

Well, this footnote was actually addressed not to our target reader, not to Sophie, but to an expert who may happen to see it. We will rediscuss these nontrivial issues in the last chapter.

(a)   **(b)**   **(c)**

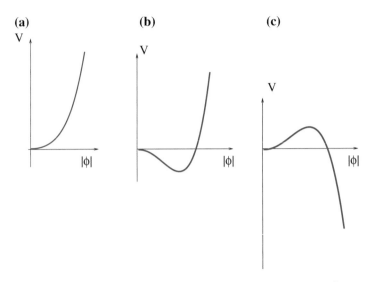

Fig. 7.2 The scalar potentials in (7.44) for different choices of parameters. **(a)** $m^2, \lambda > 0$: stable classical vacuum at $\phi = 0$; **(b)** $\lambda > 0, m^2 < 0$: stable classical vacuum at $|\phi|^2 = 2|m^2|/\lambda$; **(c)** $\lambda < 0$: no stable vacuum.

$\Pi^*(x)$ by the functional differential operators[13]

$$\Pi(x) \rightarrow -i\frac{\delta}{\delta\phi(x)},$$

$$\Pi^*(x) \rightarrow -i\frac{\delta}{\delta\phi^*(x)}. \qquad (7.47)$$

We can now pay our old debt and *derive* the Hamiltonians (4.36), (4.37) discussed earlier. To this end, we suppress the interaction term in (7.44), put the system in a finite box of length L and trade the continuous set of variables $\{\phi(x)\}$ for the discrete set of the Fourier coefficients as in (4.35). We obtain[14]

---

[13] When acting on the Lagrangian density $\mathcal{L}[\phi(x), \phi^*(x); \partial_\mu\phi(x), \partial_\mu\phi^*(x)]$, the symbol $\delta/\delta\phi(x)$ had the meaning of a usual partial derivative. But now it is a functional derivative acting on the wave functional (4.34).

[14] The reader is welcome to verify that

$$\frac{\delta}{\delta\phi(x)}\phi(y) = \delta(x - y) \qquad (7.48)$$

by trading the sum over the modes for the momentum integral:

$$\sum_n \rightarrow \frac{V}{(2\pi)^3}\int d^3p.$$

.

$$\frac{\delta}{\delta\phi(\boldsymbol{x})} = \frac{1}{V} \sum_n \exp\left\{-\frac{2\pi i \boldsymbol{n} \cdot \boldsymbol{x}}{L}\right\} \frac{\delta}{\delta c_n}. \tag{7.49}$$

Substituting this in (7.44) and doing the spatial integral, we arrive at (4.36).

The Hamiltonian (4.36) has an integral of motion

$$\boldsymbol{P} = \frac{2\pi i}{L} \sum_n \boldsymbol{n} (c_n^* \Pi_n^\dagger - c_n \Pi_n) \tag{7.50}$$

($\Pi_n = -i\partial/\partial c_n$, $\Pi_n^\dagger = -i\partial/\partial c_n^*$). That is nothing but the 3-momentum carried by the field. We can now understand the signs in (4.43) and (4.44). The eigenvalue of $\boldsymbol{P}$ in the state (4.43) is, indeed, $2\pi\boldsymbol{n}_0/L$, while it is $-2\pi\boldsymbol{n}_0/L$ for the state (4.44). For the vacuum state (4.45), the momentum is, of course, zero.

In quantum theory, the fields $\phi(x)$ and $\phi^*(x)$ become operators. The matrix elements of these operators and their Fourier components c_n, c_n^* in Fock space (we will need them later, in Chap. 10 and 12) are non-trivial. To understand what they are, recall what happens for an ordinary oscillator (4.38). As you know, one can define there the annihilation and creation operators

$$\hat{a} = \frac{m\omega x + i\hat{p}}{\sqrt{2m\omega}}, \qquad \hat{a}^\dagger = \frac{m\omega x - i\hat{p}}{\sqrt{2m\omega}}. \tag{7.51}$$

The operator $\hat{a}^\dagger$, when acting on the oscillator state of level k, creates the state $|k+1\rangle$, while the action of the operator $\hat{a}$ gives the state $|k-1\rangle$ (and $\hat{a}|0\rangle = 0$). The coordinate operator x is proportional to the sum $\hat{a}^\dagger + \hat{a}$ and has two types of matrix elements:

$$\langle k+1|x|k\rangle = \langle k|x|k+1\rangle = \sqrt{\frac{k+1}{2m\omega}}. \tag{7.52}$$

Similarly,

the operator c_n acting on a Fock state $|\Psi\rangle$ gives a superposition of two states: (i) the state where an extra antiparticle with momentum $-2\pi\boldsymbol{n}/L$ is created and (ii) the state where a particle of momentum $2\pi\boldsymbol{n}/L$ is annihilated

(if, of course, it was present in $|\Psi\rangle$).

And *the operator c_n^* may create a particle of momentum $2\pi\boldsymbol{n}/L$ or annihilate an antiparticle of momentum $-2\pi\boldsymbol{n}/L$.*

We go forth to the electromagnetic field. Well, in some sense, we go *back*; we already derived in the previous section the Lagrangian (7.27) describing the interaction of the field with *matter*: with the current generated by pointlike charged particles. And now we will write down the expression for the proper Lagrangian of the electromagnetic field.

The basic object now is the vector potential $A_\mu(x)$. Our experience with the scalar fields tells us that the kinetic part of the Lagrangian should be quadratic in derivatives of A_μ. A nice Lorentz-invariant expression for the density is

$$\mathcal{L}^{\text{EM-field}} = -\frac{1}{4}F_{\mu\nu}F^{\mu\nu} = \frac{1}{2}(\mathbf{E}^2 - \mathbf{B}^2) \tag{7.53}$$

(the coefficient $-1/4$ is the standard convention).

The density (7.53) has a remarkable symmetry (called the *gauge symmetry*). It is invariant under the transformation

$$A_\mu \rightarrow A_\mu + \partial_\mu \chi(x), \tag{7.54}$$

where $\chi(x)$ is an *arbitrary* function of the spatial coordinates and time. Indeed,

$$F'_{\mu\nu} = \partial_\mu A'_\nu - \partial_\nu A'_\mu = \partial_\mu(A_\nu + \partial_\nu\chi) - \partial_\nu(A_\mu + \partial_\mu\chi) = \partial_\mu A_\nu - \partial_\nu A_\mu = F_{\mu\nu}.$$

Thus, in contrast to ordinary symmetries, like rotational or Lorentz symmetries, with a finite number of transformation parameters (such symmetries are usually called *global* symmetries), the gauge symmetry (7.54) is a *local* symmetry involving an infinite continuum of parameters.

In some sense, the gauge symmetry is not quite a symmetry, but rather reflects a certain redundancy in our description of the electromagnetic field in terms of the variables $A_\mu(x)$. It is the components of the tensor $F_{\mu\nu}$, the electric and magnetic fields, that have a physical meaning, not the potential $A_\mu(x)$.[15]

Anyway, it is the gauge symmetry principle that distinguishes the density (7.53) compared to another independent Lorentz-invariant structure[16]

$$(\partial_\nu A_\mu)(\partial^\nu A^\mu) \equiv -A_\mu \Box A^\mu. \tag{7.55}$$

The latter is *not* gauge-invariant.

The Lagrangian density $\sim A_\mu j^\mu$ describing the interaction of the electromagnetic field with matter is not invariant under (7.54). However, the Lagrangian (7.27) *is* invariant up to a total time derivative (thus, the action $S = \int L \, dt$ *is* invariant) — that follows from the current conservation (7.29).

The Lagrangian density (7.40) of the free scalar field includes the kinetic part $(\partial_\mu \phi^*)(\partial^\mu \phi)$ and the potential part $-m^2 \phi^* \phi$. One could also try to include a potential term for the electromagnetic field. An appropriate Lorentz invariant is easily written:

$$\mathcal{L}^{\text{EM-field}}_{\text{pot}} \stackrel{?}{=} \frac{1}{2}m^2 A_\mu A^\mu. \tag{7.56}$$

However, the expression (7.56) is not gauge invariant and is not admissible for that reason.

"But *why* is it not admissible?" the confused reader may ask. "I understand the requirement of Lorentz invariance. The Lorentz symmetry is a fundamental sym-

---

[15]Indeed, one can easily *measure* $\mathbf{E}$ and $\mathbf{B}$, but not φ and A.

[16]The Lagrangian density is defined up to a total divergence. The latter does not contribute to the action (7.35).

metry of our world, it should be there. But what is wrong with a Lorentz-invariant theory of the massive vector field with

$$\mathcal{L} = -\frac{1}{4}F_{\mu\nu}F^{\mu\nu} + \frac{1}{2}m^2 A_\mu A^\mu \quad ? \tag{7.57}$$

Yes, it is not gauge-invariant. No, it is not gauge-invariant. So what?"

The answer to this legitimate question is the following. There is nothing wrong with the *free* Lagrangian density (7.57). But the trouble strikes back when one includes interaction with matter. It turns out that a *quantum* field theory (there are no pathologies at the classical level) describing the interaction of the massive vector field with the matter, such that the action is not gauge invariant, is not *renormalizable* (we already mentioned that in Chap. 5 and will discuss it in detail in Chap. 12). Thus, it is not internally consistent.

The Lagrange equations of motion, following from the density

$$\mathcal{L} = -\frac{1}{4}F_{\mu\nu}F^{\mu\nu} - eA_\mu j^\mu , \tag{7.58}$$

are Maxwell's equations[17]

$$\partial_\mu \left(\frac{\delta\mathcal{L}}{\delta(\partial_\mu A_\nu)} \right) - \frac{\delta\mathcal{L}}{\delta A_\nu} = 0 \implies \partial_\mu F^{\mu\nu} = ej^\nu . \tag{7.59}$$

When external currents are absent, these equations describe the propagation of light, whose quanta are massless photons. On the other hand, the equations of motion, following from (7.57), read

$$(\Box + m^2)A_\nu - \partial_\nu\partial_\mu A^\mu = 0 . \tag{7.60}$$

This *Proca equation* is similar to the KFG equation.[18] The Fock states of the corresponding quantum field Hamiltonian would be massive spin-1 particles.

We now understand why the physical photon is massless.[19] This is a corollary of the gauge invariance of the Lagrangian of the electromagnetic field (7.53), which in its turn is necessary to make the theory renormalizable!

Note that only the spatial components of the vector potential enter the Lagrangian density (7.58) with time derivatives. Thus, the true dynamical variables are the

---

[17]This is a compact 4-dimensional way to represent the second pair of Maxwell's equations in (4.31). The first pair of equations in (4.31) is just a corollary of (4.32).

[18]Incidentally, this explains our sign choice in (7.56) — we wanted to obtain an equation with oscillatory and not exponentially growing solutions — cf. the footnote on p. 117.

[19]The experimental limit for the photon mass is $m_\gamma < 3 \cdot 10^{-27}$ eV. This follows from the fact that the electromagnetic fields are long-range. We are sure that such long-range magnetic fields exist in our Galaxy because the light of the stars that we observe exhibits a small circular polarization, which can only be explained by interaction of the starlight with the interstellar magnetic fields. Hence the Compton wavelength of the photon cannot be much less than the size of the Galaxy.

$A(t, x)$ and not the $A_0(t, x)$. These latter play the role of *Lagrange multipliers*. The presence of the Lagrange multipliers in $\mathcal{L}$ reflects the gauge redundancy of the description of the system with the variables A_μ that we mentioned above.

The canonical momenta corresponding to the dynamical variables $A_i(t, x)$ are

$$P_i = \frac{\delta \mathcal{L}}{\delta \dot{A}_i} = F^{0i} = -E_i . \tag{7.61}$$

The canonical Hamiltonian of the electromagnetic field is

$$H = \frac{1}{2} \int dx (P^2 + B^2) = \frac{1}{2} \int dx (E^2 + B^2) . \tag{7.62}$$

Chapter 8
Cross Sections and Amplitudes

This is mostly a quantum mechanical chapter. First, we recall some basic facts about kinematics and quantum dynamics of the scattering processes.[1] In the second half of the chapter we introduce Feynman's path integrals. That section is not a necessary prerequisite to understand the subsequent chapters — our *Entrées* as well as *Desserts* will be cooked in a vegetarian manner, without path integrals.

Or rather the results that do in fact need path integrals in their derivation will be just announced and not derived. But it would be a pity not to give the reader a chance to learn this fascinating concept at the basic level.

8.1 Kinematics

Suppose there is a uniform flow of mosquitos flying in the positive x direction. The flux j characterizes the number of mosquitos passing unit area in the orthogonal yz-plane per unit time. Suppose a frog sits at the origin. It eats the passing mosquitos. We assume further that the frog does not miss any mosquito that passes sufficiently close to the origin (within the range l — the length of frog's tongue; we will assume that this length is much larger than the characteristic frog size and that the frog does not jump). On the other hand, the mosquitos that pass the plane $x = 0$ at distances from the origin larger than l fly by undisturbed (Fig. 8.1).

---

[1] The latter subject does not represent a part of the common undergraduate curriculum, but it is well described in many textbooks on quantum mechanics, in particular, in the momumental and all-encompassing Landau and Lifshitz book or in Tannoudji's pedagogical book. Besides these textbooks, I recommend the concise and very clearly written *Notes on Quantum Mechanics* by Fermi. These are notes of his lectures that he delivered back in 1954 at the University of Chicago. Scattering theory is discussed there in Sects. 23 and 33. It is just several pages, but they are worth reading. Of course, there is always an alternative: just stay with us!

© Springer International Publishing AG 2017
A. Smilga, *Digestible Quantum Field Theory*,
DOI 10.1007/978-3-319-59922-9_8

Fig. 8.1 Inelastic scattering
on an absorption center.

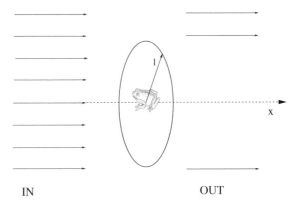

IN   OUT

A clever frog can easily calculate the number of mosquitos that it can expect to
enjoy per second. It is just

$$n = j(\pi l^2) \equiv j\sigma. \tag{8.1}$$

The value $\sigma = \pi l^2$ is the *effective cross section* for mosquito absorption. This *inelastic
scattering process* is the simplest physical situation where the notion of the effective
cross section appears.

In the model described above, the frog was active and the mosquitos passive. We
now replace the live frog by a malakhite model. At the same time, we endow the
mosquitos with some intelligence: seeing a frog, but not being able to distinguish a
model from the live frog, they try to avoid coming too close to the origin and change
the direction of their flight. Those mosquitos whose original distance from the x axis
was small, will deviate more, and those whose distance was large, will deviate less.
We are now dealing with *elastic* scattering: the mosquitos are not absorbed and their
total number is an integral of motion (Fig. 8.2).

We now place the mosquito counters far enough at the periphery of our device
and ask: how many mosquitos will pass per unit time into an element $d\Omega$ of the
solid angle? Obviously, this number is proportional to the flux j and to the solid
angle element $d\Omega$. The coefficient of proportionality has the dimension of area and
is called the *differential* cross section $d\sigma/d\Omega$:

$$n(d\Omega) = j\frac{d\sigma}{d\Omega} d\Omega. \tag{8.2}$$

The integral

$$\sigma = \int \frac{d\sigma}{d\Omega} d\Omega \tag{8.3}$$

is the *total* cross section of elastic scattering of mosquitos by the malakhite frog.

Frogs and mosquitos are classical objects, and it is the classical theory of scattering
that we have developed so far. As was mentioned in Chap. 5, it was the classical theory

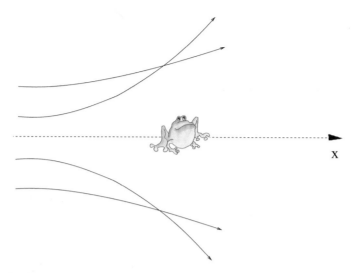

Fig. 8.2 Elastic scattering.

that Rutherford used to derive his formula (5.5) and to discover the atomic nucleus. But quantum-mechanical analysis gives the same result in this case!

The elastic cross section characterizes the probability for a particle to be scattered. In quantum mechanics, probabilities are derived from complex wave functions, $P \sim |\Psi|^2$. By the same token, the cross sections are derived from complex *scattering amplitudes*. The scattering amplitude is the amplitude of probability for the particle to come from spatial infinity, approach the scattering centre and to go to infinity again, but along a different direction.

It is clear from this definition that, to find the amplitude *directly*, one has to solve the time-dependent Schrödinger equation for the evolution of a wave packet with the appropriate boundary conditions at $t = -\infty$ and $t = +\infty$. Even better, one should use Feynman's path integral approach to calculate transition amplitudes at finite times and then take the limit $t \to \pm\infty$. However, we have not discussed path integrals yet. We will do so in the end of this chapter, but this is going to be an *optional* discussion.

For the benefit of the reader, we will now play conservative rednecks and explain how the scattering amplitude is defined and calculated in the *traditional* approach, in which one only needs to analyze the solutions of the *stationary* Schrödinger equation. This method is simpler, and it is this way that scattering is treated in most textbooks on quantum mechanics.

In the presence of the potential $V(x)$, the exact solution of the Schrödinger equation

$$(\triangle + k^2)\Psi = \frac{2mV}{\hbar^2}\Psi \tag{8.4}$$

$(k^2 = 2mE/\hbar^2)$ is complicated and can hardly ever be presented in an analytical form. In the next section, we will solve the problem *perturbatively*, assuming the potential to be small. We will also assume it to have a finite range a, being concentrated near the origin. Thus, at large distances r from the origin, the right-hand side in Eq. (8.4) is small, and the wave function represents an approximate solution of the *free* Schrödinger equation. We choose it in the form

$$\Psi = e^{ikx} + \frac{e^{ikr}}{r} f(\theta, \varphi). \tag{8.5}$$

Obviously, $(\triangle + k^2)e^{ikx} = 0$. The action of the operator $(\triangle + k^2)$ on the second term does not exactly give zero, but the result of this action is suppressed at large distances, $kr \gg 1$. Indeed, in spherical coordinates

$$\triangle\Psi = \frac{1}{r}\frac{\partial^2}{(\partial r)^2}(r\Psi) + \frac{1}{r^2}\left[\frac{1}{\sin\theta}\frac{\partial}{\partial\theta}\left(\sin\theta\frac{\partial}{\partial\theta}\right) + \frac{1}{\sin^2\theta}\frac{\partial^2}{(\partial\varphi)^2}\right]\Psi. \tag{8.6}$$

When acting on e^{ikr}/r, the radial part of the Laplacian together with the term k^2 gives zero, while the action of the angular Laplacian part gives a function that falls off as $\sim 1/r^3$ at large r.

The interpretation is the following. The term e^{ikx} describes the stationary flow of incoming quantum particles, which, following the mosquitos' example, approach the origin along the negative x axis. And the second term is an outgoing spherical wave describing the stationary flow of scattered particles.[2]

An important remark is now in order. Strictly speaking, the formula (8.5) is wrong. The term e^{ikx} describes the *initial* particles, and the second term — *final* particles. These two contributions correspond in fact to two completely different time moments (plus and minus infinity) and cannot interfere, as they do, if we take Eq. (8.5) at face value. To make the formula sensible, one should multiply the plane wave e^{ikx} by a profile function, like $\theta(R^2 - y^2 - z^2)$. This means that we collimate the ingoing beam by letting it pass through an orifice of radius R — something that an experimentalist does anyway!

We should be careful, however, to keep the aperture not too small, otherwise diffraction effects come into play and make a nuisance. More specifically, one should keep R much larger than the range a of the potential and much larger than the de Broglie wavelength $\lambda = 2\pi/k$ of the scattered particle, but much smaller than the distance r between the detector and the origin. In that case, diffraction is not important, and placing the detector at any angle other than $\theta = 0$ and $\theta = \pi$, we will observe only the scattered particles without the interference from the ingoing flow (Fig. 8.3).

The wave function (8.5) carries no dimension. The scattering amplitude $f(\theta, \varphi)$ has the dimension of length. The differential cross section is expressed via the ampli-

---

[2] That is the way we reduced the time-dependent scattering problem to a stationary one: by considering not a single particle, but the stationary flows of the ingoing and outgoing particles.

Fig. 8.3 Scattering of
quantum particles. The
vertical bars represent a
collimated ingoing plane
wave. The spreading circles
— outgoing spherical waves.
The small blob in the centre
represents the scattering
region.

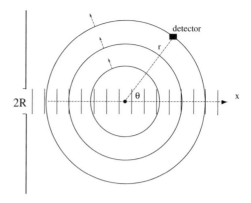

tude by a simple formula

$$\frac{d\sigma}{d\Omega} = |f|^2. \qquad (8.7)$$

Indeed, the flux density (5.2) of the ingoing plane wave is just the velocity, $j_{in} = v$.[3]
The number of ingoing particles passing unit area per unit time is proportional to this
quantity. For the scattered particles, the flux density is $j_{out} = v|f|^2/r^2$. The number
of particles passing into a solid angle element $d\Omega$ is proportional to j_{out} multiplied
by the area element $r^2 d\Omega$. Bearing in mind the definition (8.2), we deduce (8.7).

Generically, the amplitude depends on two angles, but in the important case of
scattering by a central potential, the problem is axially symmetric and the amplitude
depends only on the polar angle θ.

8.2 Born Approximation and Beyond

We are now in a position to *calculate* the amplitude. As was announced, we will solve
the problem perturbatively, assuming that the presence of the potential $V(x)$ brings
about small corrections to the wave function e^{ikx} that describes the ingoing plane
wave, and presenting the amplitude as a series in V. We start with the Schrödinger
equation (8.4) and represent its solution as a formal series:

$$\Psi = \Psi^{(0)} + \Psi^{(1)} + \Psi^{(2)} + \cdots, \qquad (8.8)$$

---

[3]Oops, it has the wrong dimension. This is due to our sloppiness — we have written the wave
function (8.5) without the normalization factor $1/\sqrt{V}$, where V is a fictitious finite spatial volume
(see the discussion on p. 36). If we took that into account, the flux density (5.2) would be divided
by the volume and acquire the correct dimension, $[j] = m^{-2} \cdot s^{-1}$. But this is all irrelevant for us,
because the cross section is defined via the *ratio* of the outgoing and ingoing fluxes. The unphysical
factor $1/V$ cancels out.

where $\Psi^{(0)} = e^{ikx}$ is the ingoing wave, $\Psi^{(1)}$ is the correction of the first order in V, $\Psi^{(2)}$ is the correction of the second order and so on. Equation (8.4) is equivalent to the infinite chain of equations

$$(\triangle + k^2)\Psi^{(1)} = \frac{2mV}{\hbar^2}\Psi^{(0)},$$

$$(\triangle + k^2)\Psi^{(2)} = \frac{2mV}{\hbar^2}\Psi^{(1)},$$

$$\cdots \qquad\qquad\qquad\qquad (8.9)$$

Our goal is to find the solution of all these equations at distances much larger than the characteristic size of the interaction region, $r \gg a$, and to compare the correction $\Psi^{(1)} + \Psi^{(2)} + \cdots$ with the second term in (8.5). This will give us the scattering amplitude.

As the first step, we determine $\Psi^{(1)}$. The equation

$$(\triangle + k^2)\Psi^{(1)} = \frac{2mV}{\hbar^2}\Psi^{(0)} \qquad\qquad (8.10)$$

is nothing but the *Helmholtz equation*, one of the basic equations of mathematical physics. Its general solution describing the outgoing wave at spatial infinity is known:

$$\Psi^{(1)}(\boldsymbol{x}) = -\frac{m}{2\pi\hbar^2} \int d\boldsymbol{y}\, \Psi^{(0)}(\boldsymbol{y})V(\boldsymbol{y}) \frac{e^{ik|\boldsymbol{x}-\boldsymbol{y}|}}{|\boldsymbol{x}-\boldsymbol{y}|}. \qquad (8.11)$$

Let us now make a digression and briefly explain here how this formula is derived. To begin with, we determine what is called the *fundamental solution*, the solution to the equation[4]

$$(\triangle + k^2)G(\boldsymbol{x}) = -\delta(\boldsymbol{x}). \qquad\qquad (8.12)$$

It reads[5]

$$G(\boldsymbol{x}) = \frac{e^{ikr}}{4\pi r}, \qquad r = |\boldsymbol{x}|. \qquad\qquad (8.13)$$

Indeed, bearing in mind (8.6), $(\triangle + k^2)G$ vanishes when $r \neq 0$. The solution is singular at the origin. In its immediate vicinity, one can replace $e^{ikr} \to 1$ and also forget about the term k^2 in the left-hand side of (8.12). Then the solution

$$G(\text{small } \boldsymbol{x}) \approx \frac{1}{4\pi r} \qquad\qquad (8.14)$$

---

[4]The negative sign on the right-hand side is a convention.
[5]Note that there are strictly speaking two fundamental solutions: the solution (8.13) and the solution $\propto e^{-ikr}$ representing a wave converging to the centre. We select the former.

is none other than the Coulomb potential of the unit pointlike charge placed at the origin, and its Laplacian indeed gives the delta-function.

We are interested here in real k, but Eq. (8.13) represents a fundamental solution of (8.12) also for *imaginary* $k = i\mu$.[6] We recognize the expression (5.16) for the Yukawa potential, stripped of the "physical" factor $-g^2$. This is by no means accidental. By the same token that the Coulomb potential represents a static solution of Maxwell's equations with pointlike charge, the Yukawa potential represents a static fundamental solution to the inhomogenous KFG equation

$$(\Box + \mu^2)\phi = \delta(x) \tag{8.15}$$

or to the equation $(\Delta - \mu^2)\phi = -\delta(x)$, the solution we have just derived.[7]

The general solution to the equation (8.10) is a convolution of the fundamental solution (8.13) with the source density on the right-hand side of (8.13). We thus derive (8.11).

The reader must have already seen a similar formula in a course of electromagnetism for the general solution of the Poisson equation

$$\Delta\varphi = -\rho(x) \implies \varphi(x) = \frac{1}{4\pi}\int dy\, \frac{\rho(y)}{|x-y|}. \tag{8.16}$$

We now go back to Eq. (8.11). When $r \gg a$ and hence $|x| \gg |y|$, we can make the approximations:

$$\frac{1}{|x-y|} \to \frac{1}{r},$$
$$e^{ik|x-y|} \to e^{ikr}e^{-ikn\cdot y}, \tag{8.17}$$

where $n = x/|x|$ is the direction of scattering. We also write

$$\Psi^{(0)}(y) = e^{ikn_0\cdot y}, \tag{8.18}$$

where n_0 is the direction of the ingoing wave. We derive

$$\Psi^{(1)} = -\frac{m}{2\pi\hbar^2}\frac{e^{ikr}}{r}\int dy\, e^{-iq\cdot y}\, V(y), \tag{8.19}$$

where $q = k(n - n_0) = \Delta p/\hbar$, Δp being the momentum transfer.

---

[6]The equation $(\Delta - \mu^2)\Psi = source$ is sometimes called the *screened Poisson equation*.

[7]The part *Chef's Secrets*, which we are in now, serves two purposes in our book. First, it gives necessary prerequisites for a reader who wishes to digest our *Entrées* and our *Desserts*. And second, we explain here *some* (not all) important facts and notions of classical and quantum mechanics that were served in the previous *Hors d'Oeuvres* part (especially, in Chap. 5) without a proper derivation. The Yukawa potential is one of these notions.

We have obtained a simple result: the scattering amplitude depends in this approximation only on the momentum transfer and is proportional to the Fourier transform of the potential:

$$f^{(1)}(q) = -\frac{m}{2\pi\hbar^2} \int dy\, e^{-iq\cdot y}\, V(y) \equiv -\frac{m}{2\pi\hbar^2} V_q. \tag{8.20}$$

For a central field, $V(y) = V(|y|)$, this amplitude depends only on $|q| = 2k\sin(\theta/2)$.

Equation (8.20) gives the scattering amplitude in the *Born approximation*. It is instructive to calculate it for the Yukawa potential (5.16). In natural units, we obtain

$$f = \frac{mg^2}{2\pi(q^2 + \mu^2)}, \tag{8.21}$$

which coincides with (5.15) multiplied by the appropriate factor and reduces to (5.4) in the limit $\mu \to 0$.

Let us now evaluate $\Psi^{(2)}$, the second-order correction to the wave function. We can write

$$\Psi^{(2)}(x) = -\frac{m}{2\pi\hbar^2} \int dy\, \Psi^{(1)}(y)V(y)\frac{e^{ik|x-y|}}{|x-y|}$$
$$= \left(\frac{m}{2\pi\hbar^2}\right)^2 \int dz\,dy\, \Psi^{(0)}(z)V(z)\frac{e^{ik|y-z|}}{|y-z|}V(y)\frac{e^{ik|x-y|}}{|x-y|}. \tag{8.22}$$

We substitute here $\Psi^{(0)}(z) = e^{ikn_0\cdot z}$, make the approximations (8.17) and represent $V(y)$ and $V(z)$ as the Fourier integrals:

$$V(z) = \frac{1}{(2\pi)^3}\int dq_1\, V_{q_1} e^{iq_1\cdot z}, \qquad V(y) = \frac{1}{(2\pi)^3}\int dq_2\, V_{q_2} e^{iq_2\cdot y}. \tag{8.23}$$

We obtain the following contribution to the amplitude:

$$f^{(2)} = \left(\frac{m}{2\pi\hbar^2}\right)^2 \frac{1}{(2\pi)^6}.$$
$$\int dz\,dy\,dq_1\,dq_2\, V_{q_1} V_{q_2} e^{ikn_0 z - ikny + iq_1 z + iq_2 y}\frac{e^{ik|y-z|}}{|y-z|}. \tag{8.24}$$

The fundamental solution of the Helmholtz equation entering (8.24) can also be represented as a Fourier integral:

$$\frac{e^{ik|y-z|}}{|y-z|} = \frac{1}{2\pi^2}\int ds\, \frac{e^{is\cdot(y-z)}}{s^2 - k^2}. \tag{8.25}$$

This result follows from (5.16) if we substitute $-ik$ for μ there.

Our watchful reader should protest at this point. How come the right-hand side of Eq. (8.25) is even in k, while the left-hand side is not. Something must be wrong there!

The reader is right. The point is that the integrand in (8.25) involves a pole at $s^2 = k^2$, and the integral is not defined until one specifies how this pole is handled.

We will not explain this here,[8] but the identity (8.25) is correct if the integral on the right-hand side is understood as a limit

$$\lim_{\epsilon \to +0} \int ds \, \frac{e^{is \cdot (y-z)}}{s^2 - (k + i\epsilon)^2}, \qquad (8.26)$$

where the notation $\epsilon \to +0$ means that ϵ tends to zero staying real and positive. The limit $\epsilon \to -0$ of the same integral would result in the opposite sign of k on the left-hand side and give a converging rather than spreading spherical wave.

We can thus rewrite (8.25) as

$$\frac{e^{ik|y-z|}}{|y - z|} = \frac{1}{2\pi^2} \int ds \, \frac{e^{is \cdot (y-z)}}{s^2 - k^2 - i0}, \qquad (8.27)$$

with $-i0$ standing for an infinitesimal negative imaginary quantity. Substituting this into (8.24), doing the integrals over $dydz$ and then over ds, we finally derive:

$$f^{(2)} = \left(\frac{m}{2\pi\hbar^2}\right)^2 \int dq_1 dq_2 \, \delta(q_1 + q_2 - q) \frac{V_{q_1} V_{q_2}}{2\pi^2[(k_{\text{in}} + q_1)^2 - k_{\text{out}}^2 - i0]}, \qquad (8.28)$$

where k_{in} and k_{out} are the initial and the final wave vectors and $q = k_{\text{out}} - k_{\text{in}}$. Unlike the leading Born amplitude (8.20), the correction $f^{(2)}$ depends not only on q, but also on k_{in}.

For perturbation theory to make sense, the correction should be small compared to the leading term. This gives a constraint for the potential; it should not be too large. Bearing in mind that the characteristic q in (8.20) are of order $1/a$, we derive the following conditions:

$$V \ll \frac{\hbar^2 k}{ma}, \qquad \text{if } ka \gg 1,$$

$$V \ll \frac{\hbar^2}{ma^2} \qquad \text{otherwise}. \qquad (8.29)$$

The higher-order corrections to the amplitude are derived quite analogously. The result for the scattering amplitude can be represented graphically as in Fig. 8.4.

---

[8]We draw moral support for our arrogant negligence from Valery Bryusov, a Russian poet of the Silver Age. He once noted: "Есть великое счастье — познав, утаить!" (To conceal what you perceived — is there a greater joy?). Actually, a clever reader familiar with the Cauchy's theorem of complex analysis can share this joy with the author. S/he can calculate the integral (8.26), and then conceal the derivation from the others!

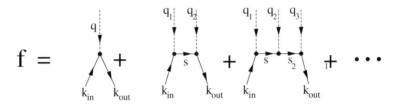

Fig. 8.4 Graphical representation for the nonrelativistic scattering amplitude.

Each graph in Fig. 8.4 encodes a particular analytic expression. The rules by which these expressions can be written are the following:

- Each blob provides the factor V_{q_i}, the Fourier component of the potential with the corresponding wave vector.
- There is no integral in the first-order *tree* graph. All other graphs have a *loop* nature and imply integration over the vectors q_i, with the sum $\sum_i q_i = q$ kept fixed:

$$f^{(n)} \sim \int (2\pi)^3 \delta\left(\sum_{i=1}^{n} q_i - q\right) \prod_{i=1}^{n} \frac{dq_i}{(2\pi)^3} \cdots .$$

- The momenta (the wave vectors) are conserved at each interaction point: the sum of the momenta entering each vertex is zero. For instance, in the second-order graph, $q_1 + k_{in} - s = 0$ and $q_2 + s - k_{out} = 0$.
- The horizontal lines are the nonrelativistic *propagators* depending on the corresponding momenta. Each such line gives a factor like

$$G(p_i) = \frac{1}{E - \frac{p_i^2}{2m} + i0}, \qquad (i = 1, \ldots, n-1) \qquad (8.30)$$

in the amplitude ($p_i = \hbar s_i$ and $E = \hbar^2 k_{in}^2/(2m) = \hbar^2 k_{out}^2/(2m)$ is the energy of the ingoing particle).
- Each graph is multiplied by a common overall factor $-m/(2\pi\hbar^2)$.

We have accurately derived thereby the *nonrelativistic diagram technique* for the scattering amplitude. Let us make several explanatory remarks.

- The reader who is familiar with ordinary perturbation theory in quantum mechanics can recognize in the propagator (8.30) a characteristic factor

$$\sim \frac{1}{E_n - E_m},$$

where E_n are the energy eigenvalues of the unperturbed Hamiltonian. Such factors appear in all perturbative formulas.

Of course, this is not at all surprising. The methods we used to derive (8.20), (8.28) and the higher-order contributions represented in Fig. 8.4 *are* the methods of stationary perturbation theory for the Hamiltonian $H = H_0 + V$.

- A heuristic interpretation for Eq. (8.28) describing the second-order correction is the following.

  First, the state $|in\rangle$ describing the free particle with momentum $\hbar k_{in}$ goes over to the intermediate state $|n\rangle$ with momentum p. The amplitude of this transition is given by the matrix element $\langle n|V|in\rangle$, which coincides in our normalization with V_{q_1}.

  Then the system lives in the intermediate state $|n\rangle$ for a time $\sim \hbar/(E_{in} - E_n)$. This "time uncertainty", associated with the energy nonconservation $\Delta E = E_{in} - E_n$, enters the amplitude as a factor.

  Finally, the system passes from the state $|n\rangle$ to the final state $|out\rangle$ with momentum $\hbar k_{out}$. The amplitude is multiplied by the matrix element $\langle out|V|n\rangle = V_{q_2}$. One should, of course, sum over all intermediate states, that is, integrate over the intermediate momenta.

- The nonrelativistic propagator (8.30) is the Fourier transform of the fundamental solution (8.13) of the Helmholtz equation multiplied by a proper normalization factor. And the Helmholtz equation is nothing but the Schrödinger equation where the potential term stays on the right, playing the role of a source. It is instructive to recall that in *momentum* space this equation acquires the form

$$\left(E - \frac{p^2}{2m}\right)\Psi_p = source.\tag{8.31}$$

It looks not as a partial differential equation any more, but as a trivial algebraic equation with the wave function multiplied by a momentum-dependent factor. The propagator (8.30) is simply the inverse of this factor.

All these observations might be useful, but they are not really needed in the nonrelativistic case. After all, we solved the problem exactly. In such cases, comments and explanations are optional. But later, when we discuss the relativistic diagram technique, we will not derive it rigorously, but, following our basic strategic idea, will only announce its rules, substantiating them by heuristic reasoning and nonrelativistic analogies. In this section we tried to provide a kind of firm ground for these analogies and reasoning.

8.3 Path Integrals

In the previous section, we derived the expression for the quantum mechanical scattering amplitude in Fig. 8.4 by solving the stationary Schrödinger equation. In QFT this approach is not convenient. To evaluate the QFT scattering amplitudes, people use other methods: *(i)* the *operator formalism* or *(ii)* *Feynman's path integral approach.*

The operator approach is more traditional. In fact, it amounts to describing a QFT system in the so-called *interaction picture* — a variant of the *Heisenberg picture* where the operators depend on time while the wave function does not.[9]

Feynman's method is more recent. At the dawn of QED, in the forties of the last century, it was not clear yet that these two approaches were equivalent. Schwinger (who developed to perfection the operator approach in QED) could not understand Feynman, and Feynman could not understand Schwinger. The equivalence of the two methods was proven in 1948 by Dyson. Even though they are equivalent, we learned by experience that the path integral approach is more powerful (it is also in some sense simpler). It is now exclusively used to treat *complicated* field theory systems, like non-Abelian gauge theories. As was said before, we will not describe in our book how it really works in QFT. But it is not so difficult to explain what is a path integral in ordinary nonrelativistic quantum mechanics. And that is what we will do. Not even speaking of applications — it is simply beautiful!

Consider a quantum system described by the time-dependent Schrödinger equation (4.8). Let the Hamiltonian involve only one dynamical variable q and its canonical momentum; the generalization for systems with several degrees of freedom is obvious. If we fix the initial conditions for $\Psi(q, t_0)$, the formal solution of Eq. (4.8) reads

$$\Psi(q, t) = \hat{U}(t - t_0)\Psi(q, t_0), \tag{8.32}$$

where

$$\hat{U}(t - t_0) = \exp\left\{-\frac{i}{\hbar}(t - t_0)\hat{H}\right\} \tag{8.33}$$

is the *evolution operator*. We now consider the *kernel* of the evolution operator, that is, the matrix element

$$\mathcal{K}(q^{(1)}, q^{(0)}; t_1 - t_0) = \langle q^{(1)}|\hat{U}(t_1 - t_0)|q^{(0)}\rangle \equiv \langle q^{(1)}, t_1|q^{(0)}, t_0\rangle \tag{8.34}$$

$(t_1 - t_0 \equiv \Delta t > 0)$. It describes the probability amplitude that the system will find itself at the position $q^{(1)}$ at $t = t_1$ provided it was located at $q = q^{(0)}$ at $t = t_0$. In terms of (8.34), the solution (8.32) is expressed as

$$\Psi(q^{(1)}, t_1) = \int dq^{(0)}\, \mathcal{K}(q^{(1)}, q^{(0)}; t_1 - t_0)\, \Psi(q^{(0)}, t_0). \tag{8.35}$$

The quantity $\mathcal{K}$ plays a fundamental role in the whole approach. The spectral decomposition of the kernel of the evolution operator reads

---

[9]In the interaction picture, wave functions do depend on time, but this dependence is not determined by the full Hamiltonian H, but only by H_0 — the part of H describing free noninteracting fields.

$$K(q^{(1)}, q^{(0)};\ \Delta t) = \sum_k \Psi_k(q^{(1)})\Psi_k^*(q^{(0)})e^{-\frac{i}{\hbar}E_k\Delta t}\,, \qquad (8.36)$$

where $\Psi_k(q)$ are the eigenstates of $\hat{H}$, with eigenvalues E_k.

I do not think that mathematicians will like to read this book,[10] but if the reader-physicist showed Eq. (8.36) to a mathematician, the latter would immediately recognize in $\mathcal{K}$ the fundamental solution of the time-dependent Schrödinger equation. For hyperbolic (time-dependent) equations, the fundamental solutions are defined in a somewhat different way compared to the solutions of elliptic (stationary) equations. $\mathcal{K}$ is the fundamental solution because:

- It satisfies

$$\left(i\hbar\frac{\partial}{\partial t} - \hat{H}^{(1)}\right)\mathcal{K}(q^{(1)}, q^{(0)};\ \Delta t) = 0\,, \qquad (8.37)$$

where the operator $H^{(1)}$ acts on the first argument in $\mathcal{K}$. Unlike (8.12), it is a *homogeneous* equation.
- The initial condition is

$$\mathcal{K}(q^{(1)}, q^{(0)};\ 0) = \delta(q^{(1)} - q^{(0)})\,.$$

In simple cases, the kernel can be found explicitly. For a free particle of mass m,

$$K(\boldsymbol{q}_1, \boldsymbol{q}_0;\ \Delta t) = \left(\frac{m}{2\pi\hbar\Delta t}\right)^{3/2}\exp\left\{\frac{im(\boldsymbol{q}_1 - \boldsymbol{q}_0)^2}{2\hbar\Delta t}\right\}\,. \qquad (8.38)$$

The reader may notice the similarity of this expression to the fundamental solution

$$\mathcal{K} = \left(\frac{1}{4\pi D\Delta\tau}\right)^{3/2}\exp\left\{-\frac{(\boldsymbol{q}_1 - \boldsymbol{q}_0)^2}{4D\Delta\tau}\right\} \qquad (8.39)$$

of the *diffusion equation*,

$$\frac{\partial n}{\partial \tau} = D\Delta n\,. \qquad (8.40)$$

This is not surprising. The diffusion equation is obtained from the free Schrödinger equation by the substitutions $t \to -i\tau$, $m \to \hbar/(2D)$. And its fundamental solution, the *heat kernel* (8.39),[11] is obtained from (8.38) by the same substitution.

---

[10]The attitude of my son was typical. Following my request, he read the first two parts and said that, even though he might have understood something, it was not pleasant reading for him — too heuristic and not at all rigorous. Well, this is, indeed, the way the book is written...

[11]It describes the spatial distribution at later times of an admixture introduced at $t = 0$ at the point $\boldsymbol{q}_0$.

An exact analytical expression for the kernel can also be derived for the Hamiltonian of the harmonic oscillator.[12] But how to solve the problem in the general case?

The wonderful fact is that, for any quantum system, the kernel $\mathcal{K}$ can be evaluated as a path integral

$$\mathcal{K}(q^{(1)}, q^{(0)}; \Delta t) = \int \prod_t dq(t) \exp \left\{ \frac{i}{\hbar} \int_{t_0}^{t_1} L(\dot{q}, q)dt \right\}, \qquad (8.41)$$

with the boundary conditions $q(t_0) = q^{(0)}$, $q(t_1) = q^{(1)}$.

The symbol $\int \prod_t dq(t)$ means that we integrate over all possible trajectories, that is, over all the coordinate values at each intermediate time moment. Thus, we have to perform an infinite continuum of integrations!

When one encounters an infinity, the only way to attribute a meaning to the symbol ∞ is to understand it as the limit of a very large number. For instance, an ordinary integral represents an infinite sum of infinitesimal contributions. This was difficult for Archimedes, but we now know how to define it rigorously: as a limit $n \to \infty$ of the sum of the large number n of very small terms. Likewise, the path integral is defined as a limit of a multidimensional integral when the number of dimensions tends to infinity.

To be more explicit:

- In order to attribute a meaning to the symbol (8.41), we divide the time interval into a large number n of small time slices, $\epsilon = \Delta t / n$. We understand $\int \prod_t dq(t)$ as

$$\int \prod_t dq(t) = \lim_{n \to \infty} C_n \prod_{j=1}^{n-1} dq_j,$$

  where

$$q_j = q(t_0 + j\epsilon), \qquad 1 \leq j \leq n - 1$$

  are the intermediate positions over which we integrate (on the other hand, $q_0 \equiv q^{(0)}$ and $q_n \equiv q^{(1)}$ are fixed). Now, C_n is an n-dependent constant. For each particular quantum system, one can evaluate C_n, but we do not need it.

- The integral $\int_{t_0}^{t_1} L(\dot{q}, q)dt$ is also replaced by a finite sum. The contribution of each time slice is[13]

$$\epsilon L \left(\frac{q_j - q_{j-1}}{\epsilon}, q_{j-1} \right).$$

For example, for one-dimensional potential motion with the Lagrangian

---

[12]The reader can find it in the classical book by Richard Feynman and Albert Hibbs, *Quantum Mechanics and Path Integrals.*

[13]For a sufficiently smooth trajectory, one can represent the derivative $\dot{q}[t_0 + (j-1)\epsilon]$ by a finite difference $(q_j - q_{j-1})/\epsilon$.

$$L = \frac{\dot{q}^2}{2} - V(q), \tag{8.42}$$

the transition amplitude is defined as

$$\mathcal{K}(q^{(1)}, q^{(0)}; t_1 - t_0) = \lim_{n \to \infty} C_n \int \prod_{j=1}^{n-1} dq_j \, \exp\left\{\frac{i}{\hbar}\left[\frac{(q^{(1)} - q_{n-1})^2}{2\epsilon}\right.\right.$$
$$\left.\left. + \cdots + \frac{(q_1 - q^{(0)})^2}{2\epsilon} - \epsilon V(q_{n-1}) - \cdots - \epsilon V(q^{(0)})\right]\right\}. \tag{8.43}$$

We will now explain where the representation (8.41) comes from. We will do so for the simple Lagrangian (8.42), where the kernel is spelled out as in (8.43). The general case is treated in Feynman's book as well as in any modern textbook in quantum field theory.

To simplify the notation, we set $t_0 = 0$, $t_1 = \Delta t \to t$. We note then that the transition amplitude $\langle q_1, t | q_0, 0 \rangle$ can be represented as

$$\langle q^{(1)}, t | q^{(0)}, 0 \rangle = \int_{-\infty}^{\infty} dq_* \left\langle q^{(1)}, t \left| q_*, \frac{t}{2} \right\rangle \left\langle q_*, \frac{t}{2} \right| q^{(0)}, 0 \right\rangle. \tag{8.44}$$

This immediately follows from the spectral decomposition (8.36) and the orthogonality condition for the functions $\Psi_k(q_*)$. But then we can subdivide the time interval t not in two halves, but in a large number n of small time slices $\epsilon = t/n$ to derive

$$\langle q^{(1)}, t | q^{(0)}, 0 \rangle = \prod_{j=1}^{n-1} \int_{-\infty}^{\infty} dq_j \, \langle q^{(1)}, t | q_{n-1}, t - \epsilon \rangle \cdots \langle q_1, \epsilon | q^{(0)}, 0 \rangle. \tag{8.45}$$

To reduce it to (8.43), one only has to show that, for small ϵ,

$$\langle q_1, \epsilon | q^{(0)}, 0 \rangle \approx \sqrt{\frac{m}{2\pi\hbar\epsilon}} \exp\left\{\frac{im(q_1 - q^{(0)})^2}{2\hbar\epsilon}\right\} \exp\left\{-\frac{i}{\hbar}\epsilon V(q^{(0)})\right\}, \tag{8.46}$$

etc. And this follows from (8.33): the first two factors in (8.46) come from the kinetic part of the Hamiltonian, it is nothing but the free kernel (8.38), and the third factor — from the potential part.

Many quantum mechanical phenomena can be understood much better, when bearing in mind the fundamental *optics-mechanics analogy*.[14] This analogy is very deep and encompasses many aspects of the two theories. For instance, a particle in mechanics corresponds to a wave packet in optics, a classical trajectory in mechanics to a ray in optics, the potential energy in mechanics to the refraction index in optics.

---

[14]Fermi's *Notes on Quantum Mechanics* start with explaining that, and this is one of the reasons why I recommended reading his book.

Finally, Fermat's principle of least time in optics — to Maupertuis' principle (a variant of the principle of least action) in mechanics.

The expression (8.41) for the quantum transition amplitude also has a close analogy in optics — the Huygens–Fresnel principle. The latter says that the amplitude of the electromagnetic wave emitted at point A and detected at point B represents an integral over all possible paths of rays between A and B, similar to (8.41). The only difference is that the action $\int L dt$ measured in units of $\hbar$ is replaced in optics by the *phase*

$$\Phi = \frac{2\pi l}{\lambda}, \tag{8.47}$$

associated with the particular ray path (here l is the optical path length, that is, the time that the light takes to travel along the path multiplied by the speed of light in vacuum, and λ is the vacuum wavelength).

In the *semiclassical limit*, when the phases (8.47) (and also the relevant phase differences) are large, wave optics goes over to geometric optics. Indeed, the Huygens-Fresnel path integral,

$$\text{wave amplitude} \sim \int_{\text{paths}} e^{i\Phi[\text{path}]}, \tag{8.48}$$

is mainly saturated in this case by the paths close to the geometric ray path realizing the minimum[15] of the functional $\Phi[\text{path}]$, i.e. the minimum of the travel time. At the minimum the variation $\delta\Phi$ is zero. This means that, for the bunch of paths in the vicinity of the minimum, the values of $\Phi[\text{path}]$ are close, and there is constructive interference between different such paths, giving a large contribution to the integral (8.48).

The same applies to mechanics. When the action expressed in units of $\hbar$ is large, the exponent in the path integral (8.41) oscillates wildly. Most trajectories interfere destructively except those trajectories that are close to the classical trajectory which realizes the minimum of the functional $S[q(t)]$. These trajectories interfere constructively, giving the main contribution to the path integral.

That is why in ordinary life, when all the characteristic actions are very large compared to $\hbar$, we do not feel quantum jerks in the motion of particles, their "*zitterbewegung*". The world is then classical and orderly and nice...

The path integral has been used in physics for more than half a century. But not in mathematics — mathematicians still do not like it and consider it not to be a rigorously defined object. To be more precise, they have nothing against the *Euclidean* path integral where the time is considered to be imaginary, $t = -i\tau$. In that case, the function to be integrated in (8.41) goes over to

---

[15] A maximum would also work, but $\Phi[\text{path}]$ has no maximum. For exotic paths straying far away from the points A and B, the optical path length can be arbitrarily large.

$$\exp\left\{-\frac{1}{\hbar}\int d\tau\left[\left(\frac{dq}{d\tau}\right)^2 + V(q)\right]\right\}. \tag{8.49}$$

If the potential grows at large q, the contribution of the trajectories with wildly deviating $q(\tau)$ is exponentially suppressed. Mathematicians call such an integrand a *Wiener measure* and can work with it. But they cannot work with real time functional integrals with oscillating complex measure. More precisely, they cannot prove the existence of the limit of a finite-dimensional integral (8.43), when the number of points goes to infinity and the time interval ϵ between the adjacent points goes to zero.

I am not myself a mathematician and cannot even try to prove it. The only thing that I can say is that, for simple-minded but nontrivial quantum mechanical problems like the anharmonic oscillator, the real-time path integrals were calculated *numerically* and the existence of the limit $\epsilon \to 0$ was established *experimentally*, if you will. Unfortunately, I know only a few scientific papers where such real time calculations were performed. There is not much research activity in this direction, which is a pity.

In this section, we discussed only path integrals in quantum mechanics. But one can also define and calculate path integrals in field theories. As was mentioned, people use them to derive the Feynman rules for perturbative calculations in complicated theories. But many efforts were also devoted to numerical calculations of path integrals in QCD. (I am talking now of Euclidean path integrals with Wiener measure. Calculating numerically Minkowski real-time path integrals in QCD is a task beyond the reach of modern computer technology.) To perform such calculations, one should not only subdivide time into small intervals, but do the same for space. We arrive at the *lattice path integral* defined on a 4-dimensional Euclidean cubic lattice.

There is no rigorous proof, but people believe that, for *non-Abelian* gauge theories, these lattice path integral have a well-defined continuum limit when the lattice spacing tends to zero. This faith is based on experience and also on the fact that the effective coupling constant in such theories decreases at small distances due to asymptotic freedom [see Eq. (5.17) and discussion thereof in Chap. 5]. On the other hand, there is no reason to expect that the continuum limit exists for theories like QED, where the effective coupling constant grows at small distances. In other words, such theories are defined only *perturbatively* — if the coupling constant is small, one can do perturbative calculations, and one can do so only in a limited range of energies.

We will come back to this question at the end of Chap. 12, where we will talk about imperfections of the Standard Model and of our present understanding of the structure of the World.

Part IV
Entrées

Chapter 9
Fermion Fields

In Chaps. 4 and 7 we presented some explicit expressions for field theory Hamiltonians and Lagrangians. But we did so only for the bosonic fields: the scalar fields describing spinless particles and the electromagnetic field that describes the photons carrying unit spin.

We know, however, that the Universe also contains the fermions: leptons and quarks that carry spin 1/2. The reader also knows (if not from other sources, then from our brief remarks in Chap. 3) that bosons and fermions have very different properties. The bosons enjoy staying together. Thus, many photons occupying the same quantum state form a coherent laser beam. On the other hand, coherent electron waves do not exist because two electrons (not to speak of *many*) cannot occupy one and the same quantum state.[1] Physicists say that the photons and electrons have different *statistics*: the photons obey the so-called Bose–Einstein statistics requiring the wave function of a many-photon state to be symmetric under photon interchange, while the electrons obey the Fermi–Dirac statistics: the multielectron wave function is necessarily antisymmetric under exchange of any pair of particles:

$$\Psi(\gamma_1, \gamma_2, \ldots) = \Psi(\gamma_2, \gamma_1, \ldots),$$
$$\Psi(e_1, e_2, \ldots) = -\Psi(e_2, e_1, \ldots). \tag{9.1}$$

In the standard courses of nonrelativistic quantum mechanics, one only shows that the supplementary conditions (9.1) *can* be imposed, that they do not contradict the principles of quantum mechanics and that, for example, an antisymmetric multielectron wave function stays antisymmetric during time evolution. But the *origin* of these conditions remains obscure. One can only understand where they really come from in the framework of relativistic field theory, and we will do so in this chapter.

---

[1] As the reader understands, it is this Pauli exclusion principle that is responsible for the rich and nontrivial atomic structure involving a set of electron shells.

© Springer International Publishing AG 2017
A. Smilga, *Digestible Quantum Field Theory*,
DOI 10.1007/978-3-319-59922-9_9

9.1 First Try

We start by writing down an expression for the relativistic Lagrangian describing fields of spin $\frac{1}{2}$ that is invariant under Lorentz transformations. Spinor representations of the Lorentz group were discussed in Chap. 6. We learned there that the group $SO(3, 1)$ is not simple and represents in some sense a product of two $SU(2)$ groups. The representations are thus labelled by two numbers (j_L, j_R), each of which can be integer or half-integer. There are two different spinor representations, the left-handed spinors ξ_α carrying the undotted indices $\alpha = 1, 2$ and belonging to the representation $\left(\frac{1}{2}, 0\right)$ and the right-handed spinors $\eta_{\dot\alpha}$ carrying the dotted indices and belonging to the representation $\left(0, \frac{1}{2}\right)$.

Take one of them, say, the left-handed spinor field $\xi_\alpha(x)$.[2] As is clear from (6.57), the complex conjugated left-handed spinor $(\xi_\alpha)^*$ belongs to the representation $\left(0, \frac{1}{2}\right)$ and can be denoted $\xi_{\dot\alpha}^*$. The simplest Lorentz-invariant expression involving the derivative operator and the field ξ_α is

$$\mathcal{L}_{\text{kin}}^{\text{ferm}} = i\xi^\alpha \partial_{\alpha\dot\beta} \xi^{*\dot\beta} \equiv i\xi^\alpha (\sigma^\mu)_{\alpha\dot\beta} \partial_\mu \xi^{*\dot\beta}, \tag{9.2}$$

with the notation σ^μ introduced in (2.11) and (6.60), and spinor indices being raised and lowered according to (6.54). By construction, the Lagrangian (9.2) is invariant under $SO(3, 1) = SU_L(2) \times SU_R(2)$. It describes free massless left-handed fermions and their antiparticles. The Lagrangian density (9.2) should have the dimension m^4. This fixes the canonical dimension of the fermion field, $[\xi] = m^{3/2}$ (we recall that the canonical dimension of bosonic scalar and vector fields is $[\phi] = [A_\mu] = m$).

Let us treat (9.2) in the same way as we did for the Lagrangian (7.40) describing the free scalar field in Chaps. 4 and 7. We introduce a finite spatial box of size L and impose periodic boundary conditions on $\xi_\alpha(\mathbf{x}, t)$:

$$\xi_\alpha(x + L, y, z) = \xi_\alpha(x, y + L, z) = \xi_\alpha(x, y, z + L) = \xi_\alpha(x, y, z). \tag{9.3}$$

We expand $\xi_\alpha(\mathbf{x})$ in a Fourier series

$$\xi_\alpha(\mathbf{x}) = \sum_{\mathbf{n}} \xi_\alpha^{\mathbf{n}} \exp\left\{\frac{2\pi i \mathbf{n} \cdot \mathbf{x}}{L}\right\}, \tag{9.4}$$

substitute it in (9.2) and integrate over $\int d\mathbf{x}$. We obtain[3]

$$L = V \sum_{\mathbf{n}} \left[i\xi_{\mathbf{n}}^\alpha \delta_{\alpha\dot\beta} \frac{d}{dt}\left(\xi_{\mathbf{n}}^{*\dot\beta}\right) + \frac{2\pi n_j}{L} \xi_{\mathbf{n}}^\alpha (\sigma_j)_{\alpha\dot\beta} \xi_{\mathbf{n}}^{*\dot\beta} \right]. \tag{9.5}$$

---

[2] We will understand why such a field is called "left-handed" a little bit later.
[3] Hopefully, the same notation L for the Lagrangian and for the length of the box will not result in any confusion.

In the quadratic approximation, the different modes do not interact and can be treated separately, as we did for bosons. The difference is that the bosonic Lagrangian (7.40) involved the squares of time derivatives, while the Lagrangian (9.5) has only terms linear in derivatives. This unusual feature does not represent, as such, a major difficulty.

To simplify things, set $V = L^3 = 1$ and consider a term in (9.5) with some definite n. One can choose e.g. $n_j = (0, 0, 1)$. The contribution to the Lagrangian reads

$$L_{(001)} = i(\xi^1 \dot{\xi}^{*1} + \xi^2 \dot{\xi}^{*2}) + 2\pi(\xi^1 \xi^{*1} - \xi^2 \xi^{*2}). \tag{9.6}$$

Here and later the dots stand for time derivatives. On the other hand, the dots over the indices have been suppressed — they do not have much meaning, if we concentrate on a particular mode so that Lorentz invariance is lost.

The Lagrangian (9.6) represents a sum of two independent contributions: *(i)* for the component ξ^1 and its conjugate and *(ii)* for the component ξ^2 and its conjugate. Consider first the part involving the components ξ^2 and ξ^{*2}. We see that the variables ξ^2 and ξ^{*2} are canonically conjugate to each other:

$$\Pi_{\xi^{*2}} = i\xi^2, \qquad \Pi_{\xi^2} = -i\xi^{*2}, \tag{9.7}$$

so that the Poisson bracket [defined in (7.12)] is $\{\xi^{*2}, \xi^2\} = i$. The contribution to the canonical Hamiltonian is

$$H^{(2)}_{(001)} = 2\pi\xi^2\xi^{*2}. \tag{9.8}$$

Our intelligent reader may recognize in the expression (9.8) the Hamiltonian of the usual harmonic oscillator, but expressed via the holomorphic variables:

$$H^{\text{osc}} = \frac{p^2 + \omega^2 q^2}{2} = \omega a^* a, \tag{9.9}$$

where $\omega = 2\pi$ ($\omega = 2\pi/L$ for the box of length L) and[4]

$$a = \frac{\omega q + ip}{\sqrt{2\omega}}, \qquad a^* = \frac{\omega q - ip}{\sqrt{2\omega}}. \tag{9.10}$$

The Poisson bracket is $\{a, a^*\} = i$ and hence the variable ξ^2 in (9.8) plays the role of a^* and the variable ξ^{*2} the role of a.

Following Heisenberg, we know how to quantize this Hamiltonian. We attribute the role of the coordinate to a^* and the role of the momentum Π_{a^*} to $-ia$. In quantum theory,

$$\Pi_{a^*} = -ia \longrightarrow -i\frac{\partial}{\partial a^*} \tag{9.11}$$

---

[4]Please do not mix up (9.10) with the variables (4.39), which were not holomorphic, but simply complex coordinates of the two-dimensional oscillator (4.37).

The classical Hamiltonian is a product of the coordinate and momentum. To write its quantum counterpart, we should resolve the ordering ambiguity, to specify in what order a^* and the differential operator (9.11) should go. Not that it is tremendously important — the different choices amount to shifting the energy by a constant — but the most aesthetic choice is

$$\hat{H} = \frac{1}{2}\omega\left(a^*\frac{\partial}{\partial a^*} + \frac{\partial}{\partial a^*}a^*\right). \tag{9.12}$$

This gives the standard quantum oscillator Hamiltonian $\hat{H} = (\hat{p}^2 + \omega^2 q^2)/2$. The normalized eigenfunctions of (9.12) are

$$\Psi_k(a^*) = \frac{(a^*)^k}{\sqrt{k!}}. \tag{9.13}$$

The energy eigenvalues are, as you know,

$$E_k = \omega\left(k + \frac{1}{2}\right). \tag{9.14}$$

The result that we obtained looks quite similar to what we derived earlier for the scalar field in Sect. 4.7. The state characterized by the quantum number k can be associated with the presence of k particles in the state with momentum $\boldsymbol{p} = \frac{2\pi}{L}(0, 0, 1)$ and negative helicity,

$$\frac{1}{2}\frac{\boldsymbol{\sigma}\cdot\boldsymbol{n}}{|\boldsymbol{n}|}\begin{pmatrix}0\\1\end{pmatrix} = -\frac{1}{2}\begin{pmatrix}0\\1\end{pmatrix}$$

[cf. Eq. (5.23)].

Consider, however, the second part in the Lagrangian (9.6) involving the positive-helicity components ξ^1 and ξ^{*1}. It is the same as the first part, but the potential term, giving the Hamiltonian, now has the opposite sign:

$$H^{(1)}_{(001)} = -2\pi\xi^1\xi^{*1}. \tag{9.15}$$

The classical Hamiltonian (9.15) is negative and the spectrum of its quantum counterpart includes negative energies. Restoring the dimensional factor $1/L$ in the energy, we finally obtain

$$E_{(001)} = \frac{2\pi}{L}(k_2 - k_1) \tag{9.16}$$

with integer k_1, k_2.

This spectrum has no bottom, a ground state is absent! This means that the whole philosophy of quantum field theory we are used to (the ground state of the field

Hamiltonian is the physical vacuum, the excited states are physical particles) breaks down. We are in trouble.[5]

9.2 Grassmann Waves in the Dirac Sea

This trouble was first detected by Dirac, who was surprised to discover that his relativistic equation describing one-electron states[6] admits solutions with both signs of energy — positive and negative. His suggestion to cope with this setback was the following:

- The electron is a massive particle and, when external fields are absent, the energies of the levels are

$$E_p = \pm\sqrt{p^2 + m^2}\,. \tag{9.17}$$

The spectrum thus involves a gap.
- We accept the Pauli exclusion principle saying that each such level can be occupied by only one particle.
- Dirac then defines the physical vacuum state as the state where all the negative-energy levels below the gap are occupied, while the positive-energy levels are empty (Fig. 9.1a). This vacuum has an infinite negative energy, but, as was mentioned earlier, in a theory not involving gravity, the absolute value of the energy is not a physically observable quantity. The ground state of the scalar-field Hamiltonian has an infinite positive energy [see Eq. (4.42)]. The ground state of the fermion field Hamiltonian has an infinite negative energy. Both are OK.

There are then two different types of elementary excitations of the multi-particle Hamiltonian. First, there are the states where all the negative-energy levels and in addition a positive-energy level are occupied (Fig. 9.1b). Such states represent physical electrons. Second, there are the states where all positive-energy levels are empty and one of the negative-energy levels is empty too, all others being still occupied. One fish has been extracted out of the sea leaving a hole in its place. (Fig. 9.1c).

The physics of the Dirac sea bears a considerable resemblance to the physics of semiconductors. The latter involve an analogue of the Dirac sea — the *valence band* of electrons including the electrons in the outer atomic shells. There are two types of charge carriers in semiconductors: the electrons in the so-called *conduction band* (their energy exceeds the energy of the valence electrons; the conduction band is empty at zero temperature) and the *holes* — unoccupied levels in the valence band.

---

[5]In Sect. 16.4.2, we will argue that the ghosts (excitations with negative energies) might be not so troublesome in some special cases, but we are not ready yet to digest this interesting speculation.

[6]We are going to discuss now the multiparticle states of the field Hamiltonian and at the same time the one-particle states representing the solutions of the Dirac equation. (The reader may be frustrated: while talking a lot about the Dirac equation, we have not written it down yet. Well, I promise to do so by the end of this chapter!) To make a distinction between them, we will still call the former "states", while the one-particle states will be further referred to as "levels".

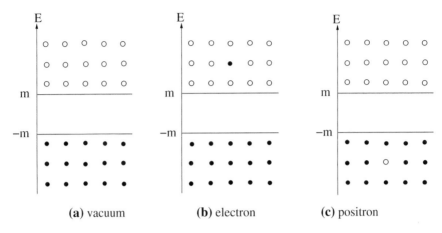

Fig. 9.1 The Dirac sea. *Black* and *white circles* represent filled and empty levels, respectively.

Imagine now that an electric field is switched on. The state with all bound levels filled up and no electrons in the conduction band just does not react to the field, and there is no electric current flow. But if conduction electrons are present, they are free to move — the conduction band is never crowded, and the electrons do not have a lot of neighbours to hinder their walk. For a hole, a neighbouring valence band electron can move to its place driven by the field, and occupy the level that was empty before. But then the adjacent level (from where the electron came from) is emptied. For an external observer this looks as the motion of the hole. It is rather obvious that the conduction electrons and the holes move in opposite directions. The electrons are negatively charged, but the holes are positively charged.

And going back to the holes in the Dirac sea — they are nothing but the positrons![7] Note that the process of e^+e^- annihilation acquires a nice interpretation in this picture. In the same way that an atomic electron occupying an excited level in an outer shell can fall down to an empty level in an inner shell, with the emission of a photon, so a free electron in the "conduction zone" can find a hole in the Dirac sea and fall down there. Kinematics does not allow for the transition energy to be released in this case as a single photon. But two (or three or more) photons can well be produced. An observer sees that the initial electron and the initial positron (the hole) disappeared, and their energy (including the energy of the gap $2mc^2$, which can be also interpreted as the sum of the electron and positron rest energies) is released

---

[7]The existence of the positrons was a major prediction of the Dirac theory, made in 1928–1929. In 1932, they were definitely discovered in cosmic rays by Carl Anderson. (One can mention also the experiment of Dmitry Skobeltzyn, who saw back in 1929 some "strange" tracks in his Wilson camera which looked similar to electron tracks, but curved to the opposite side in a magnetic field, compared to the electron tracks. That meant that the particles which produced these tracks carried positive charge. However, Skobeltzyn could not measure the energy of these particles with his methods and could not say much about their nature.)

in the form of radiation. This is quite similar to electron-hole recombination in a semiconductor.

It looks fine, but not quite. We still do not know how to reproduce and confirm this heuristic Dirac picture in the framework of the exact field theory analysis. The basic question is where does the Pauli exclusion principle come from? The formula (9.16), derived in the previous section, bears no trace of it!

This problem is not at all simple. Its almost satisfactory resolution was due to Markus Fierz (in 1939) and Wolfgang Pauli (in 1940). In the framework of the Heisenberg operator approach to quantum mechanics, they observed that, while the spectrum of the quantum field theory Hamiltonian describing particles of half-integer spin is bottomless if one proceeds in the usual way, the spectrum acquires a ground state if instead one postulates nontrivial *anticommutators* between the canonical field momenta and coordinates.

But the question of where these nontrivial anticommutators come from (standard quantum mechanics involves only nontrivial commutators, like $[\hat{p}, x] = -i\hbar$) was not answered at that time.

Full understanding came only in the nineteen sixties, when Berezin understood that, in the *classical* mechanics of fermion fields, the dynamical variables are not ordinary real or complex numbers, but Grassmann anticommuting numbers. Their mathematical theory was given in Chap. 6.

Consider the classical Lagrangian

$$L = -i\dot{\xi}\xi^\dagger + \omega\xi\xi^\dagger , \tag{9.18}$$

where now ξ and $\xi^\dagger$ anticommute:[8]

$$(\xi)^2 = (\xi^\dagger)^2 = 0, \qquad \xi\xi^\dagger + \xi^\dagger\xi = 0 . \tag{9.19}$$

The equations of motion are

$$-i\dot{\xi} + \omega\xi = 0, \qquad i\dot{\xi}^\dagger + \omega\xi^\dagger = 0 . \tag{9.20}$$

The variables ξ and $\xi^\dagger$ "parametrize" phase space (whatever the word "parametrize" means for a Grassmann coordinate). The Poisson bracket of two functions $f(\xi^\dagger, \xi)$, $g(\xi^\dagger, \xi)$ is now defined as

$$\{f, g\} = -i\left(\frac{\partial f}{\partial \xi^\dagger}\frac{\partial g}{\partial \xi} + \frac{\partial f}{\partial \xi}\frac{\partial g}{\partial \xi^\dagger}\right), \tag{9.21}$$

with the Grassmann derivative operator defined in Sect. 6.4. Note that the order of the factors on the right-hand side of (9.21) is important!

---

[8] We recall that we use the *dagger* rather than *star* symbol to denote the complex conjugation of Grassmann variables.

It is the Poisson bracket (9.21) which now enters the relation (7.12) describing the time evolution of a phase-space function,

$$\frac{df}{dt} = \{H, f\}. \tag{9.22}$$

with

$$H = \omega \xi^\dagger \xi. \tag{9.23}$$

A generic expression for the Poisson bracket in a system involving several usual real commuting phase-space variables (p_j, q_j) and several complex anticommuting phase-space variables $(\xi_\alpha, \xi_\alpha^\dagger)$ reads

$$\{f, g\} = \sum_j \left(\frac{\partial f}{\partial p_j} \frac{\partial g}{\partial q_j} - \frac{\partial f}{\partial q_i} \frac{\partial g}{\partial p_j} \right) - i \sum_\alpha \left(\frac{\partial f}{\partial \xi_\alpha^\dagger} \frac{\partial g}{\partial \xi_\alpha} + \frac{\partial f}{\partial \xi_\alpha} \frac{\partial g}{\partial \xi_\alpha^\dagger} \right). \tag{9.24}$$

We can observe that

- If at least one of the functions f, g represents an even element of the Grassmann algebra generated by $\xi_\alpha, \xi_\alpha^\dagger$, the expression (9.24) is antisymmetric under exchange $f \leftrightarrow g$. When going from classical to quantum mechanics, the Poisson bracket goes in this case to the commutator of two operators:

$$\{f, g\} \to i[\hat{f}, \hat{g}]. \tag{9.25}$$

- When both f and g are odd anticommuting elements of the Grassmann algebra, the expression (9.24) is symmetric under exchange $f \leftrightarrow g$ and goes over to the anticommutator under quantization:

$$\{f, g\} \to -i\{\hat{f}, \hat{g}\}_+. \tag{9.26}$$

Thus, the basic reason why the canonical commutators in the theory of bosonic fields are replaced by the canonical anticommutators in the theory of fermionic fields is the Grassmann nature of the fermionic variables in classical theory. We derive in particular that[9]

$$\{\hat{\xi}_\alpha, \hat{\xi}_\beta^\dagger\}_+ = \delta_{\alpha\beta} \qquad \{\hat{\xi}_\alpha, \hat{\xi}_\beta\}_+ = 0, \qquad \{\hat{\xi}_\alpha^\dagger, \hat{\xi}_\beta^\dagger\}_+ = 0. \tag{9.27}$$

For the toy model discussed above, the only nontrivial anticommutator is $\{\hat{\xi}^\dagger, \hat{\xi}\}_+ = 1$. A natural representation of this quantum algebra is $\hat{\xi} = \xi$ and $\hat{\xi}^\dagger = \partial/\partial\xi$.[10] The Hilbert space for the quantum counterpart of (9.23) (anti)symmetrized

---

[9]If the reader wants to learn a new fancy word, the algebra (9.27) is called a *Clifford algebra*. Thus, "Grassmann" becomes "Clifford" under quantization.

[10]Under this convention, related to the sign choice in (9.26), the classical canonical momentum $\Pi_\xi = -i\xi^\dagger$ derived from the Lagrangian (9.18), goes over to $\hat{\Pi}_\xi = -i\partial/\partial\xi$.

as in (9.12),

$$\hat{H} = \frac{\omega}{2} \left(\frac{\partial}{\partial \xi} \xi - \xi \frac{\partial}{\partial \xi} \right),$$ (9.28)

involves the wave functions $\Psi(\xi)$. These functions have a very simple form:

$$\Psi(\xi) = a + b\xi.$$ (9.29)

All the higher-order terms in the Taylor expansion of $\Psi(\xi)$ vanish due to the property $\xi^2 = 0$!

The Hamiltonian (9.28) has two eigenfunctions: $\Psi(\xi) = 1$ with the eigenvalue $\omega/2$ and $\Psi(\xi) = \xi$ with the eigenvalue $-\omega/2$. Thus, in contrast to the usual harmonic oscillator, our *Grassmann oscillator* does not have an infinite tower of equidistant states, but only two states. This is where the Pauli principle follows from!

Note that the Hamiltonian (9.28) also admits a matrix representation, which might be more palatable than the Grassmannian one for a student familiar with ordinary nonrelativistic quantum mechanics. One can trade the wave functions (9.29) of a holomorphic Grassmann variable ξ for the spinors

$$\Psi = \begin{pmatrix} a \\ b \end{pmatrix}.$$ (9.30)

Then the operators $\hat{\xi}$ and $\hat{\xi}^{\dagger}$ are represented by the matrices $\hat{\xi} \equiv \sigma_- = (\sigma_1 - i\sigma_2)/2$ and $\hat{\xi}^{\dagger} \equiv \sigma^+ = (\sigma_1 + i\sigma_2)/2$. The quantum Hamiltonian is then $\omega\sigma_3/2$.

We will discuss this matrix interpretation of Grassmann Hamiltonians again, while enjoying our dessert in Chap. 14 devoted to supersymmetric quantum mechanical and field-theory systems. In particular, we will show that the matrix Pauli Hamiltonian, describing the electron motion in a magnetic field, admits a Grassmannian interpretation. And in the case when the direction of the magnetic field at all spatial points is the same, we are dealing with a supersymmetric problem. For the time being, we only want to note that:

- The spinor in (9.30) has nothing to do with the usual spinors transformed by usual spatial rotations.
- The Hamiltonian of the fermion field depends on an infinite set of variables, the modes in the Fourier expansion (9.4). Each mode (representing a Grassmann variable) can be presented in the spinor form, but then the Hamiltonian becomes an infinite-dimensional matrix, which is not so convenient. The Grassmannian description is simpler.

We now take the Lagrangian (9.6), assuming both ξ^1 and ξ^2 to have a Grassmann nature. The quantum Hamiltonian (with the restored dimensional factor $1/L$) is

$$\hat{H} = \frac{\pi}{L} \left(\frac{\partial}{\partial \xi^1} \xi^1 - \xi^1 \frac{\partial}{\partial \xi^1} - \frac{\partial}{\partial \xi^2} \xi^2 + \xi^2 \frac{\partial}{\partial \xi^2} \right).$$ (9.31)

There are four states in the spectrum:

$$(A) \; \Psi = \xi^1 \quad \text{with} \quad E = -\frac{2\pi}{L} \; ;$$
$$(B) \; \Psi = 1 \; \text{and} \; (C) \; \Psi = \xi^1 \xi^2 \quad \text{with} \quad E = 0 \; ;$$
$$(D) \; \Psi = \xi^2 \quad \text{with} \quad E = \frac{2\pi}{L} \; . \tag{9.32}$$

In contrast to (9.16), the spectrum (9.32) has a bottom, and the trouble mentioned at the end of the previous section is no longer there.

The variables ξ^α in (9.31) describe the Fourier mode with $n = (0, 0, 1)$, but we can do the same analysis for any other choice of n. Again, we obtain four states in the spectrum:

$$(A) \; \Psi = \xi_n^+ \quad \text{with} \quad E = -\frac{2\pi |n|}{L} \; ;$$
$$(B) \; \Psi = 1 \; \text{and} \; (C) \; \Psi = \xi_n^+ \xi_n^- \quad \text{with} \quad E = 0 \; ;$$
$$(D) \; \Psi = \xi_n^- \quad \text{with} \quad E = \frac{2\pi |n|}{L} \; , \tag{9.33}$$

where ξ_n^+ is the right-handed spinor satisfying $(\sigma \cdot n) \, \xi_n^+ = |n| \xi_n^+$ and ξ_n^- is the left-handed spinor satisfying $(\sigma \cdot n) \, \xi_n^- = -|n| \xi_n^-$.

Bearing in mind our discussion of the Dirac sea, the physical interpretation of (9.33) is rather transparent. The left-handed modes ξ_n^- have positive energies. In the vacuum, these levels are empty. When a level ξ_n^- is occupied, this means the presence of the physical left-handed particle with momentum $p = 2\pi n / L$. On the other hand, the right-handed modes ξ_n^+ all have negative energies and belong to the sea. In the vacuum, all these levels are occupied. An excited state of the Hamiltonian with an empty sea level describes the physical right-handed antiparticle with momentum $p = -2\pi n / L$.

In other words, the Lagrangian (9.2) with Grassmann fields $\xi^\alpha(x)$ can be used to describe free massless left-handed *neutrinos* and right-handed antineutrinos. *This is the reason why we called the relativistic spinor* $\xi_\alpha = \epsilon_{\alpha\beta} \xi^\beta$ *left-handed and its conjugate right-handed* (though one has to have in mind, of course, that the attributions "left" and "right", "particle" and "antiparticle" to the physical states depend on the conventions.)

We wish, however, also to be able to describe electrons, quarks and other massive charged fermions. How can we do that?

9.3 Electromagnetic Interactions □ Dirac Equation

The important characteristic of electrons and quarks is their electric charge, by which they interact with the electromagnetic field A_μ. Let us try to write a nontrivial interacting Lagrangian including $\xi_\alpha(x)$ and $A_\mu(x)$, which would still be invariant under the gauge transformation (7.54), supplemented by an appropriate transformation of the charged field ξ_α.

The reader has already met the words "gauge transformations" and "gauge invariance" earlier in this book. First (e.g. in Chap. 5), we mentioned it without really explaining what it means. In Chap. 7 we wrote the explicit formula (7.54) describing the gauge transformations of the photon field and discussed the gauge invariance of the free Maxwell Lagrangian. Now we present for the first time a non-trivial interacting gauge-invariant field-theory Lagrangian.[11]

We emphasize that gauge invariance is, along with the Lorentz invariance, one of the basic guiding principles for building quantum field theories. The Lagrangian of the Standard Model, which we eventually want to learn and understand, enjoys a complicated non-Abelian version of this invariance.

But all this comes later. For the time being, we only try to generalize (9.2) by including the interaction with A_μ and write

$$\mathcal{L} = -\frac{1}{4}F_{\mu\nu}F^{\mu\nu} + i\xi^\alpha(\sigma^\mu)_{\alpha\dot\beta}(\partial_\mu - ieA_\mu)\xi^{\dagger\dot\beta}. \tag{9.34}$$

This Lagrangian is not quadratic any more; it involves a nontrivial trilinear interaction term.

The Lagrangian (9.34) is invariant under the transformations

$$A_\mu(x) \rightarrow A_\mu(x) + \partial_\mu\chi(x),$$
$$\xi_\alpha(x) \rightarrow e^{-ie\chi(x)}\xi_\alpha(x). \tag{9.35}$$

The second line in (9.35) describes the precise way in which the transformation law (7.54) now has to be supplemented. The operator $\partial_\mu - ieA_\mu$ or rather

$$\mathcal{D}_\mu = \partial_\mu + ieA_\mu, \tag{9.36}$$

the operator that acts on ξ^α in the expression obtained by integrating the action functional $\int d^4x\mathcal{L}$ by parts, is called the *covariant derivative*.[12]

The Lagrangian (9.34) is not, however, realistic. It fails for two reasons. One of them is purely theoretical. It turns out that the Lagrangian (9.34) is gauge-invariant only at the classical level. One cannot keep this invariance in quantum theory due to the so-called *chiral anomaly* — a phenomenon which is beyond the scope of this book. And the breaking of gauge invariance makes the theory non-renormalizable and inconsistent (cf. the discussion in Chap. 5).

---

[11] We wrote earlier the interaction Lagrangian (7.27) and (7.28), but it had a mixed nature, describing the interaction of the field $A_\mu(x)$ with the charged particles.

[12] A reader who might have studied general relativity would be familiar with the notion of covariant derivative in Riemannian geometry. In contrast to the ordinary derivative, the covariant derivative of a tensor is a tensor. Likewise, the field $\partial_\mu\xi^\alpha$ does not transform under gauge transformations in the same simple way as ξ^α. But $\mathcal{D}_\mu\xi^\alpha$ does!

And another reason is phenomenological. We are still missing two degrees of freedom. For each momentum, the Lagrangian (9.34) describes only two fermion states, and we need four: the left- and right-handed electrons and the left- and right-handed positrons.

To double the degrees of freedom, we need to introduce, along with $\xi_\alpha(x)$, *another* similar fermion field $\eta_\alpha(x)$. We further assume that the electric charge of η_α is *opposite* to the electric charge of ξ_α. The fermion part of the Lagrangian reads

$$\mathcal{L}^{\text{massless}}_{\xi\eta} = i\xi^\alpha(\sigma^\mu)_{\alpha\dot\beta}(\partial_\mu - ieA_\mu)\xi^{\dagger\dot\beta} + i\eta^\alpha(\sigma^\mu)_{\alpha\dot\beta}(\partial_\mu + ieA_\mu)\eta^{\dagger\dot\beta}. \quad (9.37)$$

It is invariant under

$$\begin{aligned}
A_\mu &\to A_\mu + \partial_\mu\chi, \\
\xi &\to e^{-ie\chi}\xi, \\
\eta &\to e^{ie\chi}\eta.
\end{aligned} \quad (9.38)$$

The quantization of the field ξ_α gives the particles e^-_L and e^+_R. The quantization of the field η_α gives the particles e^+_L and e^-_R.

With two different fermion fields of opposite charge, we can also add to the Lagrangian a gauge-invariant and Lorentz-invariant potential term which gives a mass to the electron:

$$\mathcal{L}_m = m\left(\xi_\alpha\eta^\alpha + \eta^{\dagger\dot\alpha}\xi^\dagger_{\dot\alpha}\right). \quad (9.39)$$

Note that one cannot write a mass term for a single charged fermion field ξ_α, the structure $\xi_\alpha\xi^\alpha$ (it is called a *Majorana mass term*) is Lorentz-invariant, but not gauge-invariant! On the other hand, one can well endow with a Majorana mass a single *neutral* fermion, not participating in gauge interactions and not changed by gauge transformations. In particular, it is not excluded now that the masses of the *right-handed* neutrinos have a Majorana nature. We will come back to this question in Chap. 12, when discussing the physics of the Standard Model.

The Lagrangian of quantum electrodynamics involving the electromagnetic and the electron-positron fields represents the sum of the fermion kinetic and interaction term (9.37), the fermion mass term (9.39) and the Maxwell Lagrangian (7.53).

Traditionally, the fermion part of the Lagrangian is written in a somewhat more compact way. The technicalities that follow are not *absolutely* necessary for understanding what is a fermion field, but we present them for two reasons: first, we wish to make contact with other textbooks on quantum field theory that the reader might consult, and second, this compact description will be useful for us in the subsequent chapters when we discuss Feynman graphs involving fermions and build up the Lagrangian of the Standard Model.

The two left-handed two-component fields ξ_α, η_α can be unified to form a 4-component *Dirac bispinor*

$$\psi = \begin{pmatrix} \eta^{\dagger\dot\alpha} \\ \xi_\alpha \end{pmatrix}. \quad (9.40)$$

Note that the upper and the lower components of this bispinor carry the same charge e, so that ψ is gauge-transformed as $\psi \to e^{-ie\chi}\psi$.

We also introduce four *Dirac matrices*

$$\gamma^{\mu} = \begin{pmatrix} 0 & \bar{\sigma}^{\mu} \\ \sigma^{\mu} & 0 \end{pmatrix}, \tag{9.41}$$

where $\bar{\sigma}^{\mu} = (\mathbb{1}, -\boldsymbol{\sigma})$.

They are actually 4×4 matrices presented in 2×2 block form. If we expand them, we obtain

$$\gamma^0 = \begin{pmatrix} 0 & 0 & 1 & 0 \\ 0 & 0 & 0 & 1 \\ 1 & 0 & 0 & 0 \\ 0 & 1 & 0 & 0 \end{pmatrix}, \quad \text{etc.}$$

The Dirac matrices satisfy the nice anticommutation relations

$$\gamma^{\mu}\gamma^{\nu} + \gamma^{\nu}\gamma^{\mu} = 2\eta^{\mu\nu}\mathbb{1}_{4\times 4}. \tag{9.42}$$

Consider now the matrix

$$\gamma^5 = i\gamma^0\gamma^1\gamma^2\gamma^3 = \begin{pmatrix} \mathbb{1} & 0 \\ 0 & -\mathbb{1} \end{pmatrix} \tag{9.43}$$

which anticommutes with all γ^{μ}. It plays a special role, allowing one to build the projectors onto the fields with definite *chirality*. When acting on ψ, they cook up left-handed and right-handed electron bispinor fields:[13]

$$\psi_L = \frac{1-\gamma^5}{2}\psi = \begin{pmatrix} 0 \\ \xi \end{pmatrix}, \quad \psi_R = \frac{1+\gamma^5}{2}\psi = \begin{pmatrix} \eta^{\dagger} \\ 0 \end{pmatrix}. \tag{9.44}$$

We also define *Dirac conjugation* according to $\bar{\psi} = \psi^{\dagger}\gamma^0$, $\psi^{\dagger}$ being a Hermitian conjugated spinor. Explicitly,

$$\bar{\psi} = \left(\xi^{\dagger}_{\dot{\alpha}}, \eta^{\alpha}\right). \tag{9.45}$$

It is easy to see that

$$\bar{\psi}\psi = \xi^{\dagger}_{\dot{\alpha}}\eta^{\dagger\dot{\alpha}} + \eta^{\alpha}\xi_{\alpha} \tag{9.46}$$

---

[13] Very often, people do not distinguish between the notions of chirality and helicity. It looks natural, bearing in mind that, when translated from Greek, chirality means "handedness", and we call the particles carrying negative helicity left-handed and positive helicity right-handed. But we want to make this distinction. For us, helicity is a property of a *particle state*, as defined in (5.23), while chirality is the property of a *field*. We call the fields with undotted indices *left chiral* and the fields with dotted indices *right chiral*. Then, as was discussed above, the left chiral field ξ^{α} entering (9.40) describes both left-handed electrons and right-handed positrons, while the right chiral field $\eta^{\dagger\dot{\alpha}}$ describes the right-handed electrons and left-handed positrons.

is a Lorentz scalar. One can further observe that $\bar{\psi}\gamma^5\psi$ is a pseudoscalar (changes sign under spatial reflection), $\bar{\psi}\gamma^\mu\psi$ is a vector and $\bar{\psi}\gamma^\mu\gamma^5\psi$ is an axial vector.

As promised, the full QED Lagrangian acquires a compact form in this notation:

$$\mathcal{L} = -\frac{1}{4}F_{\mu\nu}F^{\mu\nu} + i\bar{\psi}\gamma^\mu(\partial_\mu + ieA_\mu)\psi - m\bar{\psi}\psi\,. \tag{9.47}$$

It is invariant under

$$\begin{aligned} A_\mu &\to A_\mu + \partial_\mu\chi\,, \\ \psi &\to e^{-ie\chi}\psi\,. \end{aligned} \tag{9.48}$$

Finally, we fulfill our earlier promise and write down the *Dirac equation*. It is obtained by varying the Lagrangian (9.47) with respect to $\bar{\psi}$ and reads

$$i\gamma^\mu(\partial_\mu + ieA_\mu)\psi = m\psi\,. \tag{9.49}$$

The variation with respect to ψ gives the conjugate equation:

$$-i\bar{\psi}(\overleftarrow{\partial_\mu} - ieA_\mu)\gamma^\mu = m\bar{\psi}\,. \tag{9.50}$$

Equation (9.49) plays the same role for the fermion particles as the KFG equation does for the scalar particles and admits the same double interpretation, discussed in Chap. 4. We derived it as a classical field equation. But when the external field A_μ is not so strong, it can also be interpreted in the Schrödinger spirit, as the wave equation for a relativistic fermion particle.

The solutions of the *free* Dirac equation (without A_μ) represent plane waves with definite momentum p:

$$\psi(x) = Cu(p)e^{-i\epsilon t + ip\cdot x} \tag{9.51}$$

As was discussed, there are four different solutions with two helicities and with positive and negative energies, $\epsilon = \pm\sqrt{p^2 + m^2}$. For each such solution, one can find the corresponding explicit expression for the bispinor $u(p)$. We will not really *use* these expressions in the following, but to give an idea of what they look like, we are writing here the positive-energy solutions:[14]

$$u(p) = \frac{1}{\sqrt{2}}\begin{pmatrix} \sqrt{\epsilon + m}\,w + \sqrt{\epsilon - m}\,(n\cdot\sigma)\,w \\ \sqrt{\epsilon + m}\,w - \sqrt{\epsilon - m}\,(n\cdot\sigma)\,w \end{pmatrix}\,, \tag{9.53}$$

---

[14]To avoid confusion, note that we write the solutions in the *spinor representation* [of the Clifford algebra (9.42)], where the γ matrices have the form (9.41), and not in the *standard representation* with

$$\gamma^0 = \begin{pmatrix} 1 & 0 \\ 0 & -1 \end{pmatrix}, \quad \gamma = \begin{pmatrix} 0 & \sigma \\ -\sigma & 0 \end{pmatrix}, \tag{9.52}$$

as they are given in most textbooks.

where w is an ordinary 2-component spinor, which can have two different helicities, $\epsilon = p_0$, $\mathbf{n} = \mathbf{p}/|\mathbf{p}|$. The bispinor u is normalized according to $\bar{u}u = 2m$. The constant C in (9.51) is then habitually chosen as $C = 1/\sqrt{2\epsilon V}$, where V is the spatial volume where the theory is defined. Then $\psi(x)$ might be interpreted as the spinorial wave function of a *single* particle. The dimension of ψ is $[\psi] = m^{3/2}$, whereas the dimension of u is $[u] = m^{1/2}$. The object 9.53 is convenient because it does not depend on V, but only on momentum, and it is $u(p)$ along with $u(-p)$ and $\bar{u}(\pm p)$ which enter the relativistic scattering amplitudes to be discussed in the next chapter.

The variation of the Lagrangian (9.47) with respect to A_μ gives the equation

$$\partial_\nu F^{\nu\mu} = e\bar{\psi}\gamma^\mu\psi. \tag{9.54}$$

This is the same as (7.59), only instead of the electromagnetic current j^μ associated with the particles and given in (7.28), we obtain the expression

$$j^\mu = \bar{\psi}\gamma^\mu\psi. \tag{9.55}$$

In classical theory, this is a certain bosonic field, an even element of the Grassmann algebra generated by $\psi(x)$ and $\bar{\psi}(x)$. It satisfies, as it should, the conservation law

$$\partial_\mu j^\mu = 0, \tag{9.56}$$

which immediately follows from (9.54). In momentum space, this condition acquires the form

$$(p' - p)_\mu \bar{u}(p')\gamma^\mu u(p) = 0. \tag{9.57}$$

In quantum theory, the current (9.55) becomes an operator. To find the transition and scattering amplitudes, one needs to calculate the matrix elements of the Hamiltonian, which includes the current operator, between one-particle states. Then $\psi(x)$ is interpreted as the relativistic wave function, and (9.55) is nothing but a relativistic generalization of (5.2). One can also make contact with the classical current (7.28) of the localized pointlike particles. Indeed, the reader can check that, after substituting the explicit solutions (9.51), (9.53) into (9.55), one obtains

$$\langle j^\mu \rangle_p = \frac{1}{V}\frac{p^\mu}{\epsilon} = \frac{1}{V}(1, \mathbf{v}). \tag{9.58}$$

But that was the average over a delocalized plane-wave solution. If we calculated the average over a localized wave pocket, the factor $1/V$ would be traded for a δ function as in (7.28).

We address the reader to any standard textbook for further details.

Chapter 10
Feynman Graphs

In some cases we derive things accurately, but in some other cases we do not, deciding that an accurate derivation would be too technical for our semi-popular book. The latter applies in particular to the Feynman diagram technique. Its rigorous derivation is a rather challenging endeavour, which we do not dare to venture here. In Chap. 5 we gave some historical comments, which showed to the reader that Feynman himself did not derive it accurately and did not understand it quite clearly at the end of the nineteen forties. He made a remarkable guess, and its justification came later!

The logic of this chapter is the following. In Chap. 8 we derived the analytic expressions for the amplitude of nonrelativistic elastic scattering and represented them in graphic form in Fig. 8.4. We will now reformulate these results in Feynman's way and then we will ask the reader to follow Feynman's example, stretch his intution and imagination to the extreme and *accept* that the rules of the Feynman technique that I will give without derivation are reasonable and trustworthy.

But before doing that, we should treat kinematical issues. In relativistic quantum field theory, they are not trivial.

10.1   Kinematics

In Chap. 8 we discussed only the kinematics of scattering on an external scattering centre in nonrelativistic quantum mechanics. In relativistic quantum field theory we are interested, besides such scattering, in processes where the number of particles is changed. The processes of most practical importance are *(i) decays*, where a parent particle with 4-momentum P disappears and several particles with 4-momenta p'_a are created in its place, and *(ii) collisions*, where two parent particles with momenta p_1 and p_2 collide, giving rise to several particles with momenta p'_a. In the first case, we are interested in the *differential probability* to create a given set of particles with given momenta. In the second case, we are interested in the differential cross sections,

© Springer International Publishing AG 2017
A. Smilga, *Digestible Quantum Field Theory*,
DOI 10.1007/978-3-319-59922-9_10

but $d\sigma$ may now depend not only on the solid angle element $d\Omega$ (as in the case when only two particles are born), but on other relevant kinematical parameters.

As in the nonrelativistic case, the differential probability and the differential cross section depend on the scattering amplitude. We will not, however, define the latter carefully here, instead sending the reader to the standard textbooks. We only say that the *invariant amplitude* M_{fi} can be defined for any process.[1] The QFT analogues of Eq. (8.7) are somewhat more complicated. We give them without derivation, mostly for reference purposes.

- For the decay,

$$dw = (2\pi)^4 \delta^{(4)} \left(\sum_a p_a' - P \right) |M_{fi}|^2 \frac{1}{2E} \prod_a \frac{d\mathbf{p}_a'}{(2\pi)^3 2\epsilon_a'}, \qquad (10.1)$$

where E is the energy of the decaying particle. Note that the factor $(d^3 p_a')/(2\epsilon_a')$ can be presented as $d^4 p_a' \delta(p_a'^2 - m_a^2) \theta(\epsilon_a')$ and is a Lorentz scalar. Hence the only frame dependence of dw and $w = \int dw$ is due to the factor $1/E$. In the frame where the decaying particle moves, the probability is suppressed by the factor[2] $\sqrt{1 - v^2}$. Correspondingly, the lifetime of the particle $\tau = w^{-1}$ increases.

- For the collisions,

$$d\sigma = (2\pi)^4 \delta^{(4)} \left(\sum_a p_a' - p_1 - p_2 \right) |M_{fi}|^2 \frac{1}{4I} \prod_a \frac{d\mathbf{p}_a'}{(2\pi)^3 2\epsilon_a'}, \qquad (10.2)$$

where I is the *invariant flux*,

$$I = \sqrt{(p_1 \cdot p_2)^2 - m_1^2 m_2^2}. \qquad (10.3)$$

The integral $\sigma = \int d\sigma$ gives the Lorentz-invariant total cross section.

A particular interesting case is the scattering $2 \to 2$, with only two particles in the final state. The amplitude then depends on three kinematic parameters, which may be chosen as the total energy $E = \epsilon_1 + \epsilon_2 = \epsilon_1' + \epsilon_2'$ and the scattering angles θ, φ in the centre-of-mass system of the colliding particles.[3] In this system the

---

[1] M_{fi} is invariant under Lorentz transformations. The indices i and f stand for the initial and final states, their "reversed" order reflecting Dirac's *bra* and *ket* vectors convention [cf. Eqs. (4.50) and (4.51)]. We also want to warn the reader that different textbooks (to which we have just referred him) use different sign conventions for the amplitude. It is not so important, but whenever we write explicit formulas, we use the convention adopted in the books *Quantum Electrodynamics* by Berestetsky, Lifshitz and Pitaevsky and *An Introduction to Quantum Field Theory* by Peskin and Schroeder.

[2] In this chapter we revert to the natural unit system, $\hbar = c = 1$.

[3] For scalar particles the amplitude depends only on the momenta $p_1, \ldots, p_4$, from which one can construct only two independent invariants: $s = (p_1 + p_2)^2$ and $t = (p_3 - p_1)^2$. These are equivalent to E and θ in the centre-of-mass system.

expression for the cross section reads

$$d\sigma = \frac{1}{64\pi^2} |M_{fi}|^2 \frac{|\boldsymbol{p}'|}{|\boldsymbol{p}| E^2} d\Omega', \qquad (10.4)$$

where $\boldsymbol{p}$ is the momentum of the colliding particles, $\boldsymbol{p}'$ is the momentum of the produced particles (their norms need not be the same, consider for instance the process $e^+ e^- \to 2\gamma$) and E is the total centre-of-mass energy. Note that the relativistic scattering amplitude M_{fi} carries no dimension in this case.

For elastic scattering in the limit when the mass of one of the particles is very large, this particle represents in fact a static scattering centre and the amplitude $M_{fi}^{2 \to 2}$ divided by the heavy mass coincides (up to a factor of 2) with the invariant amplitude of the scattering of the second light particle in an external static potential $M_{fi}^{\text{pot.}}$. If the velocity of the light particle is small, this static scattering amplitude reduces to the nonrelativistic amplitude $f(\theta, \varphi)$ discussed in Chap. 8:

$$M_{fi}^{\text{pot.}} = 4\pi f(\theta, \phi), \qquad (10.5)$$

so that

$$d\sigma = \frac{|M_{fi}^{\text{pot.}}|^2}{16\pi^2} d\Omega. \qquad (10.6)$$

10.2 Potential Scattering

We are now ready to explain how the scattering amplitudes are actually calculated. To this end, we invite the reader to read again the middle section in Chap. 8, where the rules of the non-relativistic diagram technique were derived. Consider again the first correction to the Born scattering amplitude. Its analytic expression is given by (8.28), and it is illustrated by the middle graph in Fig. 8.4. The interpretation of this exact result is as follows:

1. The particle comes in from infinity and is scattered by the static potential, which transfers to the particle the momentum $\boldsymbol{q}_1$. The strength of this interaction is the Fourier component of the potential $V_{\boldsymbol{q}_1}$.
2. After scattering the particle acquires the momentum $\boldsymbol{s} = \boldsymbol{k}_{\text{in}} + \boldsymbol{q}_1$. Its kinetic energy $E^* = (\boldsymbol{k}_{\text{in}} + \boldsymbol{q}_1)^2/2m$ is not the same as the energy of the ingoing particle, i.e. the energy is not conserved in this scattering.
3. In quantum mechanics, nonconservation of energy is admissible for a brief moment $\tau \sim \frac{\hbar}{\Delta E}$. And very soon the particle is scattered on the potential again. This time, the momentum transfer is $\boldsymbol{q}_2$. This event in the particle's biography leaves a trace in the records as the factor $V_{\boldsymbol{q}_2}$ in the integrand in (8.28). After the second scattering, the particle acquires the momentum $\boldsymbol{k}_{\text{out}} = \boldsymbol{k}_{\text{in}} + \boldsymbol{q}_1 + \boldsymbol{q}_2$ such that

the absolute values $|\mathbf{k}_{in}|$ and $|\mathbf{k}_{out}|$ coincide. The corresponding energies coincide too. The particle can now go peacefully off to infinity, which it does.

4. The fact that the energy is not conserved between two scatterings shows up in the factor (8.30),

$$\frac{1}{E - E^*} = \frac{2m}{k_{in}^2 - (\mathbf{k}_{in} + \mathbf{q}_1)^2},$$

where E is the (initial and final) energy of the particle, and E^* — its energy in the intermediate state. As was explained, this factor can be interpreted as the fundamental solution (alias, *Green's function*) of the Schrödinger equation (8.31).

5. We are interested in the scattering amplitude corresponding to a definite momentum transfer $\mathbf{q}$. But the momentum transfers $\mathbf{q}_1$, $\mathbf{q}_2$ in individual scattering acts are not defined. We should therefore integrate over $d\mathbf{q}_1 d\mathbf{q}_2$ with the constraint $\mathbf{q}_1 + \mathbf{q}_2 = \mathbf{q}$, implemented by $\delta(\mathbf{q}_1 + \mathbf{q}_2 - \mathbf{q})$.

Actually, to get all numerical factors correct, we should rather integrate over

$$\int \frac{d\mathbf{q}_1}{(2\pi)^3} \frac{d\mathbf{q}_2}{(2\pi)^3} (2\pi)^3 \delta(\mathbf{q}_1 + \mathbf{q}_2 - \mathbf{q})$$

and multiply everything by the factor $-m/2\pi$, the same as in the Born leading-order contribution (8.20).

Let us now give a slightly different interpretation of the same formula. We assume that the energy *is* conserved at each interaction vertex, so that the particle keeps its energy E also in the intermediate state between two collisions. But then the usual dispersion relation between the energy and momentum,

$$E = \frac{p^2}{2m},\tag{10.7}$$

is not satisfied any more.

People say that the particle in the intermediate state is *off mass shell*, in contrast to the initial and final particles, for which (10.7) holds and which are thereby *on mass shell*.

Suppose now that we are interested in the potential scattering of a relativistic scalar particle. Its amplitude has the same graphic representation as in the nonrelativistic case. The difference is that the relevant wave equation is now the inhomogenous KFG equation

$$(\Box + m^2)\Psi = \text{source},\tag{10.8}$$

rather than the Schrödinger equation, and we should replace the nonrelativistic propagator by the relativistic one:

$$\frac{1}{E - \frac{p^2}{2m}} \longrightarrow \frac{2m}{p_0^2 - p^2 - m^2} = \frac{2m}{p^2 - m^2},\tag{10.9}$$

where $p_0 = m + E$ and the factor $2m$ in the right-hand side is written to match the left-hand side in the nonrelativistic limit. It is convenient, however, to define the relativistic scalar propagator without such a factor:

$$G(p) = \frac{1}{p^2 - m^2}.$$ (10.10)

We will come back to the discussion of this issue a little later.

10.3 Scalar Field Theory

What happens when an external potential is absent, but the field Lagrangian describing the scalar field is not quadratic, but involves a quartic interation term, like in (4.47)? A single particle has nothing to be scattered on, but if the initial state involves two different particles, they can scatter on each other. To the lowest order in λ, the scattering amplitude is described by the Feynman diagram which was already depicted in Fig. 4.4, but to alleviate reader's task and not forcing him to flip the book backwards *every* time, we redraw it here, now paying attention to the sign in the vertex (Fig. 10.1).

The corresponding algebraic expression for the invariant scattering amplitude is very simple. It is just a number:

$$M_{fi}^{\text{tree}} = -\lambda.$$ (10.11)

It is the same number that presents itself in the interaction Lagrangian

$$\mathcal{L}_{\text{int}} = -\frac{\lambda}{4}\phi^2(\phi^*)^2,$$ (10.12)

but the factor $\frac{1}{4}$ disappeared. It did so due to combinatorics. There are two ways to associate the initial particles 1, 2 with the two factors $\phi\phi$ in (10.12). Likewise, there are two ways to associate the final particles 3, 4 with the two factors $\phi^*\phi^*$ in (10.12). This remark (known as *Wick's theorem*) is illustrated in Fig. 10.2.

The diagram in Fig. 10.1 describes the leading (tree) contribution to the scattering amplitude. There are three diagrams describing the one-loop corrections. They are shown in Fig. 10.3. Each such diagram has a loop involving two virtual particles. For example, the graph in Fig. 10.3a involves the particles with 4-momenta q and $q + p_1 + p_2$. A virtual particle enters the analytic expression for the amplitude with its relativistic propagator (10.10). The momentum q is not fixed, and we should integrate over it. In contrast to the nonrelativistic scattering amplitude, the integration has to be done over all four components of the loop momentum with the factor $(2\pi)^4$ rather than $(2\pi)^3$ downstairs. We obtain as a result

Fig. 10.1 The scalar $2 \to 2$
scattering amplitude to the
lowest order.

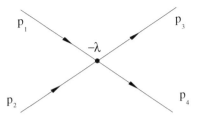

Fig. 10.2 The amplitude
$12 \to 34$ and combinatorics.
The factor $-\lambda/4$ in the
interaction Lagrangian
should be multiplied by
$2 \times 2 = 4$.

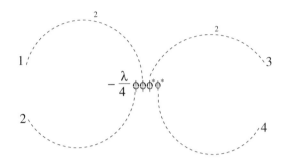

$$M_{\text{Fig. 10.3a}} = -i\lambda^2 \int \frac{d^4q}{(2\pi)^4} \frac{1}{q^2 - m^2} \frac{1}{(q + p_1 + p_2)^2 - m^2}. \qquad (10.13)$$

The integral goes over four momenta components, and, for large q, we have the factor $\sim q^4$ downstairs. Thus, the integral diverges logarithmically in the ultraviolet. This divergence is of the same nature as the divergence that shows up in the physical electron charge, discussed in Chap. 5 and to be discussed again later in this chapter.

The expression (10.13) and similar expressions below are written with the correct coefficients, but we do not want to abuse the attention of the reader and explain quite meticulously where the correct sign and i factors come from. At the end of this chapter we will present without an accurate derivation the exact Feynman rules for processes involving scalar particles, fermions and photons.

But we want to make a remark concerning the presence of the factor $2m$ in (10.9). Actually, we use it as a pretext to acquaint the reader with another important field theory.

10.4 Scalar Electrodynamics

Consider the Lagrangian

$$\mathcal{L} = -\frac{1}{4} F_{\mu\nu} F^{\mu\nu} + (\mathcal{D}^\mu \phi)^* \, \mathcal{D}_\mu \phi, \qquad (10.14)$$

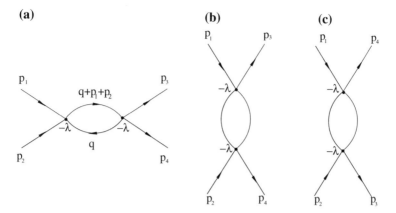

Fig. 10.3 One-loop contributions to the amplitude.

where $\mathcal{D}_\mu = \partial_\mu + ieA_\mu$ as in (9.36). This Lagrangian describes charged scalar particles (they exist in nature, for example — the π^+ and π^- mesons) and their interaction with the electromagnetic field. It is invariant under the gauge transformations

$$A_\mu \rightarrow A_\mu + \partial_\mu \chi \,,$$
$$\phi \rightarrow e^{-ie\chi}\phi \,. \tag{10.15}$$

The Lagrangian (10.14) involves the cubic and quartic interaction terms:

$$\mathcal{L}_{\text{int}}^{\text{scalar QED}} = ieA_\mu[(\partial^\mu \phi^*)\phi - \phi^*(\partial^\mu \phi)] + e^2 A_\mu A^\mu \phi^* \phi \,. \tag{10.16}$$

Consider first the quartic term. It looks similar to the interaction term (10.12) in the scalar theory. It generates vertices where two scalar lines and two photon lines meet, as in Fig. 10.4b. These vertices[4] enter with the factor $2e^2\eta^{\mu\nu}$ (the 2 comes from combinatorics), with the indices μ, ν reflecting the presence of the photon lines — photons (real or virtual) can have different polarizations.

The cubic term generates vertices where only three lines meet (see Fig. 10.4a). Note, however, the presence of the derivatives in the Lagrangian that act on the scalar fields. When the latter represent plane waves, the derivatives give 4-momenta of the corresponding states and, in the graphical representation, the momenta of the corresponding lines. Please accept without proof that it is true also for virtual particles. The derivatives act both on the left and on the right, and we obtain the factor $-e(p + p')^\mu$ associated with every cubic vertex.

Consider now the process when a scalar particle scatters in a static external electromagnetic field. To the leading order, the scattering amplitude is described by the graph in Fig. 10.4a, where $q = (0, \boldsymbol{q})$. The analytic expression for the amplitude is obtained by multiplying the factor $-e(p+p')^\mu$ in the vertex by the Fourier component

---

[4]Poetically minded theorists christened them *seagulls*.

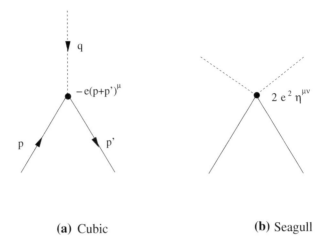

(a) Cubic   (b) Seagull

Fig. 10.4 Elementary vertices in scalar QED.

of the vector potential:

$$M_{fi}^{(1)} = -e(p + p')^{\mu} A_{\mu}(\boldsymbol{q}) . \tag{10.17}$$

In a pure static electric field, only the temporal component of the vector potential is present, $A_{\mu}(\boldsymbol{q}) = \delta_{\mu 0} \Phi(\boldsymbol{q})$. If the velocities of the ingoing and the outgoing particles are small, we can replace $(p + p')^0$ by $2m$ and write

$$M_{fi}^{(1)} = -2me\Phi(\boldsymbol{q}) = -2m V_q . \tag{10.18}$$

Dividing this by 4π, as dictated by (10.5), we reproduce the nonrelativistic Born amplitude (8.20).

To order $\sim e^2$, the scattering amplitude represents the sum of the contributions of the seagull graph in Fig. 10.4b and of the graph in Fig. 10.5. In the nonrelativistic case, the analytic expression for the latter may be written as

$$M_{fi}^{(2)} = -\int \frac{d\boldsymbol{q}_1}{(2\pi)^3} 2me\Phi(\boldsymbol{q}_1) \frac{1}{(p + q_1)^2 - m^2} 2me\Phi(\boldsymbol{q} - \boldsymbol{q}_1), \tag{10.19}$$

where one can discern the factors $2me\Phi(\boldsymbol{q}_1)$ and $2me\Phi(\boldsymbol{q}_2)$ coming from the two vertices and the relativistic propagator (10.10) for the virtual intermediate particle. Dividing this by 4π, we reproduce the result (8.28) with the nonrelativistic propagator (10.9).

In a diagram like Fig. 10.5, the dashed lines may stand not only for an external static potential, but also for real on-mass-shell photons with $q_1^2 = q_2^2 = 0$. We are then dealing with the scattering of photons on scalar charged particles, $\gamma\pi^+ \to \gamma\pi^+$ or $\gamma\pi^- \to \gamma\pi^-$. The same diagrams describe the annihilation $\pi^+\pi^- \to \gamma\gamma$ and the

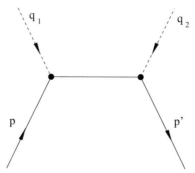

Fig. 10.5 Two consecutive scatterings.

inverse process, $\gamma\gamma \to \pi^+\pi^-$, depending on the signs of the temporal components of the momenta $q_{1,2}$, p, p' — an ingoing particle with negative energy is nothing but an outgoing antiparticle!

The vector potential of a real electromagnetic wave involves a polarization vector $\epsilon_\mu(q)$ as a factor. The wave can have two transverse polarizations and hence there are two such polarization vectors $\epsilon_\mu^{(1,2)}(q)$. In the reference system where $q = (0, 0, |\mathbf{q}|)$,

$$\epsilon_\mu^{(1)}(q) = (0;\, 1, 0, 0)\,, \qquad \epsilon_\mu^{(2)}(q) = (0;\, 0, 1, 0)\,. \tag{10.20}$$

The dashed lines in the Feynman diagrams for scattering of real photons are spelled out as the factors (10.20) in the amplitudes.[5]

Another important process is $\pi\pi$ electromagnetic scattering. Two diagrams contributing to the amplitude are depicted in Fig. 10.6.

These diagrams involve the exchange of a virtual photon. Hiding the subtleties under the carpet for a while (we will fetch out some of them by the end of this chapter), its propagator can be chosen as the Green's function of the inhomogeneous d'Alembert equation

$$\Box A_\mu = \text{source}\,. \tag{10.21}$$

This gives

$$D_{\mu\nu}(q) = \frac{\eta_{\mu\nu}}{q^2}\,. \tag{10.22}$$

---

[5]To be more exact, for the outgoing photons one should take complex conjugate polarization vectors. This rule is closely related to our convention in (4.43) and (4.44) and the italicized remark on p. 119 For linearly polarized photons, $\epsilon_\mu(q)$ is real and this is irrelevant, but the polarization vectors of circularly polarized photons are complex, and we (or rather not *we*, but the people who want to perform actual exact calculations) should be careful at this point.

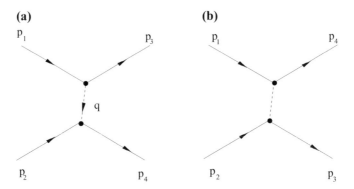

Fig. 10.6 $\pi\pi$ scattering.

Thus, the total expression for the amplitude of $\pi\pi$ scattering, given by the sum of two diagrams in Fig. 10.6, reads

$$M_{fi}^{\gamma\pi\to\gamma\pi} = e^2 \left[\frac{(p_1 + p_3) \cdot (p_2 + p_4)}{(p_1 - p_3)^2} + \frac{(p_1 + p_4) \cdot (p_2 + p_3)}{(p_1 - p_4)^2} \right]. \quad (10.23)$$

10.5   Spinor Electrodynamics

The π mesons are not elementary particles, but complicated objects made of quarks and antiquarks. Thus, the Lagrangian (10.14) describes reasonably well the dynamics of their electromagnetic interaction only for low-frequency fields, with wavelength much larger than the pion size. On the other hand, the electron is an elementary particle. Or, better to say, it can be treated as an elementary particle over a wide range of energies — up to at least $\sim 100-200\,\text{GeV}$.[6]

Due to this and also due to the fact that the electrons are stable and that there are a lot of electrons in the Universe, while the pions are only created for a brief moment in accelerators, spinor electrodynamics, the theory of electron-photon interactions, is more relevant than the scalar one for describing and understanding Nature.

The Lagrangian of QED was written in (9.47). It involves, besides the free quadratic part, only one cubic interaction term, and the Feynman graphs describing QED processes involve only triple vertices.

---

[6]For still higher energies, one penetrates the electron core and starts to feel the presence of the Higgs field in the electron structure — see Chap. 12.

I cannot help recalling at this point that Vladimir Lenin, the eminent Russian revolutionary, once distracted himself by writing a philosophical essay (it was at a time when revolutionary movement was on the decline, and Lenin was not so busy in his main job). I read it and I must say that it was not so stupid. Among other things, he wrote there that "the electron is as inexhaustable as the atom". He might have been right at this point. It is a real pity that this man did not choose a less violent path in life. He could have become a good physicist …

Fig. 10.7 Electron
scattering in an external
field.

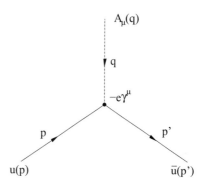

The diagram giving the amplitude of electron scattering by an external field in the lowest order is drawn in Fig. 10.7.

The analytic expression for this amplitude is

$$M_{fi} = -e\bar{u}(p')\gamma^{\mu}u(p)A_{\mu}(q),\qquad(10.24)$$

where the factor $A_{\mu}(q)$ is the external vector potential, $-e\gamma^{\mu}$ is the interaction vertex following from the Lagrangian (9.47), $u(p)$ is the Dirac bispinor describing the free ingoing electron — it enters the plane wave solution (9.51) of the Dirac equation and was explicitly written in (9.53). $\bar{u}(p')$ is related to $u(p')$ by Dirac conjugation. It describes the outgoing electron. The factors $u(p)$ and $\bar{u}(p)$ play for the electrons exactly the same role as the polarization vectors (10.20) play for the photons.

The cross section is proportional to $|M_{fi}|^2$. If we are interested in the cross section for the scattering of unpolarized electrons, we should sum over the polarizations of the initial and the final electron. For the Coulomb external static field created by the charge Ze, we arrive in the nonrelativistic limit at the Rutherford formula [cf. Eq. (5.5)]:

$$\frac{d\sigma}{d\Omega} = \frac{(2mZ\alpha)^2}{q^4}.\qquad(10.25)$$

The amplitude of the process of e^-e^- (Møller) scattering was drawn in Fig. 5.3. The diagrams look identical to the diagrams in Fig. 10.6 for the charged pion scattering, but the vertices now include the structures $-e\gamma^{\mu}$, while the external lines are spelled out as the bispinors $u(p_1)$, $u(p_2)$, $\bar{u}(p_3)$, $\bar{u}(p_4)$. The photon propagator is given by the expression (10.22), as earlier.

We consider now the process of photon-electron scattering (*Compton scattering*. There are two diagrams, drawn in Fig. 10.8.

We already know that the dashed lines representing real photons give the factors $\epsilon_{\mu}(q)$ and $\epsilon_{\nu}^{*}(q')$ in the analytic expressions for the amplitude. The vertices involve the structures $-e\gamma^{\mu}$ and $-e\gamma^{\nu}$. The external solid lines describe the ingoing and outgoing electrons and give the factors $\bar{u}(p')$ and $u(p)$.

(a)   **(b)**

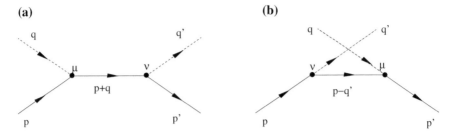

Fig. 10.8 Compton scattering.

But what analytic expression corresponds to the internal solid virtual electron line?

Well, we already know that the virtual photon line gives the Green's function of the d'Alembert equation (10.21) and that the virtual scalar line gives the Green's function of the KFG equation (10.8). Then it is easy to conjecture (and this conjecture is confirmed by the exact analysis) that the virtual fermion line gives the Green's function of the Dirac equation.

In momentum space, the inhomogeneous (including interactions) Dirac equation is

$$(\hat{p} - m)\Psi = \text{source}, \tag{10.26}$$

where we introduced the notation $\hat{p} = p_\mu \gamma^\mu$. The fermion propagator is

$$G_F(p) = \frac{1}{\hat{p} - m} = \frac{\hat{p} + m}{p^2 - m^2}. \tag{10.27}$$

Assembling together all the structures and counting all the signs accurately (by a yet undisclosed rule!), we can write the explicit expression for the Compton amplitude, given by the sum of two graphs in Fig. 10.8:

$$M_{fi} = -e^2 \epsilon_\mu^{(s)}(q) \epsilon_\nu^{(s')*}(q') \bar{u}^{(\sigma')}(p') \left[\gamma^\nu \frac{\hat{p} + \hat{q} + m}{(p+q)^2 - m^2} \gamma^\mu \right.$$
$$\left. + \gamma^\mu \frac{\hat{p} - \hat{q}' + m}{(p-q')^2 - m^2} \gamma^\nu \right] u^{(\sigma)}(p). \tag{10.28}$$

Here the variables $s = \pm 1$ and $s' = \pm 1$ are the helicities of the ingoing and the outgoing photons, and $\sigma = \pm 1/2$ and $\sigma' = \pm 1/2$ are the helicities of the ingoing and the outgoing electrons.

In Chap. 12, devoted to the electroweak theory, we will discuss the graphs involving virtual massive vector particles. To this end, we need to know their propagator. The latter is the Fourier image of the Green's function of the Proca equation (7.60)

Fig. 10.9 Schwinger correction to μ_e.

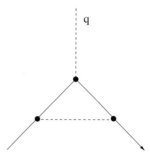

and reads[7]

$$D_{\mu\nu}^{\text{Proca}}(q) = \frac{\eta_{\mu\nu} - q_\mu q_\nu / m^2}{q^2 - m^2}. \tag{10.29}$$

10.6 Loops

The diagram in Fig. 10.7 may describe scattering not only in an external electric, but also in an external magnetic field. Calculating the amplitude in the latter case, we can distinguish there a term that describes the interaction of the field with the intrinsic electron magnetic moment predicted by Dirac's theory,

$$\mu_e = \frac{e\hbar}{2m_e c} = -9.274\ldots \cdot 10^{-21} \frac{\text{erg}}{\text{gauss}}. \tag{10.30}$$

However, when making precise measurements, people observed that the actual electron magnetic moment does not exactly coincide with (10.30), but its absolute value is a little bit larger. This deviation can be explained if we recall that the diagram in Fig. 10.7 only gives the scattering amplitude in the lowest order in perturbation theory. There are corrections, described by loop diagrams (we already discussed them in Chap. 5). The lowest nontrivial correction to the electron magnetic moment comes from the graph in Fig. 10.9. The exact calculation[8] gives

$$\mu_e = \frac{e\hbar}{2m_e c} \left[1 + \frac{\alpha}{2\pi} + O(\alpha^2) \right]. \tag{10.31}$$

---

[7] You may check that

$$[(m^2 - q^2)\delta_\mu^\nu + q_\mu q^\nu] D_{\nu\lambda}^{\text{Proca}}(q) = -\eta_{\mu\lambda}.$$

[8] It was done by Schwinger in 1948. Schwinger was very proud of this result. The formula (10.31) is engraved in his tombstone.

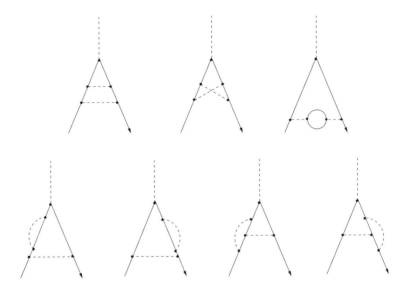

Fig. 10.10 Two-loop corrections to μ_e.

The corrections of higher order in α are also known. For sure, the higher the order, the more difficult are the calculations. There are seven two-loop graphs, depicted in Fig. 10.10. At the 3-loop level, one has to evaluate 72 graphs, at the 4-loop level — 891 graphs, at the 5-loop level — 12672 graphs…And all that was done: done analytically up to the level $\sim \alpha^3$ and numerically at the next two levels. As a result, our theoretical knowledge of the electron magnetic moment is incredibly precise.

People usually quote the result for the correction to the 1 in Eq. (10.31), $a = \alpha/(2\pi) + \ldots$, giving the *anomalous* magnetic moment. The theoretical prediction is

$$a^{\text{theor}} = (11\,596\,521\,816 \pm 8) \cdot 10^{-13} . \tag{10.32}$$

The experimental measurements of this quantity are also incredibly precise. We know that

$$a^{\text{exp}} = (11\,596\,521\,807 \pm 3) \cdot 10^{-13} . \tag{10.33}$$

You would not accuse me of wishful thinking and unjustified exaggeration, if I call the agreement between theory and experiment satisfactory in this case.

As you see, the experimental error is somewhat smaller than the theoretical one. The main source of the latter are not the still-unknown 6-loop corrections, but our imprecise knowledge of the fine-structure constant α. Incidentally, if we believe in the theory, we can *improve* this knowledge just on the basis of the experimental result (10.33).

Fig. 10.11 Photon
polarization operator at one
loop.

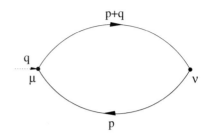

Going back to the corrections of order $\sim \alpha$, the graph in Fig. 10.9 is not the only
one-loop graph that one can draw. Another important one-loop graph was depicted
in Fig. 5.7. This graph does not contribute to the electron magnetic moment, but it
determines, as was pointed out in Chap. 5 the renormalization of the electron charge.
In Chap. 5 we claimed that this graph involves an ultraviolet logarithmic divergence,
but did not explain why. Now we can try to do so.

As was mentioned before, loop diagrams involve integrals over virtual 4-momenta.
Consider the block in the graph in Fig. 5.7 involving the fermion loop. This block (or
rather the corresponding analytic expression) represents the one-loop contribution
to what is called the *photon polarization operator* $\mathcal{P}^{\mu\nu}(q)$. The block is redrawn in
Fig. 10.11. Note that, by definition, $\mathcal{P}^{\mu\nu}(q)$ involves only the fermion loop, but not
the photon propagators that are attached to this loop in Fig. 5.7.

The Feynman rules outlined above and spelled out precisely at the end of the
chapter dictate the following analytic expression for $\mathcal{P}^{\mu\nu}(q)$:

$$\mathcal{P}^{\mu\nu}(q) = ie^2 \int \frac{d^4 p}{(2\pi)^4} \, \mathrm{Tr} \left\{ \gamma^\mu \frac{1}{\hat{p} - m} \gamma^\nu \frac{1}{\hat{p} + \hat{q} - m} \right\}. \tag{10.34}$$

At first sight, this graph involves not just a logarithmic, but a *quadratic* ultraviolet
divergence at all momenta q including $q = 0$:

$$\mathcal{P}^{\mu\nu}(q) \sim e^2 \int \frac{d^4 p}{p^2} \sim e^2 \Lambda^2,$$

where Λ is the ultraviolet cutoff — the artificially introduced upper limit in the
momenta integral. However, if we introduce this cutoff *accurately*, respecting both
Lorentz and gauge invariances, (we will not try to do such an accurate *ultraviolet
regularization* here), one can be convinced that the quadratic divergence actually
cancels and the result of integration is

$$\mathcal{P}^{\mu\nu}(q) = \left(\eta^{\mu\nu} - \frac{q^\mu q^\nu}{q^2} \right) \mathcal{P}(q^2), \tag{10.35}$$

with $\mathcal{P}(0) = 0$, so that $\mathcal{P}^{\mu\nu}(q)$ is regular at $q = 0$. Then $\mathcal{P}(q^2)$ only involves the
logarithmic ultraviolet divergence.

Fig. 10.12 Exact photon propagator.

This result is not surprising. A nonzero quadratically divergent $P(0)$ would mean a nonzero quadratically divergent photon mass [cf. Eq. (10.37) below]! And such a nonsense is not allowed by gauge invariance.

Note that the polarization operator satisfies the transversality condition $q_\mu P^{\mu\nu} = 0$. That is not accidental and actually follows from the current conservation law (9.56) (heuristically: the derivative ∂_μ goes over to q_μ in momentum space).

The scalar function $P(q^2)$ depends, besides q^2, on the electron mass m and the ultraviolet cutoff Λ. The exact expression for $P(q^2)$ can be found, but it is not particularly transparent. In the asymptotics, however, it becomes rather nice:

$$P(q^2) \approx -\frac{e^2}{12\pi^2} q^2 \ln\frac{\Lambda^2}{q^2}, \qquad |q^2| \gg m^2,$$

$$P(q^2) \approx -\frac{e^2}{12\pi^2} q^2 \ln\frac{\Lambda^2}{m^2}, \qquad |q^2| \ll m^2. \tag{10.36}$$

Our next task is to derive the expression (5.12) for the physical electron charge. The point is that a fermion loop in Fig. 10.11 can be inserted in *any* photon virtual line in a Feynman graph. And not only one such loop, but any number of loops. Consider a set of graphs in Fig. 10.12. The corresponding analytic expression is

$$\mathcal{D}_{\mu\nu}(q) = \frac{\eta_{\mu\nu}}{q^2} + \frac{\eta_{\mu\rho}}{q^2}\left(\eta^{\rho\sigma} - \frac{q^\rho q^\sigma}{q^2}\right) P(q^2)\frac{\eta_{\sigma\nu}}{q^2} + \cdots$$

$$= \frac{\eta_{\mu\nu}}{q^2 - P(q^2)} - q_\mu q_\nu \frac{P(q^2)}{q^4[q^2 - P(q^2)]}. \tag{10.37}$$

Let us call it the *exact photon propagator*.

A remarkable fact is that the second term on the right-hand side of (10.37) is completely irrelevant! It is, again, a corollary of the current conservation law. Indeed, consider as an example the graphs in Fig. (5.3) for electron scattering. If we replace the tree propagator (10.22) by the exact propagator (10.37) in the virtual photon line, the term $\propto q_\mu q_\nu$ in the propagator is multiplied by the fermion brackets $\bar{u}(p_3)\gamma^\mu u(p_1)$ and $\bar{u}(p_4)\gamma^\nu u(p_2)$. And this product vanishes due to (9.57).

This is exactly the subtlety that we have now fetched out from under the carpet, as promised. One can always add to the photon propagator a term $\propto q_\mu q_\nu$, and all the results for the physical amplitudes stay intact. To add or not to add such terms and, if they are added, with what coefficient, is a matter of taste and convenience. This freedom is actually a direct consequence of the gauge freedom (9.48).

We now go back to (10.37). Disregard there the term $\propto q_\mu q_\nu$ and assume q^2 to be small. We are left with the first term $\propto \eta_{\mu\nu}$, where we should substitute for $P(q^2)$

Fig. 10.13 Electron polarization operator.

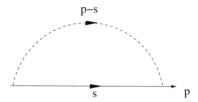

the second line in (10.36). We obtain[9]

$$\mathcal{D}_{\mu\nu}(q) = D_{\mu\nu}(q) \frac{1}{1 + \frac{e_0^2}{6\pi^2} \ln \frac{\Lambda}{m}} + \text{irrelevant terms}, \qquad (10.38)$$

i.e. the exact propagators coincide in this limit with the tree propagators, multiplied by a certain factor. And this boils down to multiplying e_0^2 (associated with each virtual photon line) by the same factor. The result (5.12) is thus reproduced.

Not only does the physical electron charge differ from the value entering the Lagrangian. The same is true for the electron mass. By the same token that for the photon, one can define the *electron polarization operator* and the exact electron propagator. The lowest-order 1-loop graph giving a contribution to the former is drawn in Fig. 10.13. This graph involves a logarithmic ultraviolet divergence.

The physical electron mass is given by the pole of the exact propagator. At the 1-loop level,

$$m_{\text{phys}} = m_0 \left(1 + \frac{3e_0^2}{8\pi^2} \ln \frac{\Lambda}{m_0} \right), \qquad (10.39)$$

where m_0 is the bare mass. Taking also in account higher loops and keeping only the *leading logarithmic* terms $\propto (e^2 \ln \frac{\Lambda}{m})^n$ [that is what we actually did when deriving (5.12)], one can derive

$$m_{\text{phys}} = m_0 \left(\frac{e_0^2}{e_{\text{phys}}^2} \right)^{9/4}. \qquad (10.40)$$

The physical mass is larger than the bare one.

We conclude this chapter by accurately formulating the Feynman rules, with which any diagram in scalar theory and in scalar and spinor electrodynamics can be exactly calculated. Frankly, to give such an accurate formulation was not in my preliminary plans: as was repeatedly stated, we do not have the ambition to really *learn* how to calculate graphs. Still, having written this chapter, I imagined the sufferings of a *very* dedicated reader, who wanted to reproduce the expressions for the different

---

[9] We have restored the notation e_0, used in Chap. 5, for the bare charge appearing in the Lagrangian.

scattering amplitudes that I wrote, but could not do so, and changed my mind. The following page is for him. All others can safely skip it.

1. The analytic expressions for the vertices following from the Lagrangian and given in Figs. 10.1, 10.4 and 10.7 should all be actually multiplied by i.[10]
2. A scalar virtual line gives the factor $i D(p) = i/(p^2 - m^2)$. No nontrivial factors are associated with external scalar lines.
3. A photon virtual line gives the factor $-i D_{\mu\nu}(q) = -i\eta_{\mu\nu}/q^2$.
   An ingoing photon line provides the polarization factor $\epsilon_\mu(q_{in})$; an outgoing photon line provides the polarization factor $\epsilon_\mu^*(q_{out})$.
   A static external electromagnetic field gives the factor $2\pi\delta(q_0)A_\mu(q)$, where $A_\mu(q)$ is the Fourier transform of the field.
4. The virtual line of a vector massive particle provides the factor (10.29) multiplied by $-i$.
5. An electron virtual line gives the factor $i G_F(p) = i/(\hat{p} - m)$.
   An ingoing electron gives the polarization factor $u(p_{in})$; an outgoing electron gives the factor $\bar{u}(p_{out})$.
   An ingoing positron gives the polarization factor $\bar{u}(-p_{in})$ and an outgoing positron the factor $u(-p_{out})$. $(p_0 > 0!)$[11]
6. If certain internal virtual 4-momenta $s_\mu^{(j)}$ or the momenta associated with an external field are not determined by the kinematics, one should perform an integration over $\prod_j d^4 s^{(j)}/(2\pi)^4$.
7. Each fermion loop provides an extra minus sign.[12]
8. After performing all the calculations, we obtain the analytic expression for the scattering amplitude M_{fi} multiplied by i.[13]
9. For completeness, we also mention that, in the case when a loop involves n identical bosonic particles and the diagram is symmetric under exchange of the corresponding virtual lines, the extra combinatorial factor $1/n!$ should be introduced. This follows from Wick's counting in the spirit of Fig. 10.2.

---

[10]When accurately deriving the diagram technique in the framework of the path integral approach, this follows from the fact that the Minkowski path integrals involve $\exp\{i \int \mathcal{L} d^4 x\}$ in the integrand, rather than $\exp\{\int \mathcal{L} d^4 x\}$ [cf. Eq. (8.41)].

[11]A convenient arrow policy is then: *arrows outwards* for the outgoing electrons and ingoing positrons and *arrows inwards* for the ingoing electrons and outgoing positrons — cf. the remark at the end of Sect. 5.1.

[12]This follows from the Grassmann nature of the fermion fields. Do not forget this minus if you want to reproduce (10.34)!

[13]This i has the same origin as the extra factor i for the vertices mentioned above. Note that, to the first order, these two factors cancel out and one can deduce the analytical expressions for the amplitudes in Figs. 10.1, 10.4 and 10.7 directly from the Lagrangian, which we actually did.

Chapter 11
Quantum Chromodynamics

The two long chapters that follow are central to our book. After all our preparations, we are finally going to do it: to *present* the quantum field theories which describe the physics of the strong interactions and the physics of the weak interactions, and we will try to explain many of their properties mentioned before, in Chaps. 3 and 5. This chapter is devoted to the theory of the strong interactions — quantum chromodynamics (QCD in short).

QCD has many features in common with spinor electrodynamics. Both theories involve fundamental fermions (electrons in QED and quarks in QCD) and fundamental vector particles (photons in QED and gluons in QCD). The difference is that both quarks and gluons carry colour. Quarks belong to the fundamental representation of the $SU(3)$ colour group; one can say that there are quarks of three different colours (and of six different flavours). Gluons belong to the adjoint representation of $SU(3)$, and there are $3^2 - 1 = 8$ different gluon colour species.

11.1 QCD Lagrangian

Our intention now is to write down the field theory Lagrangian for QCD. We will take as our starting point the QED Lagrangian (9.47) and show how it can be generalized to the multicolour non-Abelian case.

We concentrate on the fermion part of (9.47), which reads

$$\mathcal{L}_{\text{ferm}}^{\text{QED}} = i\bar{\psi}\gamma^\mu \mathcal{D}_\mu \psi - m\bar{\psi}\psi, \tag{11.1}$$

where

$$\mathcal{D}_\mu = \partial_\mu + ieA_\mu \tag{11.2}$$

is the covariant derivative. The expression (11.1) is invariant under the Abelian gauge transformations (9.48), which we write here again:

© Springer International Publishing AG 2017
A. Smilga, *Digestible Quantum Field Theory*,
DOI 10.1007/978-3-319-59922-9_11

$$A_\mu \to A_\mu + \partial_\mu \chi \,,$$
$$\psi \to e^{-ie\chi} \psi \,. \tag{11.3}$$

The transformations (11.3) form a group. This group is Abelian because the result of two consecutive transformations does not depend on their order. The group of gauge transformations is *infinite-dimensional*. There is a distinct parameter $\chi(x)$ at each space-time point. People say (and we also have already mentioned it — see p. 120) that the transformations (11.3) are *local* to distinguish from the *global* symmetry transformations with only a finite number of transformation parameters not depending on x and t. The global counterparts of (11.3) are the transformations where the charged field ψ is multiplied by a universal phase factor while the neutral field A_μ does not transform. We recognize the one-parameter Abelian $U(1)$ group. It is also natural to attribute the name $U(1)$ to the infinite-dimensional gauge group (11.3).

An important observation is that the structure $\mathcal{D}_\mu \psi$ is gauge-transformed in the same way as ψ:

$$\mathcal{D}_\mu \psi \to e^{-ie\chi} \mathcal{D}_\mu \psi \tag{11.4}$$

[see also the footnote after (9.36)].

Let now replace the electron field ψ by the quark field q_j carrying the colour index, $j = 1, 2, 3$. It is clear that the mass term in (11.1), replaced by $-m\bar{q}^j q_j$, is now invariant under local non-Abelian $SU(3)$ transformations,

$$q(x) \to \Omega(x)q(x), \qquad \bar{q}(x) \to \bar{q}(x)\Omega^\dagger(x) \,, \tag{11.5}$$

with $\Omega(x) \in SU(3)$.

We now go over to the first term in (11.1). It is clear again that the *free* fermion kinetic term $i\bar{\psi}\gamma^\mu \partial_\mu \psi$ is easily generalized to $i\bar{q}^j \gamma^\mu \partial_\mu q_j$. This structure is not invariant, however, with respect to local gauge transformations (11.5); it is only invariant under global $SU(3)$ transformations where Ω is one and the same at all space-time points.

To write the interaction term, we replace (11.2) by[1]

$$\mathcal{D}_\mu = \delta^k_j \partial_\mu - ig(t^a)^k_j A^a_\mu \equiv \mathbb{1}\partial_\mu - ig\hat{A}_\mu \,, \tag{11.6}$$

where t^a are the eight generators of $SU(3)$ defined in (6.43) and g is the "strong charge" — a dimensionless constant determining how strong the strong interaction is. The matrices $\hat{A}_\mu$ transform under $SU(3)$ as in (6.51), $\hat{A}_\mu \to \Omega\hat{A}_\mu\Omega^\dagger$. The structure

$$\mathcal{L}^{QCD}_{ferm} = i\bar{q}\gamma^\mu\left(\partial_\mu - ig\hat{A}_\mu\right)q \,, \tag{11.7}$$

---

[1] According to the usually adopted convention, the sign of the charge in (11.6) is opposite compared to (11.2). This mess can be traced back to the fact that the electron charge is negative, $e = -|e|$. Unfortunately, when Benjamin Franklin defined which charge is positive and which is negative $\sim$260 years ago, he knew nothing about electrons.

(we do not display the colour indices any more) is then invariant under global $SU(3)$ symmetry transformations

$$q(x) \rightarrow \Omega q(x), \qquad \bar{q}(x) \rightarrow \bar{q}(x)\Omega^\dagger, \qquad \hat{A}_\mu(x) \rightarrow \Omega \hat{A}_\mu(x)\Omega^\dagger. \qquad (11.8)$$

What we now want to do is to accept the expression (11.7) as the *true* contribution in the QCD Lagrangian including the kinetic quark term and the term describing quark interactions with the gluon field and ask the following question: assume that the quark fields transform *locally* as in (11.5); how should the matrix vector field $\hat{A}_\mu$ simultaneously transform for the Lagrangian (11.7) to stay invariant? The equivalent question is: how $\hat{A}_\mu$ should transform for the derivative (11.6) to be really covariant, i.e. for the structure $\mathcal{D}_\mu q(x)$ to transform as $\mathcal{D}_\mu q(x) \rightarrow \Omega(x)\mathcal{D}_\mu q(x)$, in the same way as $q(x)$. Yet another reformulation is to find a transformation law for $\hat{A}_\mu$ such that the differential operator (11.6) transforms as

$$\mathcal{D}_\mu \rightarrow \Omega \mathcal{D}_\mu \Omega^\dagger. \qquad (11.9)$$

The answer to these questions is

$$\hat{A}_\mu \rightarrow \Omega \hat{A}_\mu \Omega^\dagger - \frac{i}{g}(\partial_\mu \Omega)\Omega^\dagger \qquad (11.10)$$

(we leave that for the reader to check). The laws (11.5) and (11.10) constitute together the local gauge symmetry with respect to which (11.7) is invariant.

It is instructive to see how these laws go over to the Abelian laws (11.3) for the group $U(1)$. For $U(1)$, A_μ is no longer a matrix and the first term in (11.10) is just A_μ. In the second term, we have to substitute $e^{-i e \chi}$ for Ω and $-e$ for g. Then it is easy to see that (11.10) goes over to $A_\mu \rightarrow A_\mu + \partial_\mu \chi$.

Our next task is to generalize to the non-Abelian case the Lagrangian of the electromagnetic field $-\frac{1}{4}F_{\mu\nu}^2$. To this end, we first notice that the electromagnetic field strength $F_{\mu\nu}$ can be represented as a commutator of the covariant derivatives in (11.2):

$$F_{\mu\nu} = \partial_\mu A_\nu - \partial_\nu A_\mu = -\frac{i}{e}[\mathcal{D}_\mu, \mathcal{D}_\nu]. \qquad (11.11)$$

Let us now do the same for the non-Abelian covariant derivatives (11.6) and define

$$\hat{F}_{\mu\nu} = \frac{i}{g}[\mathcal{D}_\mu^{non-Ab}, \mathcal{D}_\nu^{non-Ab}] = \partial_\mu \hat{A}_\nu - \partial_\nu \hat{A}_\mu - ig[\hat{A}_\mu, \hat{A}_\nu]. \qquad (11.12)$$

$\hat{F}_{\mu\nu}$ belongs, as does $\hat{A}_\mu$, to the adjoint representation of the group and can be written as $\hat{F}_{\mu\nu} = F_{\mu\nu}^a t^a$ with

$$F_{\mu\nu}^a = \partial_\mu A_\nu^a - \partial_\nu A_\mu^a + g f^{abc} A_\mu^b A_\nu^c, \qquad (11.13)$$

where f^{abc} are the structure constants of $SU(3)$.

The Abelian field strength (11.11) is invariant under Abelian gauge transformations. The non-Abelian field strength (11.12) is not invariant under (11.10), but transforms in a simple way:

$$\hat{F}_{\mu\nu} \rightarrow \Omega \hat{F}_{\mu\nu} \Omega^{\dagger} , \tag{11.14}$$

as immediately follows from (11.9). It is now easy to observe that the structure

$$\mathcal{L}^{\text{gauge}} = -\frac{1}{2}\text{Tr}\{\hat{F}_{\mu\nu}\hat{F}^{\mu\nu}\} = -\frac{1}{4}F^a_{\mu\nu}F^{\mu\nu\,a} \tag{11.15}$$

is invariant under the non-Abelian gauge transformations (11.10). We can say now that (11.15) is the true Lagrangian describing the dynamics of the gluon field in QCD!

As was mentioned in Chap. 5, the Lagrangian (11.15) was first written down by Yang and Mills in 1954.[2] It is interesting that the concept of a non-Abelian gauge field was simultaneously and independently developed by pure mathematicians. But if you ever discuss this subject with a mathematician, please, do not say "gauge field". You will simply not be understood. Talking about $\hat{A}_{\mu}$, call it the *connection of a principal fibre bundle*. As for $\hat{F}_{\mu\nu}$, it is evidently the *curvature form of a principal G-connection*.

The full gauge-invariant QCD Lagrangian reads

$$\mathcal{L}^{\text{QCD}} = -\frac{1}{2}\text{Tr}\{\hat{F}_{\mu\nu}\hat{F}^{\mu\nu}\} + \sum_{f=1}^{6}\left[i\bar{q}_f\gamma^{\mu}\left(\partial_{\mu} - ig\hat{A}_{\mu}\right)q_f - m_f\bar{q}_f q_f\right], \tag{11.16}$$

where the sum runs over six quark flavours. m_f are their masses quoted in Table 3.1.

11.2 Feynman Rules □ Asymptotic Freedom □ Confinement

The quark-gluon interaction described by the Lagrangian (11.16) is very similar to the electron-photon interaction. The only difference is that the interaction vertex now includes the $SU(3)$ generator t^a.

An essentially new feature is self-interaction of gluons, described by the first term in (11.16). By substituting there (11.12), we obtain the cubic term $\sim A^3$ in the Lagrangian, as well as the quartic term. In the cubic term, one of the fields enters with a derivative, giving a momentum in the 3-gluon vertex. There are no momenta in the quartic vertex, but its colour and Lorentz structure appear to be relatively

---

[2] A nice year, when your author was born...

complicated. To amuse the reader, we present below the explicit expressions for the QCD vertices and propagators.

$b\nu$ ⌇⌇⌇⌇⌇⌇ $a\mu$
$\overset{q}{}$

$\qquad$ gluon propagator: $\quad -iD_{\mu\nu}^{ab}(q) \;\; = \;\; \dfrac{-i\eta_{\mu\nu}}{q^2}\delta^{ab}\,,$

$$\tag{11.17}$$

$c\lambda$

r $\qquad$ p

$\qquad$ $a\mu$

$b\nu$ $\qquad q$

3–gluon vertex: $\quad \Gamma_{\mu\nu\lambda}^{abc}(p,q,r) \;\; = \;\; -gf^{abc}[(p-q)_\lambda\eta_{\mu\nu} +$
$\qquad\qquad\qquad\qquad\qquad (q-r)_\mu\eta_{\nu\lambda} + (r-p)_\nu\eta_{\mu\lambda}]\,,$

$$\tag{11.18}$$

$d\sigma$

$c\lambda$ ⌇⌇⌇⌇⌇⌇ $a\mu$

$b\nu$

4–gluon vertex: $\quad \Gamma_{\mu\nu\lambda\sigma}^{abcd} \;\; = \;\; -ig^2 f^{abe}f^{cde}(\eta_{\mu\lambda}\eta_{\nu\sigma} - \eta_{\mu\sigma}\eta_{\nu\lambda})$
$\qquad\qquad\qquad\qquad -ig^2 f^{ace}f^{bde}(\eta_{\mu\nu}\eta_{\sigma\lambda} - \eta_{\mu\sigma}\eta_{\nu\lambda})$
$\qquad\qquad\qquad\qquad -ig^2 f^{ade}f^{bce}(\eta_{\mu\nu}\eta_{\sigma\lambda} - \eta_{\mu\lambda}\eta_{\nu\sigma})\,,$

$\xrightarrow{\quad p \quad}$

$$\tag{11.19}$$

$\qquad$ quark propagator: $\quad iG_k^j(p) \;\; = \;\; \dfrac{i}{\hat{p} - m}\delta_k^j\,,$

$\qquad$ quark - gluon vertex: $\quad (\Gamma_\mu^a)_k^j \;\; = \;\; ig\gamma_\mu(t^a)_k^j\,.$

$$\tag{11.20}$$

(For the 3-gluon vertex, we have chosen the convention that all the momenta p, q and r are outgoing, so that $p + q + r = 0$.)

We have to admit, however, that, as often happens, we have not only amused the reader, but also cheated him. The Feynman rules (11.17)–(11.20) are sufficient to correctly calculate tree amplitudes, but are not sufficient to calculate loops.

To understand the problem, let us first go back to QED. The photon propagator describes the propagation of a virtual photon. The physical photon has two transverse polarizations (10.20). However, the propagator (10.22) also describes on an equal footing the propagation of an unphysical longitudinally polarized photon with $e_\mu^{(L)} = (0; 0, 0, 1)$ and of the "scalar" or "temporal" photon with $e_\mu^{(T)} = (1; 0, 0, 0)$. Inclusion of unphysical degrees of freedom may be troublesome, but a remarkable fact [a corollary of the current conservation law (9.56)] is that the contributions of unphysical virtual photon polarizations exactly cancel in any physical amplitude! It allows one to use the simple expression (10.22) for the propagator and not to worry about anything.

The point is that in QCD life is more complicated. If we use the propagator (11.17), the contributions of the longitudinal and temporal gluons do *not* cancel in graphs involving loops. The cancellations can only be assured, if we add graphs with loops of some extra unphysical particles, called *ghosts*. The Feynman rules in (11.17)–(11.20) should be supplemented to include the ghost propagator and the ghost interaction vertex. As we already mentioned in Sect. 5.3, this was done by Faddeev and Popov in 1967. But we will not explain here what they did. The reader should consult more serious books... We make only one remark. One can avoid the introduction of ghosts by using another form of the gluon propagator that incorporates only physical transverse polarizations of virtual gluons. This form is, however, more complicated than (11.17), and this method is technically less convenient.

With the Feynman rules in hand, one can perform perturbative calculations.

We now invite the reader to look again into the end of Sect. 5.3, where the phenomenon of asymptotic freedom and its consequences for phenomenology were discussed. In brief: calculations show that the effective strength of the interaction depends on energy. In contrast to QED, where the effective charge $e^2(\mu)$ and $\alpha(\mu) = e^2(\mu)/4\pi$ grow with the characteristic energy μ, in QCD the situation is the opposite: as the energy grows, the coupling $\alpha_s(\mu) = g^2(\mu)/4\pi$ decreases according to (5.20).[3] This gives us a small parameter and makes perturbative expansion meaningful. The most spectacular perturbative QCD phenomenon, where theoretical predictions are in beautiful agreement with experiment, is jet production in e^+e^- – annihilation [see Fig. 5.10]. There are others.

---

[3] We reproduce here this fundamental result:

$$\alpha_s(\mu) = \frac{2\pi}{\left(11 - \frac{2}{3}N_f\right) \ln \frac{\mu}{\Lambda_{QCD}}}, \tag{11.21}$$

where N_f is the number of quarks whose mass does not exceed μ and $\Lambda_{QCD} \sim 200$ MeV is the fundamental dimensionful QCD constant.

The coupling constant (11.22) is small when μ is large, but it becomes large when μ decreases. At $\mu = \Lambda_{QCD}$, it is even infinite, but the one-loop approximate formula (11.22) does not apply, of course, for such small μ. One can only say that for $\mu \lesssim .5$ GeV, perturbative calculations lose meaning.

Consider now a heavy quark-antiquark pair (one can think of a pair of t quarks of mass $m_t \approx 170$ GeV). At small distances (this corresponds to large characteristic energies) they are attracted due to the exchange of a virtual gluon, by the same token that the electron and the proton in the hydrogen atom are attracted due to the exchange of a virtual photon. This gives a Coulomb-like potential

$$V_{Q\bar{Q}}(r) = -c_F \frac{\alpha_s \left(\frac{1}{r}\right)}{r}, \qquad (11.22)$$

where the factor c_F, defined as $t^a t^a = c_F \mathbb{1}$, appears, due to the presence of the generator t^a in the vertex (11.20). Its numerical value for $SU(3)$ is $c_F = 4/3$. Mathematicians call it the *eigenvalue of the quadratic Casimir operator in the fundamental representation*. But never mind — you can check that the arithmetic is correct by using the explicit expressions (6.43) for the generators.

The result (11.22) for the potential is correct when $r \ll \Lambda_{QCD}^{-1}$, such that $\alpha_s \left(\frac{1}{r}\right)$ is small and the corrections due to exchange by several gluons or light quark-antiquark pairs are small too. When $r \gtrsim \Lambda_{QCD}^{-1}$, the Coulomb formula (11.22) is not valid any more, and it is not easy to say what one should write instead of (11.22) — perturbative evaluations are no longer possible.

A *conjecture* is that the potential $V_{Q\bar{Q}}(r)$ grows linearly at large distances:

$$V_{Q\bar{Q}}(r \gg \Lambda_{QCD}^{-1}) \sim \sigma r, \qquad (11.23)$$

where σ is a nonzero positive constant. Equation (11.22) and (11.23) together give the so-called *funnel* potential represented in Fig. 11.1.

Fig. 11.1 Quark-antiquark potential.

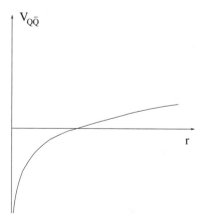

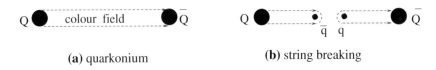

(a) quarkonium   **(b)** string breaking

Fig. 11.2 Strings in QCD.

The growth of the potential at large distances means confinement — the quarks simply cannot be separated, as this would cost an infinite energy!

Assume that the potential grows linearly, as in (11.23). Well, as was already mentioned in Chap. 5, we cannot prove it at the moment. Neither can we prove a weaker claim that $V_{Q\bar{Q}}(\infty) = \infty$, which would imply the absence of coloured states in the spectrum.[4]

But the linear growth is the simplest and the most natural assumption. A heuristic justification comes from the picture drawn in Fig. 11.2a. It implies that, at large interquark distances, the gluon *chromo*electric field $\boldsymbol{E}^a(\boldsymbol{x})$ is not spread out (as it would be for weak coupling), but is concentrated in a kind of tube, which is usually called a *string*.[5] The field carries energy. The picture in Fig. 11.2a suggests a constant linear string energy density. This density (called the *string tension*) is nothing but the constant σ in (11.23). Certain numerical estimates and phenomenological fits give

$$\sigma \approx 0.9 \, \frac{\text{GeV}}{\text{fermi}} . \tag{11.24}$$

The whole system — *quark + antiquark + connecting string made of gluon field* — carries no colour. It is a physical hadron state. This type of hadron is called

---

[4]This applies also to the somewhat more simple *pure* Yang–Mills theory, the theory without quarks where the Lagrangian only involves the first term in (11.16). At the moment we cannot prove that the physical spectrum of this theory does not involve coloured massless gluon states.

As a matter of fact, the confinement problem belongs to the set of so-called *Millennium Prize Problems* — seven difficult long-standing conceptual problems in mathematics and physics formulated by the Clay Mathematics Institute and the associated fund, which allocated $7 million in prizes for their solution. Since 2000 when the prizes were announced, one of the problems (It is known as the *Poincaré conjecture*. Please forgive me for not going into further details. This would lead us too far astray.) has been solved by Grigory Perelman, a mathematician from St. Petersburg, who opted not to receive the prize. This was a purely mathematical problem; four of the six remaining problems are also mathematical, but two others refer to physics. One of them will be awarded to a scholar who can prove the existence and uniqueness of the solution to the Navier–Stokes equation (4.26) that describes the flow of turbulent liquid — this problem has remained unresolved for almost two hundred years.

And the second physical millennium problem is about confinement. For a reader who might wish to try to solve it and simultaneously to solve to a considerable extent his/her financial problems, I give here the exact official formulation:

Prove that for any compact simple gauge group G, a non-trivial quantum Yang–Mills theory exists on $\mathbb{R}^4$ and has a mass gap $\Delta > 0$. Good luck!

[5]No, do not worry. It is not a *fundamental string* of string theory — we will try to digest this latter notion later, while enjoying our dessert. It is a low-profile phenomenological QCD string.

heavy quarkonium. Its mass is roughly twice the heavy quark mass. For t quarks, $M_{t\bar{t}} \approx M_t + M_{\bar{t}} \approx 340$ GeV, and the energy stored in the string only represents a small fraction of the total.

Suppose now that we pull the quarks further and further apart. The string stretches more and more and carries more and more energy. Evidently, this growth cannot go on forever. Pretty soon our string is *broken* in two, with the creation of an extra pair of light quarks. This is shown in Fig. 11.2b.

The emergence of extra quarks at the ends of the freshly broken string is necessary to neutralize the colour charges of two pieces. As a result, nothing keeps two new colour-neutral systems together any more. They can go in different directions and forget about each other. We are thus dealing with the *decay* of a colourless excited quarkonium state $(Q\bar{Q})^*$ into two mesons, $Q\bar{q}$ and $q\bar{Q}$. Each such meson represents a heavy quark, accompanied by its light anticomrade and a short gluon string.

11.3 November Revolution and Quarkonium

On November 12 or maybe November 13, 1974, Lev Okun, who was teaching us, graduate students of the Moscow Institute of Physics and Technology, a course on particle phenomenology, entered the class and said: "There was a sensation yesterday!". He was referring to the discovery of the J/ψ particle, simultaneously announced on 11/11/74 by two American research groups, one at the Brookhaven National Laboratory, headed by Samuel Ting, and another at the Stanford Linear Accelerator Center, headed by Burton Richter.

The BNL group studied the inclusive production of electron-positron pairs in the collisions of the proton beam with the beryllium target, the process $p + Be \rightarrow e^+e^- + X$, where X is an unspecified set of particles — that is what "inclusive" means. Ting supposed that the e^+e^- pairs were produced via the decay of a virtual photon created in the collision [see Fig. 11.3a].

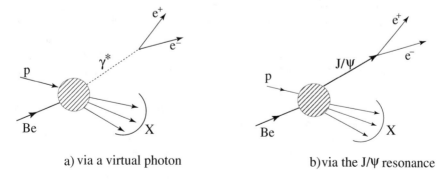

a) via a virtual photon   b) via the J/ψ resonance

Fig. 11.3 Two mechanisms of the e^+e^- pair production.

Fig. 11.4 Distribution over
invariant masses of the e^+e^-
pairs in Ting's experiment.

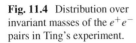

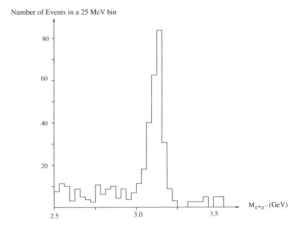

Virtual photons can have different masses $M^2_{\gamma*} = p^2_{\gamma*}$, and Ting and his colleagues expected to see a smooth distribution over the *invariant masses* of the pair, $M^2_{e^+e^-} = (p_+ + p_-)^2$. However, to their great surprise, they observed a sharp peak in the distribution near $M_{e^+e^-} \approx 3.1$ GeV. [See Fig. 11.4 — the plot drawn after the plot from their original paper.] That meant that, for this value of $M_{e^+e^-}$, the electron and positron were produced not through a virtual photon, but rather by a *resonance* mechanism via the decay of a real particle having this mass [see Fig. 11.3b]. Ting christened this particle "J".

At about the same time, physicists at SLAC, the laboratory where electrons and positrons were accelerated in the same storage ring but in opposite directions, and their collisions were then studied, succeeded in raising the energy to an impressive $E_+ = E_- \sim 1.5$ GeV and more.[6] And they saw a sharp peak in the total cross section of e^+e^- annihilation at the same energy as Ting! Richter's group christened the particle ψ.

The SLAC peak was even sharper than the BNL one. The finite width Γ of the peak in Fig. 11.4 is actually due to experimental imprecisions — one can only measure the momentum and the energy of the produced electrons and positrons and hence their invariant mass with a finite accuracy. But the accuracy with which the energies of colliding electrons and positrons can be fixed is much greater. Not in the first publication, but rather soon after that, the *intrinsic* width of the resonance[7], not related to experimental uncertainties, but arising due to the finite lifetime of the

---

[6]Impressive in a historical perspective, of course. Still larger machines were soon built. By 2000, the electron and positron energies at LEP, the e^+e^- accelerator at CERN, had reached the value $E_+ = E_- = 104.5$ GeV. Incidentally, soon after that, when the programme of experimental studies at this energy was fulfilled, LEP was shut down and later dismantled in order to make room in the tunnel for the construction of the Large Hadron Collider.

[7]*Resonances* is a common name for all particles whose lifetime is so small that it cannot be measured directly by any kind of watch, but only via a finite width of the distribution in invariant mass of the products of their decay.

J/ψ particle and the Heisenberg uncertainty principle, $\Delta E \Delta t \sim \hbar \implies \Gamma \tau \sim \hbar$, was measured at SLAC. This width turned out to be pretty small, $\Gamma_{J/\psi} \approx 90$ keV, corresponding to the lifetime $\tau_{J/\psi} \approx 7 \cdot 10^{-21}$ s.

The discoveries of the BNL and SLAC groups were announced on the same day and were submitted to Physical Review Letters with a one-day interval. That is why the particle has got the double name J/ψ, the only such case in the whole history of experimental particle studies. For their remarkable discovery, Richter and Ting were awarded the Nobel Prize in 1976.

But what *is* J/ψ? On the day it was discovered, that was not clear. In the first memorable Okun lecture (and Lev Borisovitch abandoned the regular planning of his course and talked exclusively about J/ψ for two or, I think, even three lectures, conveying to us the content of the discussions in the physicists community and the consensus that was eventually reached), he told us that it was probably an extra intermediate neutral boson, a mediator of the weak interactions. The standard theory of the weak interactions predicted the mass of the Z boson to be much larger [see Eq. (5.29)], but at that time people were not sure yet that the Standard Model was correct (and the term "Standard Model" had not yet been coined), and non-minimal variants of the theory with several Z bosons were discussed. In these versions, it was possible to have a heavy Z boson and some number of supplementary light intermediate bosons.

However, it became clear very soon that J/ψ is not a Z boson. It is a hadron, but a hadron of a special type, made not of light up, down and strange quarks, only known at that time, but rather it represents a bound state of a new relatively heavy charmed quark and a charmed antiquark. J/ψ was the first experimentally observed elementary particle representing a heavy quarkonium, a system we talked about in the preceding section.

J/ψ carries unit spin, like the photon. In analogy with *orthopositronium* — a bound state of an electron and a positron where the spins of the constituents are parallel so that the total angular momentum of the system is equal to one — one can call J/ψ *orthocharmonium*.

One can now understand why this state is so narrow. For J/ψ to decay, the $c\bar{c}$ pair should disappear (annihilate). And for heavy quarks, it is not so easy. By the same token that orthopositronium decays into three photons (the decay into two photons is not allowed by quantum numbers), the main decay channel of J/ψ is $J/\psi \rightarrow ggg$. Then energetic gluons (with energy ~ 1 GeV each) go over into light hadrons, observed by experimentalists. But this happens *later*, and if one is not interested in the details (which particular mesons or baryons are produced), but only in the total decay probability, one can consider only the underlying process $c\bar{c} \rightarrow ggg$, multiply its amplitude by the probability amplitude for the quarks c and $\bar{c}$ to meet (i.e. the wave function of the bound $(c\bar{c})$ system at the origin), and calculate the decay probability w (or the total width Γ) by integrating the differential probability (10.1). Some graphs contributing to the amplitude are drawn in Fig. 11.5.

It is exactly the same philosophy that was outlined in Chap. 5 when the total cross section of e^+e^- annihilation was discussed. The underlying fundamental process for the annihilation $e^+e^- \rightarrow$ hadrons is $e^+e^- \rightarrow q\bar{q}$, with the formation of an energetic

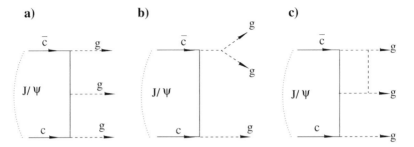

Fig. 11.5 Feynman diagrams for the decay $J/\psi \rightarrow 3g$: (**a**), (**b**) tree graphs; (**c**) a 1-loop graph.

quark pair. The underlying fundamental process for the decay of J/ψ into hadrons is $J/\psi \xrightarrow{\Psi(0)} c\bar{c} \rightarrow ggg$, with the formation of three energetic gluons. One can thus express the total width of J/ψ as

$$\Gamma_{J/\psi} = \frac{160(\pi^2 - 9)\alpha_s^3 |\Psi(0)|^2}{81 m_{J/\psi}^2} + O(\alpha_s^4). \qquad (11.25)$$

The wave function at the origin entering (11.25) can be evaluated numerically by solving the Schrödinger equation for the funnel potential in Fig. 11.1. The 1-loop correction to the width $\propto \alpha_s^4$ was also calculated, but we conceal this result from the reader. First, because we want to protect his brain from bursting due to the excess of information and, second, because the coefficient of α_s^4 in (11.25) depends on *which energy* μ the strong coupling constant is evaluated at. It is natural to calculate it for μ of the order of the mass of J/ψ. At this energy, the coupling constant (11.21) is already comparatively small, $\alpha_s(3\,\text{GeV}) \approx 0.25$, the 1-loop corrections are not too large, and the width (11.25) finds itself in good agreement with experiment.

The formula (11.25) is very similar to a formula giving the probability of orthopositronium decay. For the latter, one should substitute α for α_s, m_e for m_c, change the numerical coefficient in an appropriate way (the positronium calculation is simpler; in particular, the graph like in Fig. 11.5b is absent) and substitute the analytically known value for the positronium wave function

$$\Psi_{e^+e^-}(0) = \sqrt{\frac{m_e^3 \alpha^3}{8\pi}} \qquad (11.26)$$

Orthopositronium (resp. orthoquarkonium) is only one of many bound states which an electron and a positron (resp. a quark and an antiquark) can form. For example, there is also *parapositronium* — the ground state in the positronium spectrum. It has unit principal quantum number (lies in the innermost shell), zero orbital angular momentum and also zero total spin, so that the spins of the electron and positron are antiparallel. This state decays into two, rather than three, photons. Hence the decay probability is much larger, $\Gamma_{para} \propto \alpha^2$, compared with $\Gamma_{ortho} \propto \alpha^3$, and the lifetime is, correspondingly, much shorter.

In exactly the same way the quarks c and $\bar{c}$ can form the *paracharmonium* bound state with antiparallel spins of the constituent quarks and zero total spin and angular momentum. This state (called η_c) was observed at SLAC in 1975. Its mass is $m_{\eta_c} \approx$ 2980 MeV, somewhat less than the mass of J/ψ. The masses of these particles are different for the same reason that the binding energies of ortho- and para-positroniums are different and the energies of the triplet ortho-state $|1^3 S_1\rangle$ and the singlet para-state $|1^1 S_0\rangle$ in the hydrogen atom are different.[8] This difference in mass and energy is called the *hyperfine splitting*. In positronium and the hydrogen atom, it is due to the interaction of the magnetic moments of the constituents. In charmonium, it is due to the interaction of the *chromo*magnetic moments of the constituent quarks.

Paracharmonium does not easily decay into a $\mu^+ \mu^-$ pair. In contrast to orthopositronium, this decay mode needs the mediation of two (rather than just one) photons, which brings about a factor α in the amplitude and the factor α^2 in the probability. Also the cross section of its direct formation in $e^+ e^-$ collisions is rather tiny. But it was observed in one of the decay modes of J/ψ: $J/\psi \to \eta_c \gamma$.[9] The paracharmonium state can decay into two gluons; its width is hence proportional to α_s^2 and is much larger than the width of J/ψ, $\Gamma_{\eta_c} \approx 30$ MeV.

The orbital momentum of quarkonium states is not necessarily zero. It can be $L = 1$, giving P-states. The quark spins in these states can be parallel or antiparallel. Finally, the principal quantum number is not necessarily equal to 1. There are also excited states with $n = 2$ and $n = 3$. It is remarkable that the whole zoo (well, maybe not the whole zoo, but its considerable fraction) of charmonium states has been observed in experiment! Some of the animals of this zoo are displayed in the cells of Table 11.1.

Among the states with the principal quantum number $n = 2$, we see the S-states η_c' and ψ' representing the radial excitations of η_c and J/ψ, and also P-states with $L = 1$. There are four of them: the state h_c, where the spins of the constituent quarks are antiparallel, and the states χ_c, where the spins are parallel, giving $S = 1$. Unit spin and unit angular momentum can add to give the states with $J = 0, 1, 2$.

---

[8]We follow the standard nomenclature for the atomic levels: in the term symbols above, 1 is the principal quantum number n, S means S-state with zero orbital momentum (there also are letters P for $L = 1$, D for $L = 2$, etc.), the right subscript is the total angular momentum J and the left superscript is the multiplicity $2S + 1$, where S is the total spin (not to be confused with the notation "S" for S-levels).

We should warn the reader at this point. For historical reasons not quite clear to me, in the most of the scientific papers and reviews devoted to quarkonium and also in the *Table of Particle Properties* booklet, which some of you might have held in your hand, a slightly different convention is used, where the leftmost numeral is not the principal quantum number n, but rather the radial quantum number $n_r = n - L$. For example, what are called $1P$ levels there are $2P$ levels in our notation.

[9]A similar transition between $|1^3 S_1\rangle$ and $|1^1 S_0\rangle$ states in hydrogen gives the famous 21 cm line by which the interstellar gaz is observed.

At the level $n = 3$, we see the radially excited P-states χ'_{c0} and χ'_{c2}.[10] We also see a D-state with $L = 2$. The last line of the table exhibits the double radial excitation of J/ψ of mass 4040 MeV. Its width is much larger than the width of J/ψ and ψ'. The reason is that its mass exceeds the threshold, beyond which the string connecting c and $\bar{c}$ can be kinematically broken in two, as in Fig. 11.2, with the production of two D-mesons[11] made of the charmed quark and a light antiquark and of the charmed antiquark and a light quark. But the mass of ψ'' is rather close to the threshold [the mass of $D^0 = (c\bar{u})$ and of $D^+ = (c\bar{d})$ mesons is $m_D \approx 1870$ MeV], the width is still not too large, and the resonance structure in the total cross section of e^+e^- annihilation at the center-of-mass energy $E_{\text{c.m.}} \sim 4040$ MeV is rather pronounced.

The width of charmonium states with still larger mass grows rapidly, and very soon it no longer makes sense to talk about resonances. The spectrum becomes *continuous*.

It is now difficult to fully grasp the strength of the impact that these discoveries exerted on the physics community. This book is primary addressed to curious young people, who must have been hearing about quarks since childhood. They *know* that quarks exist. The fact that they are confined and do not appear as asymptotic states may still seem to them surprising (as it does for the author), but the existence of the rich spectrum of bound states should look rather natural. However, at the beginning of the seventies not everybody was sure that quarks were really there. Quarks had the status of hypothetical objects and, along with arguments *for* their existence, there were arguments *against*.

I remember quite well that some time in the spring of 1974, being a fourth-year student, I went to an experimental seminar in ITEP.[12] The speaker said that his data were in agreement with the quark model. Having heard that, one of the respected theorists who was attending the seminar said that it was a pity that the speaker compromised his excellent experimental work by a reference to a doubtful model with shaky foundations...

But that was before the November revolution. In the summer of 1975 when, on the one hand, part of Table 11.1 has already been experimentally revealed and, on the other hand, the theoretical discovery of asymptotic freedom in QCD has become widely known, there could be no more doubts — the quarks exist and their interactions are described by QCD!

Up to now, we were only discussing charmonium in this section. But there is also *bottomonium*, the states made of the b quark and its antiparticle. The first bottomonium state, the Υ-meson with mass 9.46 GeV, was discovered at Fermilab by the collaboration headed by Leon Lederman in 1977. At present, the bottomonium zoo is as densely populated as the charmonium one. Mesons and baryons with naked beauty (including only one b or $\bar{b}$ quark) have also been observed and studied.

---

[10] We do not see χ'_{c1}, but I am too lazy to find out why. Also, I have not put in the table a mysterious $X(3872)$ resonance, which does not have a simple spectroscopic interpretation. Perhaps it is not a biquark, but a tetraquark.

[11] Again, do not mix them up with the quarkonium D-states.

[12] The Institute for Theoretical and Experimental Physics, where we did our graduate studies.

Table 11.1 Spectrum of charmonium.

Particle	Term symbol	Mass (MeV)	Width (MeV)
η_c	1^1S_0	2984	32
J/ψ	1^3S_1	3097	0.09
h_c	2^1P_1	3525	0.7
χ_{c0}	2^3P_0	3415	10
χ_{c1}	2^3P_1	3511	0.85
χ_{c2}	2^3P_2	3556	2
η_c'	2^1S_0	3640	11
ψ'	2^3S_1	3686	0.3
$\psi(3770)$	3^3D_1	3773	27
χ_{c0}'	3^3P_0	3918	20
χ_{c2}'	3^3P_2	3927	24
ψ''	3^3S_1	4040	80

Toponium must also exist, but the energy of e^+e^- machines is not and never was high enough to see it.

11.4 Light Mesons and Baryons

The quark model was formulated in the beginning of the nineteen-sixties, well before the advent of QCD and before the discovery of charmonium and bottomonium states. Light hadrons also display features which make very probable their composite nature and the existence of smaller elementary constituents. As we have just discussed, these features are less pronounced than for heavy quarks and leave room for justified doubts, but they actually give rather serious indications of the presence of quarks, and most (though not all) theorists would have bet that quarks existed even before 1974.[13]

Many light hadron properties can be understood in the framework of the constituent quark model, which says that light hadrons are made of light quarks in basically the same way as an atom is made of a nucleus and electrons, a nuclei is made of protons and neutrons and (as we have just learned) heavy quarkonium states are made of heavy quarks.

Look again at Table 2.1. One can observe that the masses of the u and d quarks are small compared to the characteristic QCD scale: $m_u \approx 3$ MeV and $m_d \approx 6$ MeV. But these are the so-called *current* masses, the parameters entering the QCD Lagrangian (11.16). These masses show up at high energies, where the coupling is

---

[13]I remember reading a long time ago that at some chemical congress in the first half of the 19th century the participants *voted* on the existence of atoms. The atomic hypothesis won with a narrow margin.

small and the physics of the strong interaction can be described perturbatively. In the phenomenological constituent quark model describing low-energy hadron physics, each quark is dressed in a gluon cloud[14] of mass $m_{\text{cloud}} \approx 300$ MeV. The mass of a constituent quark is *roughly* the sum of the current quark mass and the mass of the cloud. We see that the constituent quarks u and d have about the same mass, the constituent d quark being fatter than the constituent u quark by only several MeV.

This approximate degeneracy shows up in approximate *isotopic symmetry* — the global $SU(2)$ symmetry with respect to the unitary transformations mixing u and d quarks. It is translated into the approximate degeneracy of hadron states. Thus, the neutron, consisting of two d quarks and one u quark, is heavier than the proton, consisting of one d quark and two u quarks, by ≈ 1.3 MeV (this includes the effects due to the quark mass difference and due to the different Coulomb energies of the composite systems). The proton and neutron form an isotopic doublet. Other isotopic multiplets will show up soon.

The strange quark mass, $m_s \approx 150$ MeV, is also not so large. Thus, it sometimes makes sense to talk about the symmetry mixing three types of quarks — the approximate *flavour $SU(3)$ symmetry*.[15]

A gold-plated flavour $SU(3)$ multiplet is the baryon *octet* [the adjoint representation of $SU(3)$]. It includes the nucleon isodoublet, the isotriplet of strange Σ *hyperons*, $|\Sigma^+\rangle = |uus\rangle$, $|\Sigma^0\rangle = |uds\rangle$ and $|\Sigma^-\rangle = |dds\rangle$, the isosinglet $\Lambda = |uds\rangle$ and the isodoublet of Ξ hyperons including two strange quarks, $|\Xi^0\rangle = |uss\rangle$ and $|\Xi^-\rangle = |dss\rangle$. This octet is graphically represented in Fig. 11.6.

The reader might be puzzled by the fact that the strangeness grows not bottom-up in this figure (which would be more natural — the number of strange quarks in the composition grows bottom-up), but top-down. Well, it is the same story as with the negative electric charge of the electron. When physicists first saw in the beginning of the fifties that some particles produced in cosmic rays have a tremendously long lifetime, they were surprised by that and thought that it was *strange*. At that time, they knew nothing about quarks and just established a phenomenological law of conservation of "strangeness" S in strong interactions. And they attributed the value $S = +1$ to the mesons K^0 and K^+ and the value $S = -1$ to the mesons $\bar{K}^0$ and K^- and to the baryons Σ and Λ. Later people understood that the strangeness thus defined is simply the number of strange *anti*quarks minus the number of strange quarks, but they did not change the established convention.

The baryons in Fig. 11.6 have spin 1/2. That means that not all spins of the constituent quarks are parallel. If two of them look up, the third spin must look down. All these baryons are relatively stable. The proton is absolutely stable. Other particles are not, but they decay due to the weak interaction with rather a long lifetime. The neutron decay is associated with the conversion of a d quark into u quark,

---

[14]Of course, when the quark model was first formulated, people knew nothing about gluons, but I am not following the historical path here.

[15]This global symmetry has nothing to do with the exact $SU(3)$ colour gauge symmetry. Please do not mix them up!

Fig. 11.6 Octet of baryons.

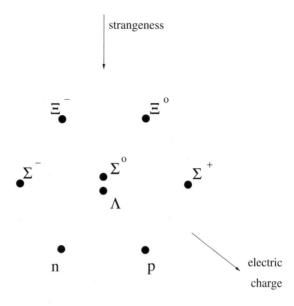

Table 11.2 The masses of the baryons in the lowest lying octet.

Particle	p	n	Λ	Σ^+	Σ^0	Σ^-	Ξ^0	Ξ^-
Mass (MeV)	938	940	1116	1189	1193	1197	1315	1322

$d \rightarrow u + e^- + \bar{\nu}_e$. The underlying process driving the decay of Σ and Λ hyperons is a similar disintegration of the strange quark, $s \rightarrow u + e^- + \bar{\nu}_e$. The particle Ξ, called sometimes the *cascade hyperon*, decays in two stages. First, only one of its strange quarks disintegrates and it gives Σ or Λ. The Σ or Λ baryon lives for some time ($\approx 10^{-10}$ s), leaving a visible track in the bubble or spark chambers, and then decays into a proton or a neutron, with the emission of a π meson or a lepton pair.

The masses of these baryons are given in Table 11.2.

A glance at the table reveals that the masses of the particles belonging to the same isomultiplet are rather close. The second observation is that the masses of Σ and Λ, including one strange quark, are larger than the nucleon masses, but smaller than the masses of the Ξ-hyperons, including two strange quarks. The third observation is that the Λ hyperon, having the same quark decomposition $|uds\rangle$ as Σ^0, is lighter than Σ^0 and its isotopic partners. This Σ-Λ splitting is due to the fact that, even though their composition is the same, the *wave function* of Λ is not the same as that of Σ^0, and the interactions between the quarks are also different.

Let us denote by m_N, m_Λ, m_Σ and m_Ξ the average values of the masses in one isotopic multiplet. Bearing in mind that $m_\Lambda \neq m_\Sigma$, we define their average mass according to

$$m_{\Sigma\Lambda} = \frac{3m_\Lambda + m_\Sigma}{4} \approx 1135\,\text{MeV}\,. \tag{11.27}$$

One can then make one more noteworthy observation: the relation

$$m_{\Sigma\Lambda} - m_N = m_\Xi - m_{\Sigma\Lambda} \tag{11.28}$$

is satisfied with tremendously good accuracy!

A heuristic explanation is simple: the baryons Σ and Λ include in their composition a strange quark, which is heavier than the u and d quarks, and hence their masses are larger than the nucleon mass. The baryon Ξ includes two strange quarks, and its mass should be still larger.

What is not explained in this way is why the mass difference (11.28) has the value ≈ 190 MeV, larger than the strange quark mass, $m_s \approx 150$ MeV. And what remained absolutely unclear is why the average Σ–Λ mass was chosen as in (11.27) and not, say, in a more naturally looking form $\tilde{m}_{\Sigma\Lambda} = (m_\Lambda + 3m_\Sigma)/4$.

Well, heuristic reasoning has its limits. The *Gell-Mann–Okubo (GMO) relation* (11.28) can be derived using group theory methods (we prefer not to give this derivation here), and that explains (11.27). A heuristic explanation of the somewhat larger than expected mass difference (11.28) can be given, however. The mass difference $m_{\Sigma\Lambda} - m_N$ is larger than the current strange quark mass because the *constituent* quark masses are different in different "environments" — for the same reason that the effective mass of an electron in a crystal does not coincide with its proper mass in vacuum.

The situation is somewhat simpler for another $SU(3)$ baryon multiplet, the *decuplet* [we explained what the decuplet representation of $SU(3)$ is in the vicinity of Eq. (6.52)]. The decuplet states are represented in Fig. 11.7.

In this case, the constituent quarks have all spins parallel, and the total baryon spin is $3/2$. The interquark interaction in the decuplet baryons differs from that in the

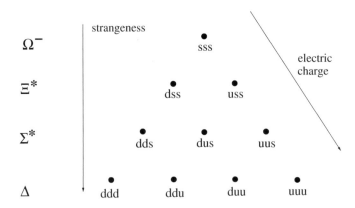

Fig. 11.7 Decuplet of baryons.

Table 11.3 Masses in the decuplet.

Particle	Δ	Σ^*	Ξ^*	Ω^-
Mass (MeV)	1232	1385	1533	1672

octet baryons, and the decuplet masses [they are given in Table 11.3, where we disregarded the small mass differences within the isotopic quartet (Δ^-, Δ^0, Δ^+, Δ^{++}), the isotriplet (Σ^{*-}, Σ^{*0}, Σ^{*+}) and the isodoublet (Ξ^{*-}, Ξ^{*0})] are larger than the octet masses (it is the same hyperfine interaction between the quarks chromomagnetic moments that makes the orthocharmonium J/ψ state heavier that η_c). As a result, almost all decuplet states have a very short lifetime $\approx 10^{-23}$ s — they decay into the octet states by the strong interaction, where the quarks are not converted into one another and only extra quark-antiquark pairs may be created. For example, the particle Δ^{++}, made of three u quarks with parallel spins, may decay, with the creation of an extra $d\bar{d}$ pair, into a proton ($|uud\rangle$) and a π^+ meson ($|u\bar{d}\rangle$). Δ particles are observed as pronounced resonances in the total cross section of πN scattering — like the resonance in Fig. 11.4, but much wider.[16]

The only exception is Ω^-. It is the lightest state including three strange quarks. It can decay only by weak interaction into Ξ hyperons (with lifetime of order $10^{-10}s$), and further Ξ decays into Σ, Λ, and the latter — into nucleons.

The different isotopic multiplets in Fig. 11.7 are practically equidistant in mass, and the mass differences between two adjacent lines in Fig. 11.7 are close to the current strange quark mass. That is how the existence of the Ω^- and its mass were predicted. This prediction was brilliantly confirmed in an experiment in 1964.

We now go over to mesons. The lowest meson states are the pions, $\pi^\pm$ of mass 140 MeV and π^0 of mass 135 MeV. They form an isotriplet. The quark decompositions are $|\pi^+\rangle = |u\bar{d}\rangle$, $|\pi^-\rangle = |d\bar{u}\rangle$, $|\pi^0\rangle = \frac{1}{\sqrt{2}}(|u\bar{u}\rangle - |d\bar{d}\rangle)$. The latter relation means that the wave function of π^0 represents a superposition of the states $|u\bar{u}\rangle$ and $|d\bar{d}\rangle$, so that the probabilities to find in π^0 a pair $u\bar{u}$ or a pair $d\bar{d}$ are equal.

In the pions, the quark spins look in opposite directions: the pions are "parastates", like η_c, and the total pion spin is zero. If we disregard their internal structure (which is justified when their energy is much less than their inverse size), the pions are described by scalar (or rather *pseudoscalar* — the pion wave function changes sign under spatial reflection) fields.

The pion isotriplet belongs to the flavour $SU(3)$ pseudoscalar meson octet, including besides the pions also two isodoublets of K mesons, ($|K^0\rangle = |d\bar{s}\rangle$, $|K^+\rangle = |u\bar{s}\rangle$)

---

[16] As the octet of baryons, in contrast to the decuplet, involves only quasi-stable long-living particles, and there is also the meson quasi-stable octet, the octet representation of $SU(3)$ is in some sense "more important" here than the decuplet one. This allowed Gell-Mann to dub the theory of hadron classification based on $SU(3)$ flavour symmetry the *Eightfold Way* (alluding to the Noble Eightfold Way Path of Buddhism).

Fig. 11.8 Pseudoscalar
meson octet.

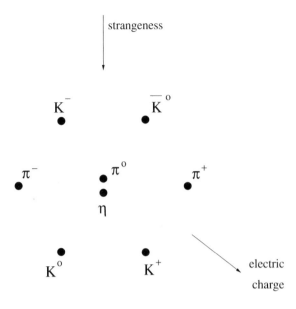

and ($|K^-\rangle = |s\bar{u}\rangle$, $|\bar{K}^0\rangle = |s\bar{d}\rangle$), and the isosinglet η with the wave function[17]
$|\eta\rangle = \frac{1}{\sqrt{6}}(|u\bar{u}\rangle + |d\bar{d}\rangle - 2|s\bar{s}\rangle)$. The meson octet is represented in Fig. 11.8.

By analogy with the baryon octet, we can derive the relationship between the
masses of the pseudoscalar mesons:

$$m_K = \frac{3m_\eta + m_\pi}{4}.$$ (11.29)

The experimental values of the masses are given in Table 11.4.

Substituting in (11.29) the average multiplet values $m_K = (m_{K^+} + m_{K^0})/2$ and
$m_\pi = (2m_{\pi^+} + m_{\pi^0})/3$, we obtain

$$496\,\text{MeV} = 445\,\text{MeV}.$$ (11.30)

The agreement is there, but it is not so fantastic. Well, this might be natural, — thought
the physicists in the sixties. The derivation of the Gell-Mann–Okubo relations is based
on the assumption that the mass splitting in the $SU(3)$ multiplet is comparatively
small and the inequality in quark masses can be treated as a perturbation. And for
the pseudoscalar octet, the masses of mesons are not close — the mass of η is almost
four times larger than the mass of π.

But then people noticed that the GMO formula better works for mesons if in
(11.29) we replace the masses by their squares. Doing that, we obtain

---

[17] In the wave functions of π^0 and η, one can recognize the $SU(3)$ generators t^3 and t^8 in (6.43). This
is not accidental, of course, and I am making an effort not to go here into more detailed explanations.

Table 11.4 Masses of the pseudoscalar mesons.

Particle	$\pi^{\pm}$	π^0	$K^{\pm}$	$K^0, \bar{K}^0$	η
Mass (MeV)	140	135	494	498	548

$$0.246\,\text{GeV}^2 \; = \; 0.230\,\text{GeV}^2\,, \tag{11.31}$$

which is better than (11.30), indeed.

When I was a student, our professors said: "Look, mesons are bosons and the meson fields satisfy the KFG equation involving m^2. And baryons are fermions, their fields satisfy the Dirac equation, involving the first power of mass. Thus, it is natural to write the GMO relation for the masses in the case of baryons and for m^2 in the case of mesons." A very heuristic argument, indeed!

Now we understand this much better. In the case of pseudoscalar mesons, the GMO formula works better for m^2 because the nature of the pseudoscalar mesons is somewhat special, compared to other hadrons. One can describe them in the framework of the constituent quark model, as we did in this section, but this model does not work so well for pseudoscalars. In fact, the pseudoscalar mesons in Fig. 11.8 are *pseudo-Goldstone* particles associated with *spontaneous breaking* of approximate *chiral symmetry*.

The meaning of these fancy new notions will be explained in the next section.

11.5 Chiral Symmetry and Its Breaking

11.5.1 Chiral Symmetry in QCD

As was mentioned, the current masses of u and d quarks are small compared to the characteristic QCD scale. It is interesting to pose the question: what would our world look like if these quarks were exactly massless?

In that case the light quark part of the QCD Lagrangian would read

$$\mathcal{L}_{\text{massless }u,d} \; = \; i\bar{q}\gamma^{\mu}(\partial_{\mu} - ig\hat{A}_{\mu})q\,, \tag{11.32}$$

where q is an isodoublet,

$$q = \begin{pmatrix} u \\ d \end{pmatrix}.$$

One can immediately observe that, besides the gauge symmetry (11.5), (11.10), the Lagrangian (11.32) enjoys the global isotopic unitary $U_V(2)$ symmetry that mixes two light quark flavours: $q \longrightarrow S_V q$. The subscript V means that this is a *vector* symmetry. And the latter means that the corresponding Noether currents are vectors — an isotopic singlet $\bar{q}\gamma^\mu q$ and an isotopic triplet $\bar{q}\gamma^\mu\sigma^A q$, σ^A being the Pauli matrices acting on the quark flavour indices.

But the actual symmetry of (11.32) is larger. To see that, we introduce the left-handed and right-handed quark fields,[18]

$$q_{L,R} = \frac{1 \mp \gamma^5}{2}q \quad \text{and hence} \quad \bar{q}_{L,R} = \bar{q}\frac{1 \pm \gamma^5}{2}, \tag{11.33}$$

and notice that the Lagrangian (11.32) can be represented as

$$\mathcal{L}_{\text{massless }u,d} = i\bar{q}_L\gamma^\mu(\partial_\mu - ig\hat{A}_\mu)q_L + i\bar{q}_R\gamma^\mu(\partial_\mu - ig\hat{A}_\mu)q_R \tag{11.34}$$

[cf. Eqs. (9.44), (9.37)].

The Lagrangian (11.34) is invariant under two different unitary rotations

$$q_L \rightarrow S_L q_L, \qquad q_R \rightarrow S_R q_R. \tag{11.35}$$

The full symmetry is thus $U_L(2) \times U_R(2)$.

The presence of separate left and right symmetries is specific for massless quarks. The fermion mass term involves both left-handed and right-handed fields:

$$\mathcal{L}_m = m\bar{q}q = m(\bar{q}_L q_R + \bar{q}_R q_L) \tag{11.36}$$

[cf. (9.39)]. As the masses of the different quark flavours are different, the true QCD Lagrangian does not have an exact isotopic symmetry. But if one imagines a world where the u and d quark masses were exactly equal but nonzero, the Lagrangian of such a theory would be invariant under the transformations (11.35) only if $S_L = S_R$. These are vector transformations. On the other hand, the transformations (11.35) with $S_L = S_R^\dagger$ are *axial* transformations, the corresponding Noether's currents being the isoscalar and isotriplet axial vectors: $\bar{q}\gamma^\mu\gamma^5 q$ and $\bar{q}\gamma^\mu\gamma^5\sigma^A q$.

Now, do not ask me why, but the isosinglet axial current is actually not conserved even if the quarks are massless. This *anomalous* nonconservation is brought about by quantum effects — we cannot keep the singlet axial symmetry of the classical Lagrangian at the quantum level and have to impose a constraint

$$\det(S_L S_R^\dagger) = 1 \tag{11.37}$$

on the unitary transformation matrices in (11.35). We are left with the global vector symmetry $U_V(2)$ and the global axial symmetry $SU_A(2)$.

A remarkable fact is that the latter symmetry is also broken in massless QCD, but this is not an explicit, but a *spontaneous* breaking.

---

[18]A good occasion to recall the definition of Dirac conjugation, $\bar{q} = q^\dagger\gamma^0$, and the property $\gamma^\mu\gamma^5 + \gamma^5\gamma^\mu = 0$.

Fig. 11.9 Double-well
potential.

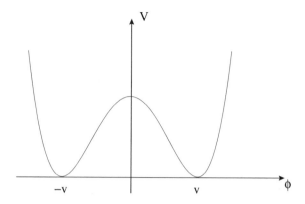

11.5.2 *Spontaneous Symmetry Breaking □ Goldstone's Theorem*

This is a very important subsection. Probably, we would not touch this subject if our goal were only to give an idea about the theory of the strong interactions and hadron physics. But the notions of spontaneous breaking of global and local symmetry, the Goldstone effect and Higgs effect, are indispensable to understand how the theory of *electroweak* interactions is formulated. Thus, we will now acquaint the reader with some interesting facts of pion physics and simultaneously provide him with the necessary prerequisites for understanding the next chapter.

To begin with, we consider the Lagrangian of a real scalar field

$$\mathcal{L} = \frac{1}{2}(\partial_\mu \phi)^2 - V(\phi) \tag{11.38}$$

with a somewhat unusual potential[19]

$$V = \frac{\lambda}{24}(\phi^2 - v^2)^2 . \tag{11.39}$$

This Lagrangian has a discrete Z_2 symmetry

$$\phi \to -\phi . \tag{11.40}$$

However, the potential (11.39) [depicted in Fig. 11.9] does not have a minimum at the symmetric point $\phi = 0$, as was the case for the systems discussed earlier. Instead, it has *two* different minima at $\phi = v$ and $\phi = -v$. Hence, it has two classical vacuum

---

[19]We follow the standard normalization convention. It would be an unnecessary distraction to explain its origin.

states. In quantum field theory living in an infinite volume, this means the existence of two degenerate ground states in the spectrum of the Hamiltonian.[20]

Note now that each of two vacua is not invariant under the symmetry transformation (11.40). They are instead transformed into each other. It is this phenomenon (the Hamiltonian is invariant under a certain symmetry transformation, but it has not one, but several ground states, each of which is *not* invariant) that is called spontaneous symmetry breaking. In the case considered, it is the discrete symmetry (11.40) which is broken.

Which vacuum is actually realized in Nature is a matter of chance. As we know, right after the Big Bang, the Universe was very hot. If it involved the real scalar fields with interaction (11.39), space would be filled at that epoch with energetic interacting real scalar particles[21] — the excited states of the Hamiltonian — and the structure of the ground states of the latter would not be important. When the Universe cooled down, the excited states would disappear and the system (11.38) would "fall down" either on the left or on the right so that the symmetry would be broken.

In fact we already discussed spontaneous breaking of a discrete symmetry in Chap. 5 [after Eq. (5.23)]. The ordinary electromagnetic interactions responsible for the structure of our bodies are left-right symmetric. But many molecules — sugars, amino acids, proteins, DNA — can have two different, left and right forms — mirror images of each other. In our bodies we have only one type of these molecules — sugars are right-handed, amino acides are left-handed, etc. This asymmetry is a matter of pure chance, determined by the type of organic molecules that were synthesized first at the dawn of the Earth's history. One can well imagine a biosphere with left-handed sugars.

One can also imagine a situation where in a certain region of space, the system fell after cooling into the left well, while in some other region of space, it fell into the right. In that case, the two regions would be separated by a *domain wall* represented in Fig. 11.10. This domain wall would have a finite width (depending on λ and v) and a finite surface energy. In our Universe such domain walls probably do not exist. But if they existed, they would be quite physical objects — one could knock on them, see them (scatter light), etc.

We now go over to systems involving spontaneous breaking of continuous symmetry. Consider the following Lagrangian of a *complex* scalar field:

---

[20]In quantum mechanics with the Lagrangian

$$L \ = \ \frac{1}{2}\dot{x}^2 - \frac{\lambda}{24}(x^2 - v^2)^2 \tag{11.41}$$

involving a single real variable x, this would not be the case. Even for a high barrier between the left and right minimum, the degeneracy between the corresponding states would be lifted due to tunneling. But in field theory with an infinite number of degrees of freedom, the tunneling amplitude is zero and the degeneracy is exact.

[21]In the minimal Standard Model, there are no fundamental *real* scalar fields, but suppose there were.

Fig. 11.10 Domain wall.

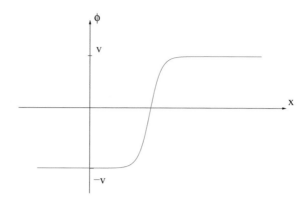

$$\mathcal{L} = (\partial^\mu \phi)^* (\partial_\mu \phi) - \frac{\lambda}{4}(\phi^* \phi - v^2)^2 \tag{11.42}$$

[cf. Eq. (7.45) and Fig. 7.2b]. This Lagrangian is invariant under a continuous $U(1)$ symmetry

$$\phi \rightarrow \phi e^{i\chi}. \tag{11.43}$$

The $3D$ plot for the potential resembles the bottom of a champagne bottle [see Fig. 11.11].[22] The minimum is realized at $|\phi| = v$, but the *phase* of the complex vacuum field ϕ is absolutely arbitrary. Thus, there are an infinite number of classical vacua. Each such vacuum is no longer invariant under (11.43) — the symmetry is spontaneously broken.

Imagine now that the phase of the vacuum field depends slowly on x:

$$\phi_{\text{vac}}(x) = v e^{i\alpha(x)/\sqrt{2}}. \tag{11.44}$$

Substitute this into (11.42). We obtain an effective Lagrangian

$$\mathcal{L} = \frac{v^2}{2}(\partial_\mu \alpha)^2. \tag{11.45}$$

This Lagrangian describes a massless real scalar field $v\alpha(x)$. In other words, the low-energy spectrum of the theory (11.42) involves free massless scalar states. This massless particle is called the *Goldstone boson*.[23]

---

[22]It is also sometimes called the *Mexican hat* potential, but I prefer champagne if you ask me. Or good Crimean sparkling wine.

[23]This name was given to honour Jeffrey Goldstone, who understood the universal nature of these modes in the quantum field theory context and proved the theorem below. But the still more universal Arnold principle also works, of course, in this particular case. The first person to discover the existence of the Goldstone modes was Yoichiro Nambu in 1960.

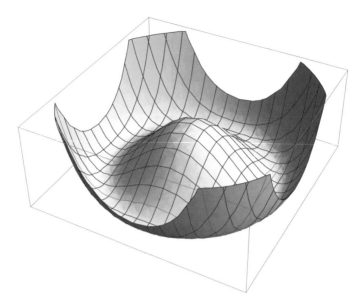

Fig. 11.11 A bottom of a bottle.

This observation can be generalized. The following powerful *Goldstone theorem* can be proven:

Suppose the Lagrangian enjoys a continuous global symmetry G, which is sponta-neously broken down to H ⊂ G. Then the spectrum of the theory involves massless particles, whose number is equal to the difference of the dimension of G and the dimension of H. The strength of interaction of these particles depends on the energy. In the limit E → 0, the Goldstone bosons decouple.

The meaning of the theorem is quite transparent. The set of all vacua coincides (modulo possible discrete factors) with the so-called *coset space* G/H.[24] The di-mension of this coset space is $dim(G/H) = dim(G) - dim(H)$. The system can walk freely along this coset space and the effective Lagrangian describing this mo-tion involves only the kinetic, but not the potential term. In field theory, the absence of a potential means the absence of a mass for the corresponding effective field.

The theorem has, indeed, a universal nature and can be illustrated in an example that has nothing to do with elementary particles. Consider a ferromagnet. As we know, when a hot piece of iron cools down below a certain temperature (called the *Curie point*), the iron is magnetized: the electron spins align along a certain distinguished direction. If no external magnetic field is present, this direction is absolutely arbitrary. Moreover, if the size of our grain of iron exceeds a certain small value of several microns, the grain would develop several domains with different

---

[24]The coset G/H is defined as a manifold obtained from the group manifold G after identifying all points belonging to every *orbit* of the subgroup H, i.e. a set of points in G related by the action of H.

directions of magnetization. We will consider for simplicity the case of a small magnetic particle representing a single domain.

A nonzero magnetization spontaneously breaks the rotational $SO(3)$ symmetry of the Hamiltonian. However, it does not break it completely: our magnet is still symmetric under rotations around the magnetization axis. This is the $SO(2) \equiv U(1)$ symmetry. The set of all possible ground states (and, as was mentioned, the direction of spontaneous magnetization is completely arbitrary) represents a $3 - 1 = 2$-dimensional coset $SO(3)/SO(2) \sim S^2$.[25] Goldstone's theorem tells that there should be two massless modes corresponding to "walking" over this sphere (this means that the direction of magnetization in different spatial points may fluctuate, and that for slow long-wavelength fluctuations, this costs no energy).

In solid state physics, one has to think in the framework of quantum field theory for low temperatures, but, for high temperatures, one should not. These magnetic Goldstone modes are then interpreted not as particles, but as classical waves (called *spin waves* or *magnons*) with a gapless dispersion law $\omega = Ak^2$. (Do not ask me where this law comes from, but it is natural that it is not linear in k — there is no Lorentz symmetry in a ferromagnet.)

The transition from the high-temperature phase with no magnetization to the low-temperature phase with spontaneous magnetization is actually a second-order *phase transition*.[26] The magnetization is an *order parameter* (a parameter that tells us into which of many possible vacua our system has fallen after spontaneous symmetry breaking occurred) associated with this transition. The Goldstone spin waves are related to fluctuations of this order parameter.

Likewise, in the field theory example (11.42) considered above, the order parameter is the value of the vacuum scalar field (or rather of its phase α). The Goldstone particle appears as the result of the quantization of the mode describing the fluctuations of $\alpha(x)$.

11.5.3 Quark Condensate □ Pseudoscalar Pseudogoldstones

We go back to QCD and consider first a world with two massless quarks. As was mentioned, the global axial symmetry $SU_A(2)$ of the Lagrangian (11.34) is broken spontaneously. As we have just learned, this means that the vacuum state of the theory is not invariant under the action of this symmetry. In the example (11.42), such non-invariance was due to a nonzero vacuum value of ϕ.

---

[25] A reader who might be confused by new fancy words ("coset space", etc.) can, however, easily understand that the set of all directions of a 3-dimensional vector can be mathematically described as a 2-sphere.

[26] We recall the definitions. A *first*-order phase transition is characterized by a discontinuous energy density — at the boiling point, it is much larger for vapour than for water. On the other hand, for *second*-order phase transitions, the energy density is continuous, but the specific heat (the derivative of the energy with respect to temperature) is discontinuous. Phase transitions associated with spontaneous symmetry breaking are mostly second-order.

But there are no scalar fields in QCD, and it is not immediately clear what is the relevant order parameter. The answer is that one can distinguish different vacua by considering a composite bifermion operator,[27] $O^{fg} = q_L^f \bar{q}_R^g$, where $f, g = 1, 2$ are the flavour indices, which we now display. The order parameter is the *vacuum expectation value* (the average value in the vacuum state) of this operator:

$$\Sigma^{fg} = \langle q_L^f \bar{q}_R^g \rangle_{\text{vac}} . \qquad (11.46)$$

A generic complex 2×2 matrix depends on 8 real parameters: a dimensionful parameter $\sim \sqrt{\text{Tr}\{\Sigma\Sigma^\dagger\}}$, giving the "extent" to which the symmetry is broken [like the parameter v in (11.44) or like the magnetization of a ferromagnetic domain] and seven "angular" parameters. The latter would correspond to the spontaneous breaking of the full chiral symmetry (11.35), involving 8 parameters, down to $U_V(1)$ — the symmetry under the common phase transformation of all the quark fields.

However, as was already mentioned at the end of Sect. 11.5.1, the $U_A(1)$ part of $U_L(2) \times U_R(2)$ is *explicitly* broken due to the anomaly. Hence it cannon be broken spontaneously. In addition, one can *prove* that the ordinary isotopic vector symmetry $SU_V(2)$ that mixes u and d Dirac spinors is *not* spontaneously broken, it is still there.[28]

That means that only the $SU_A(2)$ part of the symmetry (11.35) can be spontaneously broken, and the set of all vacua is a 3-dimensional manifold $SU(2) \equiv S^3$. The matrix (11.46) is thus proportional to a unitary matrix:

$$\Sigma^{fg} = \frac{1}{2} \Sigma \, \Omega^{fg}, \qquad \Omega \in SU(2) . \qquad (11.47)$$

A generic chiral transformation (11.35) transforms one such vacuum into another, $\Omega \to \Omega' = S_L \Omega S_R^\dagger$. It is convenient to assume in practice that, in the true physical vacuum where we live in, $\Omega^{fg} = \delta^{fg}$.

The constant Σ is called the *quark condensate*. Its experimental value is

$$\Sigma \approx (250 \text{ MeV})^3 . \qquad (11.48)$$

The fact that $\Sigma \neq 0$ is rather nontrivial. It could be otherwise, the chiral symmetry (11.35) could stay unbroken. There are theoretical arguments to believe that it *would* stay unbroken in a different world where the number of light quark flavours was larger — five, six... We live, however, in a world where the quark condensate is well formed.

According to Goldstone's theorem, there should be three massless particles associated with the fluctuations of Σ^{fg}.

---

[27] If the reader has heard about superconductors and about the condensate of electron *Cooper pairs* there, he will not be too surprised at this point!

[28] This is the so-called *Vafa–Witten theorem*. We will not give the proof here, with the hope that the reader finds this statement natural and accepts it without much discontent.

"But there are no massless particles in the QCD spectrum," a confused reader may protest, "all hadrons are massive!"

The reader is right. But there is no contradiction with Goldstone's theorem. The latter is valid only when the symmetry is exact, but the symmetry (11.35) is not exact in the real world. It is broken by the u and d quark masses, which are small, but nonzero. Thus, in QCD we are dealing with the spontaneous breaking of an *approximate* symmetry. As a result, the would-be Goldstone particles acquire a small, but nonzero, mass.[29] Such nearly massless modes, related to spontaneous breaking of an approximate global continuous symmetry, are called the *pseudo-Goldstone particles*. Colloquially — "pseudogoldstones".

And such particles are well-known in QCD. They are our old friends, the pions, whose mass, indeed, is considerably less than the mass of other hadrons. There is a nice formula (due to Murray Gell-Mann, Robert Oakes and Bruno Renner) expressing the pion mass in terms of nonzero quark masses, the quark condensate and a certain extra parameter F_π, which has the dimension of mass and can be determined from the experimental decay rates $\Gamma_{\pi \to e \nu_e}$ and $\Gamma_{\pi \to \mu \nu_\mu}$: $F_\pi \approx 93$ MeV. We give this formula without derivation:

$$F_\pi^2 m_\pi^2 = (m_u + m_d)\Sigma \,. \tag{11.49}$$

We see that, in the *chiral limit* when $m_u = m_d = 0$, pions become truly massless Goldstone particles.

I cannot resist quoting here a beautiful expression for the low-energy effective Lagrangian that describes pion physics in the chiral limit. It reads

$$\mathcal{L} = \frac{F_\pi^2}{4} \mathrm{Tr}\{\partial_\mu U \partial^\mu U^\dagger\} \,, \tag{11.50}$$

where

$$U(x) = \exp\left\{ \frac{i\pi^A(x)\sigma^A}{F_\pi} \right\} \tag{11.51}$$

are $SU(2)$ matrices. In this expression, $\pi^{A=1,2,3}(x)$ are the pion fields.

The Lagrangian (11.50) plays the same role for QCD with massless u and d quarks as the Lagrangian (11.45) for the model (11.42). Both Lagrangians carry the symmetries of the original theories. The Lagrangian (11.45) is invariant under the transformations $\alpha \to \alpha + \mathrm{const}$, which realize the $U(1)$ symmetry transformations (11.43) of the original theory. The Lagrangian (11.50) is invariant under the $U_L(2) \times U_R(2)$ transformations

$$U \to S_L U S_R^\dagger \tag{11.52}$$

---

[29]To visualize the appearance of nonzero mass, imagine that the champagne bottle in Fig. 11.11 has been tilted a little. Now it touches our dinner table at only one point, the potential has only one true minimum and both eigenvalues of the matrix of the second derivatives of the potential are positive.

that correspond to the symmetry transformations (11.35) of the classical massless QCD Lagrangian (but we recall that the singlet axial part of this group is not actually a symmetry of quantum theory due to the anomaly).

We see that in this case the Goldstone particles are not free as they were for the $U(1)$ model in (11.42), but are involved in nontrivial interactions. If we substitute (11.51) into (11.50) and expand the exponentials, we derive:

$$\mathcal{L} = \frac{1}{2}(\partial_\mu \pi^A)^2 + \frac{1}{6F_\pi^2}[(\pi^A \partial_\mu \pi^A)^2 - \pi^A \pi^A (\partial_\mu \pi^B)^2] + \cdots \qquad (11.53)$$

The quartic interaction term involves here two extra derivatives, compared with (10.12). This means that the interaction becomes weak for slowly varying fields (for small momenta), as is required by the Goldstone theorem, but it rapidly grows with energy. This behaviour is characteristic for any *nonrenormalizable* field theory, of which (11.50) represents an example. In particular, pion scattering amplitudes in the effective chiral theory behave in a very similar way to graviton scattering amplitudes. We will capitalize on this similarity when discussing perturbative quantum gravity in Chap. 16.

Up to now, we only talked about the part of the QCD Lagrangian involving the light u and d quarks. But the s quark with its mass ~150 MeV is also rather light. If we treat m_s in the same manner as we treated m_u and m_d — as a perturbation of the idealized world with three massless quarks, we can notice that the QCD Lagrangian has an approximate $U_L(3) \times U_R(3)$ chiral symmetry, reduced to $U_V(3) \times SU_A(3)$ by the anomaly. This symmetry is further spontaneously broken down to $U_V(3)$.

Of course, one may be confused here. How can a symmetry that is already explicitly broken by quark masses be further broken spontaneously? Well, we can only repeat what we just said. To study the properties of QCD, we simplify the theory by crossing out the light quark mass terms in the Lagrangian. We study the properties of such an idealized theory with true spontaneous breaking and then we switch on the mass term again, treating it as a perturbation.

As a result, we can generalize the GMOR relation (11.49) and give the expressions not only for pion masses, but also for the masses of K and η mesons. Writing these two new formulas together with (11.49), we obtain:

$$F^2 m_\pi^2 = 2\bar{m}\Sigma \,,$$
$$F^2 m_K^2 = (\bar{m} + m_s)\,\Sigma \,,$$
$$F^2 m_\eta^2 = \frac{2\bar{m} + 4m_s}{3}\,\Sigma \,, \qquad (11.54)$$

where $\bar{m} = (m_u + m_d)/2$. The constants F entering these three relations might be slightly different, but this difference is proportional to quark masses, and taking it into account would mean calculating the effects quadratic in $m_{u,d,s}$, which we actually cannot do very well.

We observe now that it is, indeed, the squares of the meson masses m_π^2, m_K^2, m_η^2, rather than m_π, m_K, m_η, that satisfy the Gell-Mann–Okubo relation (11.29), in a-greement with the heuristic argument of the theorists in the nineteen sixties.

There are many other experimentally observed properties of the pions and other pseudoscalar mesons that can be explained by their pseudogoldstone nature, but this is already beyond the scope of our book.

11.6 Hadron Masses □ Baryon Magnetic Moments

In Chap. 3, when we were discussing the fundamental interactions and the structure of the Universe, not knowing yet how the theory of strong interactions is formulated, we said that the proton mass could play the role of the fundamental strong interaction constant. The QCD Lagrangian (11.16) that we have learned in this chapter includes several parameters: besides the dimensionless "strong charge" g, there are also dimensionful quark masses. Does that mean that it is the latter that mostly determine the masses of hadrons? Does that mean that, in the chiral limit when the masses of the light quarks vanish, the proton would also stay massless?

There is no unique answer to the first question (see the discussion below), but the answer to the second one is definitely negative. As we already mentioned in Chap. 5 and repeated in this chapter, the scale symmetry of the classical massless QCD Lagrangian is broken by quantum effects.[30] The dimensionless coupling constant "runs" — it depends on the characteristic energy scale of the process in interest, as in (11.21). This leads to dimensional transmutation — a dimensionful parameter Λ_{QCD} pops out [see Eq. (5.19) for its one-loop expression], which roughly determines the scale below which the perturbative approach is no longer valid. In addition, Λ_{QCD} determines the characteristic scale of light hadron masses!

The latter heuristic conjecture can be made more precise. In the previous section, we introduced the notion of quark condensate. Its numerical value given in (11.48) can be represented as $\Sigma = c\Lambda_{QCD}^3$ with some numerical coefficient c. There is a remarkable relation (the *Ioffe formula*) that expresses the proton mass via the quark condensate:

$$m_p = \sqrt[3]{4\pi^2\Sigma}.\tag{11.55}$$

This is not an exact result, but an approximate estimate obtained in the framework of the so-called *ITEP sum rules* method, whose essentials we have to hide from the reader. If we substitute (11.48) into (11.55), we obtain the estimate $m_p \approx 850$ MeV, to be compared with the experimental value, $m_p = 938$ MeV. The accuracy of the prediction is reasonably high. What is still more important, the Ioffe formula tells us

---

[30]Theorists call this phenomenon the *conformal anomaly*. It is akin to the chiral anomaly (the explicit breaking of the isosinglet axial symmetry by quantum effects), mentioned but not explained in the previous section.

that, if the chiral symmetry were not broken, the protons would stay massless in the massless quark limit.

In other words, the phenomenon of spontaneous chiral symmetry breaking is not just a theoretical adornment allowing us to improve the agreement of the Gell-Mann–Okubo relation (11.29) with experiment. It plays a crucial role in determining the structure of nucleons, of nuclei, of stars and of our bodies!

The quark condensate together with another important characteristic of the QCD vacuum, the *gluon condensate*

$$\left\langle \frac{\alpha_s}{\pi} \mathrm{Tr}\{\hat{F}_{\mu\nu} \hat{F}^{\mu\nu}\} \right\rangle_{\mathrm{vac}} \propto \Lambda^4_{QCD}, \tag{11.56}$$

also determine the masses of other light hadrons, like ρ mesons, etc. For charmed hadrons, the largest contribution to the mass is due to the massive charmed quark, but the vacuum condensates (11.48) and (11.56) also play an important role. The masses of the hadrons made of heavy b and t quarks are largely determined by the masses of their constituents.

The ITEP sum rules method allows one to evaluate not only the masses of hadrons, but also their other properties, like decay probabilities, formfactors (distributions of electric charges inside hadrons) and magnetic moments. Let us dwell on the latter.

In Eq. (10.30) we wrote the expression for the so-called Dirac electron magnetic moment. It enters the tree scattering amplitude of an electron in an external magnetic field, described by the diagram in Fig. 10.7. In Chap. 10 we discussed loop corrections to the moment (10.30); they represented a series in the small parameter $\alpha \approx 1/137$ with calculable coefficients.

Like the electron, the proton carries electric charge and spin and has a magnetic moment. At the tree level, the proton electromagnetic interactions are described by the same diagram in Fig. 10.7, and the corresponding magnetic moment is given by the same expression (10.30) with m_p substituted for m_e. The quantity

$$\mu_N = \frac{|e|\hbar}{2m_p c} \tag{11.57}$$

is called the *nuclear magneton*.

However, the proton is a strongly interacting particle, and loop corrections to the estimate (11.57) are not small and cannot be evaluated perturbatively. Experimentally, the proton magnetic moment exceeds μ_N by the factor ≈ 2.79. Going to the neutron, it does not have a net electric charge and hence does not have a magnetic moment in the tree approximation. However, strong interaction effects bring about a nontrivial distribution of charge inside the neutron and a nonzero magnetic moment. Experimentally, $\mu_n \approx -1.91 \, \mu_N$.

Even though the nucleon magnetic moments cannot be calculated perturbatively, they can be approximately evaluated using the non-perturbative ITEP sum rules method. Boris Ioffe and myself derived the following approximate expressions:

$$\mu_p \approx \frac{8\mu_N}{3} \left(1 + \frac{2\pi^2 \Sigma}{3m_p^3} \right) \approx 3\,\mu_N$$

$$\mu_n \approx -\frac{4\mu_N}{3} \left(1 + \frac{8\pi^2 \Sigma}{3m_p^3} \right) \approx -2\,\mu_N \,. \tag{11.58}$$

The agreement with experiment is more than satisfactory.

Similar approximate formulas can be derived for other baryons from the spin $1/2$ octet in Fig. 11.6. For example, for the Λ hyperon, we obtain

$$\mu_\Lambda \approx -\frac{2\mu_N}{3} \frac{m_p}{m_\Lambda} \left(1 + \frac{8\pi^2 \Sigma}{3m_\Lambda^3} \right) \approx -0.7\,\mu_N \tag{11.59}$$

(to be compared with the experimental value, $\mu_\Lambda^{\text{exp}} \approx -0.61\,\mu_N$). The factor m_p/m_Λ in (11.59) has a kinematic origin — it corresponds to measuring μ_Λ in nuclear magnetons rather than in the more natural units $|e|\hbar/(2m_\Lambda c)$.

Chapter 12
Theory of the Electroweak Interactions

In Chap. 5 we already talked about the phenomenological Fermi theory, as well as the modern fundamental electroweak theory (the Standard Model), and described their salient features in words and pictures. Since then, we have learned a lot of new stuff (what are fermion fields, what is a field theory Lagrangian, how to calculate scattering amplitudes, what is Abelian and non-Abelian gauge invariance, etc.) and are now prepared to give more precise mathematical formulations.

We have to warn the reader. The chapter he has now started to read is more difficult than the preceding ones. Simply because there is a lot of material. I would not say that the physics of the weak interactions is more complicated than the physics of the strong interactions (probably not; a weak interaction is always weak and in all circumstances can be studied by perturbative methods, no Millennium Prize for that), but the weak interactions are more versatile, if you like; they have many faces and it might be difficult for a newcomer to get acquainted with all of them at once. Anyway, your author has set to write this book with the ambition of eventually explaining what the Standard Model is to everybody who wishes to learn it by applying moderate but not negligible efforts, and he has no choice...

12.1 Fermi Theory □ Weak Currents

In 1934 Fermi suggested the following Lagrangian describing neutron decay:

$$\mathcal{L}_F = G_F \, \bar{p}\gamma_\mu n \cdot \bar{e}\gamma^\mu \nu_e \,, \tag{12.1}$$

© Springer International Publishing AG 2017
A. Smilga, *Digestible Quantum Field Theory*,
DOI 10.1007/978-3-319-59922-9_12

where p, n, e, ν_e are the fermion fields describing the corresponding particles.[1] Then the tree amplitude of neutron decay is given by the diagram in Fig. 5.11 and reads

$$M_{n \to p e^- \bar{\nu}_e} = G_F \, \bar{u}_p(p_p)\gamma^\mu u_n(p_n) \cdot \bar{u}_e(p_e)\gamma_\mu u_\nu(-p_{\bar{\nu}_e}), \qquad (12.2)$$

where p_n, p_p, p_e and p_ν are the 4-momenta of the interacting particles and u_p, etc. are the Dirac bispinors (9.53) entering the plane wave solutions (9.51). The bispinors (9.53) have the dimension $[u] = m^{1/2}$, so that the amplitude (12.2) is dimensionless. With the expression (12.2) in hand, one can calculate the electron spectra in nuclear β-decays (Fermi did it). They are in perfect agreement with experiment.

Later it was understood that the expression (12.1) was not quite correct. In particular, the Lagrangian (12.1) is parity-even and does not include parity-breaking effects characteristic of all weak processes. The true *effective* Lagrangian describing the neutron decay reads

$$\mathcal{L}_{\text{neutron decay}} = \frac{G_F}{\sqrt{2}} \cos\theta_c \, \bar{p}\gamma_\mu(1 - g_A\gamma^5)n \cdot \bar{e}\gamma^\mu(1 - \gamma^5)\nu_e, \qquad (12.3)$$

where $g_A \approx 1.26$ is a phenomenological constant and θ_c is the so-called *Cabbibo angle*; we will explain what it is a bit later. The factor $\sqrt{2}$ by which G_F is now divided has historical origin. It was inserted to roughly match the total probability of the neutron decay following from (12.3) to the probability following from (12.1) without an essential change of the numerical value of G_F.

The Lagrangian (12.3) involves a product of two factors. The factor

$$j_\mu^{-(e)} = \bar{e}\gamma_\mu(1 - \gamma^5)\nu_e \qquad (12.4)$$

is called the *weak charged lepton current*. It is called "charged" because it carries negative electric charge.[2] The current (12.4) is akin to the *neutral* electromagnetic electron current $\bar{e}\gamma_\mu e$, but it involves two different fermion fields (so that the corresponding operator describes the transitions $\nu_e \to e$ as well as the creation of the particles $\bar{\nu}_e$ and e from the vacuum) and, which is very important, is not a pure vector, but represents a combination of a vector and an axial vector. Bearing in mind the usual convention [see the remark after Eq. (9.46)], people call (12.4) the V–A current.

The second factor,

---

[1] Fermi wrote down somewhat different formulas, but we modernized his reasoning a little bit. It is worth recalling at this point that the fermion field carries the canonical dimension $m^{3/2}$. Bearing in mind that $[\mathcal{L}] = m^4$, this gives the canonical dimension of Fermi's constant, $[G_F] = m^{-2}$. Its numerical value was given in (5.21).

[2] The precise meaning of the latter statement is the following. The current (12.4) enters the classical Lagrangian (12.3). Consider, however, the corresponding *quantum* Hamiltonian. It is an operator acting in Fock space and involving the product of the current operators. The field operator $\bar{e}(x)$ in the quantum counterpart of (12.4) may either create an electron state or annihilate a positron one (cf. the discussion on p. 119). In both cases, the electric charge of the state is changed by -1.

$$j_\mu^{+(N)} = \bar{p}\gamma_\mu(1 - g_A\gamma^5)n , \tag{12.5}$$

is the nucleon weak charged current. As we see, it is not exactly V–A, but a somewhat different combination.

The appearance of a nontrivial coefficient g_A is due to the fact that the proton and neutron are not fundamental particles and the fields $p(x)$ and $n(x)$ are not fundamental fields. One can say that the original V–A structure of the nucleon weak current (or rather of the quark one, see below) is distorted by strong interaction effects.[3] It is noteworthy that this distortion involves only the axial current. The coefficient of the vector current g_V is exactly 1, for the same reason that the proton electric charge coincides by the absolute value with the electron charge in spite of the fact that the proton participates in strong interactions, while the electron does not. It is a corollary of the current conservation law (9.56).

Fermi did not know (but we do) that the truly fundamental particles are not the nucleons, but the quarks (and also the gluons, but they do not participate either in the electromagnetic, or in weak interactions). Thus, the fundamental Lagrangian should involve not the nucleon but the quark fields. Let us try

$$\mathcal{L} = \frac{G_F}{\sqrt{2}}[j_\mu^{-(e)} j^{\mu+(q)} + j_\mu^{+(e)} j^{\mu-(q)}], \tag{12.6}$$

where $j_\mu^{+(e)} = \bar{\nu}_e\gamma_\mu(1 - \gamma^5)e$ carries charge $+1$ and is related to $j_\mu^{-(e)}$ by complex conjugation (Hermitian conjugation in quantum theory), and the quark weak currents are

$$j_\mu^{+(q)} = \bar{u}\gamma_\mu(1 - \gamma^5)d, \qquad j_\mu^{-(q)} = \bar{d}\gamma_\mu(1 - \gamma^5)u . \tag{12.7}$$

We can now observe that

• The quark currents (12.7) are V–A.
• This means that both the quark currents and the lepton currents include only left-handed fermions:

$$j_\mu^{-(e)} = 2\bar{e}_L\gamma_\mu\nu_{eL}, \qquad j_\mu^{-(q)} = 2\bar{d}_L\gamma_\mu u_L \tag{12.8}$$

[see the definitions (9.44)].
• The Lagrangian (12.6) is real: the two terms in the sum are complex conjugate to each other.

Neutron decay is described by the first term in (12.6). The underlying process is $d \to ue^-\bar{\nu}_e$. In the language of the constituent quark model, one of the neutron's d quarks goes over into a u quark, with the emission of an electron and an antineutrino. The latter fly away, while the newly born u quark stays in the premises. It finds in

---

[3] Actually, one can evaluate this distortion using the ITEP sum rules method, but this is a separate interesting story.

Fig. 12.1 Neutron decay in
the constituent quark model.

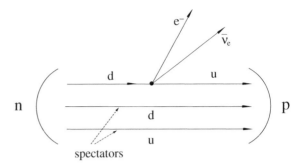

the vicinity two former roommates of the disappeared d quark (nothing happened
to them; they were pure observers of this dramatic event and are often called the
spectator quarks) and stays to live with them, forming the proton bound state.

This reasoning, illustrated in Fig. 12.1, was heuristic and visual. In this chapter,
however, we are after more rigorous and precise mathematical descriptions. To give
one, we use the quantum Hamiltonian language where the expression (12.6) represents an operator $\hat{V}$. It contributes to the Hamiltonian (with a negative sign, but let
us disregard the signs!) and can be treated as a perturbation. In the leading order, the
amplitude of neutron decay is given by the matrix element

$$M \;=\; \langle p e^- \bar{\nu}_e | \hat{V} | n \rangle \;=\; \frac{G_F}{\sqrt{2}} \, \langle p | j_\mu^{+(q)} | n \rangle \cdot \langle e^- \bar{\nu}_e | j^{\mu\,-(e)} | \mathrm{vac} \rangle \,. \tag{12.9}$$

The second factor in this expression is equal to $\bar{u}_e(p_e)\gamma^\mu(1 - \gamma^5)u_\nu(-p_{\bar{\nu}_e})$. And the
first factor is

$$\langle p | j_\mu^{+(q)} | n \rangle \;=\; \bar{u}_p(p_p)\gamma_\mu(1 - g_A\gamma^5)u_n(p_n) \,. \tag{12.10}$$

How can one derive the latter relation? Well, the bispinors $\bar{u}_p$ and u_n should appear
for kinematical reasons. The operator $j_\mu^{+(q)}$ is a mixture of a vector and an axial
vector, and its matrix element should also be a superposition of the vector $\bar{u}_p\gamma_\mu u_n$
and the axial vector $\bar{u}_p\gamma_\mu\gamma^5 u_n$. Then the only non-trivial statement is that $\bar{u}_p\gamma_\mu u_n$
enters with the coefficient $g_V = 1$. We gave arguments in its favour above.

For the time being, we have derived the expression

$$M \;=\; \frac{G_F}{\sqrt{2}} \, \bar{u}_p(p_p)\gamma_\mu(1 - g_A\gamma^5)u_n(p_n) \cdot \bar{u}_e(p_e)\gamma^\mu(1 - \gamma^5)u_\nu(-p_{\bar{\nu}_e}) \tag{12.11}$$

for the amplitude. But the true tree amplitude following from the effective Lagrangian
(12.3) has the extra factor $\cos\theta_c$. Where does that come from?

The point is that, besides the electrons, their neutrinos, u and d quarks, there
are also the fermions (μ, ν_μ, c, s) and (τ, ν_τ, t, b) belonging to the second and the
third generations. And weak interactions between quarks and leptons of different

generations are *not* excluded. For example, the Λ hyperon can decay in the same way as the neutron, giving the proton, electron and $\bar{\nu}_e$ in the final state. But the Λ hyperon is a strange hadron including a strange quark in the composition. And the proton is not. To go from Λ to p, the strange quark should disappear, and the underlying fundamental process for the Λ decay is $s \rightarrow u + e^- + \bar{\nu}_e$. The latter process is only possible if the Lagrangian includes the term describing the interaction of the strange charged current $\sim \bar{u}s$ with the ordinary lepton current (12.4).

Let us leave aside the third generation for a while and consider the model with only two generations. Then the true quark current $j_\mu^{-(q)}$ is not given by Eq. (12.7) and not, say, by

$$j_\mu^{-(q)} = \bar{d}\gamma_\mu(1 - \gamma^5)u + \bar{s}\gamma_\mu(1 - \gamma^5)c, \qquad \text{(wrong)}$$

but is rather

$$j_\mu^{-(q)} = \bar{d}\gamma_\mu(1 - \gamma^5)(\cos\theta_c\, u - \sin\theta_c\, c)$$
$$+ \bar{s}\gamma_\mu(1 - \gamma^5)(\sin\theta_c\, u + \cos\theta_c\, c). \qquad (12.13)$$

The generations are *mixed*. The Cabbibo angle θ_c is the mixing angle between the first and the second generations.

Experimentally, $\theta_c \approx 13°$. This angle is not too large. That means that the amplitudes of the decay of strange baryons involving the emission of e^- and $\bar{\nu}_e$ are several times smaller than the amplitude of the neutron β decay. And speaking of the latter, the difference between (12.11) and the amplitude following from (12.3) is not too large, $\cos\theta_c \approx 0.97$.

If we take into account all three generations, the mixing matrix, the *Cabbibo–Kobayashi–Maskawa (CKM) matrix*, includes four numerical parameters — three mixing angles and a small phase making the matrix complex. These small complexities (at the level $\approx 10^{-3}$) lead to small effects of violation of the CP symmetry.[4] There is a similar mixing matrix in the lepton sector.[5] The mixing of leptons brings about the beautiful phenomenon of neutrino oscillations. And we *will* talk about this in more detail at the end of this chapter.

The full modernized and generalized Fermi Lagrangian reads

$$\mathcal{L} = \frac{G_F}{\sqrt{2}} j_\mu^+ j^{\mu-}, \qquad (12.14)$$

where $j_\mu^\pm = j_\mu^{\pm(l)} + j_\mu^{\pm(q)}$ and the quark and lepton currents take into account the mixing effect. The quark current $j_\mu^{-(q)}$ represents a Kobayashi–Maskawa

---

[4]This symmetry and its violation was already mentioned in a telegraphic style on p. 78. We ask the reader to forgive us that we decided not go into further details concerning this extremely interesting subject in our book.

[5]It is usually called the *Pontecorvo–Maki–Nakagawa–Sakata* or PMNS matrix.

generalization of (12.13) and the lepton current has a similar structure, but with the different set of parameters.

The Lagrangian (12.14) describes not only the *semileptonic* processes like neutron decay or Λ decay, but also pure *leptonic* processes like νe and $\nu \mu$ elastic scattering or muon decay. It also describes *non-leptonic* weak decays of strange particles like $K^+ \rightarrow \pi^+ \pi^0$ or $\Lambda \rightarrow p\pi^-$, which are due to the part of the Lagrangian (12.14) when both currents j_μ^+ and j_μ^- are of quark nature. It is, of course, much easier to experimentally study weak interaction effects in non-leptonic processes where strangeness is changed and there is no strong background. But by applying some effort, one can also detect small parity nonconservation effects in processes involving only ordinary hadrons. For example, a small circular polarization of γ quanta in the process of radiative neutron capture by a proton, $n + p \rightarrow d$(euterium) $+ \gamma$, has been measured.

The Lagrangian (12.14) is far from being the end of the story. The reader remembers that the Fermi theory is effective in nature; it works only at small energies and should be replaced by the true renormalizable theory at energies approaching and exceeding the unitary limit (5.24). However, even for small energies the Lagrangian (12.14) does not describe everything. Namely, it does not describe processes with *neutral currents*, when the fundamental fermions do not change their nature during interaction. Examples of such processes: νp scattering (see Fig. 5.16) and parity nonconservation in atomic transitions were given in Chap. 5.

The neutrino weak neutral current is V–A:

$$j_\mu^{(\nu)} = \bar{\nu}\gamma_\mu(1 - \gamma^5)\nu. \tag{12.15}$$

But the electron neutral current is not; it represents a more complicated combination of the V and A currents. We will derive it soon.

We are now leaving for a while the zoo of weak phenomena that we were visiting[6] and pose a natural question: what is the *fundamental biology* of all these beasts that we have just seen? It is worthwhile recalling at this point the logic that finally led people to construct what is now called the Standard Model.

- The Fermi-like theory of the weak interactions with pointlike 4-fermion vertices is not renormalizable and does not allow one to calculate loops. It works only at small energies, $E \ll G_F^{-1/2}$, and has the status of an effective theory. To cope with non-renormalizability, one should assume that the interaction is not pointlike, but has a finite range given by the Compton wavelength of the massive vector bosons that serve as mediators (see Fig. 5.14).

- However, a naïve theory involving massive vector particles does not work. It is not gauge-invariant and, as a result, not renormalizable. We simply mentioned that in the *Hors d'Oeuvres* part, in Chap. 5, but we have now acquired enough tasty credits to understand why.

---

[6]Now I've suddenly thought of Sophie who accompanied us during our visit of the Theoretical Physics Museum in Chap. 4. Is she still with us? Is she not too tired?

Fig. 12.2 Loop graph for elastic $\nu_e e$ - scattering.

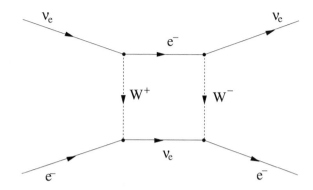

Take for example the 1-loop graph in Fig. 12.2 describing elastic $\nu_e e$-scattering. At large momenta the fermion propagators behave as $\propto 1/p$, while, unlike the photon propagators (10.22), the massive W propagators (10.29) do not fall off at all. As a result, the loop integral associated with the graph in Fig. 12.2 behaves in the ultraviolet as $\propto \int d^4 p/p^2$ and involves a quadratic UV divergence.

- However, one can still make vector bosons massive and keep the renormalizability. To this end, one should assume that the bosons acquire mass by the *Higgs mechanism* — roughly in the same way as the photons acquire mass in superconductors. We will explain what the Higgs mechanism is in the next section.
- One cannot construct the fundamental theory of the weak interactions without taking care of the electromagnetic ones. If we gave a mass to the $W^{\pm}$ by the non-Abelian Higgs mechanism, but left the photon aside, we could not keep intact the ordinary Abelian gauge invariance, and hence such a model would not be renormalizable. A consistent theory should deal with the mediators of the weak and electromagnetic interactions simultaneously.
- It should involve (at least) four such mediators: $W^{\pm}$ coupled to the charged weak currents, Z^0 coupled to the neutral weak currents and the photon. That means that the theory should involve at least four gauge fields, which corresponds to the gauge group $SU(2) \times U(1)$.

12.2 Higgs Mechanism

12.2.1 Abelian

In the previous chapter, we discussed the spontaneous breaking of a continuous symmetry and considered, as a simplest field theory example, the Lagrangian (11.43) of a complex scalar field with a double-well potential, whose minima are characterized by nonzero $|\phi|$. Let us assume now that our scalar particles carry electric charge.

Then the Lagrangian includes the electromagnetic field, and the ordinary derivatives become covariant ones:

$$\mathcal{L} = -\frac{1}{4} F_{\mu\nu} F^{\mu\nu} + (\mathcal{D}^{\mu}\phi)^* \mathcal{D}_{\mu}\phi - \frac{\lambda}{4}(\phi^*\phi - v^2)^2 , \tag{12.16}$$

with $\mathcal{D}_{\mu}\phi = (\partial_{\mu} + ieA_{\mu})\phi$.[7] In contrast to the Lagrangian (11.43), which only enjoys the global $U(1)$ symmetry, the Lagrangian (12.16) is invariant with respect to local gauge transformations

$$\phi(x) \rightarrow e^{-ie\chi(x)}\phi(x)$$
$$A_{\mu} \rightarrow A_{\mu} + \partial_{\mu}\chi(x) . \tag{12.17}$$

The classical vacuum state is characterized, as earlier, by a nonzero $|\phi| = v$. However, we cannot say any more that there are an infinite number of vacua distinguished by the phase of ϕ. Changing the phase of $\phi(x)$ amounts to changing the gauge choice. It does not bring about a new physical state. Therefore, there is now no reason to expect the existence of a massless Goldstone boson in the spectrum.

And there is none. To see that, the reader may substitute in (12.16) the vacuum field (11.45). S/he obtains, instead of (11.46),

$$\mathcal{L}_{\alpha} = \frac{v^2}{2}(\partial_{\mu}\alpha + \sqrt{2}eA_{\mu})^2 . \tag{12.18}$$

One can completely get rid of the field $\alpha(x)$ by a gauge transformation (12.17) with $\chi(x) = -\alpha(x)/(e\sqrt{2})$; $\alpha(x)$ is thus an unphysical gauge degree of freedom, indeed.

We now choose the gauge where $\phi(x)$ is real and express it as

$$\phi(x) = v + \frac{H(x)}{\sqrt{2}} . \tag{12.19}$$

Now, $H(x)$ is a physical real scalar field describing a deviation of $\phi(x)$ from its vacuum value. We may assume it to be small, $H(x) \ll v$. We substitute (12.19) in (12.16) to obtain

$$\mathcal{L} = -\frac{1}{4} F_{\mu\nu} F^{\mu\nu} + e^2 v^2 A_{\mu}^2 + \frac{1}{2}(\partial_{\mu}H)^2 - \frac{\lambda v^2}{2} H^2$$
$$+ \text{ small cubic and quartic interaction terms.} \tag{12.20}$$

The most striking feature of this expression is the second term. The photon has acquired a mass, $m_A^2 = 2e^2v^2$! This phenomenon is called the Higgs effect, or, more often, the Higgs mechanism. The real scalar field H is also massive, $m_H^2 = \lambda v^2$. It is

---

[7]Wishing to stay within the limits of the standard Latin alphabet, we use the same notation e for the charge of the scalar particle as for the electron charge. But they actually need not be the same.

nothing but the *Higgs boson* [not the Higgs boson of the Standard Model, which we have not talked about yet, but the Higgs boson in the simple Abelian model (12.16)].

In fact, a small miracle occurred. We said at the end of Chap. 7 that the Lagrangian of a free massive vector field is not gauge-invariant. Which is true. Neither is the first line in (12.20) gauge-invariant. But the *full* Lagrangian, which includes also the interaction terms that we have not explicitly written and also the terms including the gauge degree of freedom $\alpha(x)$ that we eliminated by a gauge choice, coincides with (12.16) and *is* invariant under the transformations (12.17).

It is instructive to do a little arithmetic at this point and count the physical degrees of freedom. Consider a Lagrangian like that of Eq. (12.16), but with an ordinary mass term $-m^2 \phi^* \phi$ instead of the double-well potential. The vacuum average $\langle \phi \rangle$ vanishes in this case. The photon stays massless and there are two physical asymptotic photon states with different transverse polarizations. The complex scalar field also gives two different asymptotic states.

In reality, a nonzero vacuum average is formed and the photon becomes massive. A massive vector spin 1 particle can be "stopped" and has three physical polarizations. On top of that, there is a *single* real Higgs boson state. A remarkable equality

$$2 + 2 = 3 + 1 \; ,$$

tells us that we have not made a mistake!

Thus, the essence of the Higgs mechanism is somewhat macabre. When a system that would otherwise include in the spectrum a massless Goldstone mode due to spontaneous breaking of a global continuous symmetry gets into contact with a gauge field, the latter *swallows* the Goldstone boson and becomes massive at its expense.

As was mentioned in Chap. 5, the Higgs effect was first discovered by Vitaly Ginzburg and Lev Landau, who suggested a field-theoretical model for superconductivity, where the vacuum condensate of a phenomenological scalar field (actually, the condensate of Cooper pairs) renders the photon massive. But that was a nonrelativistic model. The relativistic realization of the Higgs mechanism was worked out by Peter Higgs and independently by Robert Brout and François Englert and by Gerald Guralnik, Carl Hagen and Tom Kibble in 1964. Higgs and Englert were awarded the Nobel Prize in 2013 (Brout died in 2011) after the Higgs boson of the Standard Model had been experimentally discovered at CERN.

12.2.2 Non-Abelian

We need to give a mass not to one, but to several bosons, $W^\pm$ and Z^0. This can be done using the non-Abelian Higgs mechanism.

Consider the Lagrangian

$$\mathcal{L} = -\frac{1}{2}\mathrm{Tr}\{\hat{G}^{\mu\nu}\hat{G}_{\mu\nu}\} + (\mathcal{D}^{\mu}\phi)^* \mathcal{D}_{\mu}\phi - \frac{\lambda}{4}(\phi^*\phi - v^2)^2, \tag{12.21}$$

where ϕ belongs to the fundamental doublet representation of $SU(2)$, $\mathcal{D}_{\mu}\phi = (\partial_{\mu} - ig\hat{W}_{\mu})\phi$, $\hat{W}_{\mu}$ is the $SU(2)$ gauge field, $\phi^*\phi \equiv \phi^{*k}\phi_k$ and

$$\hat{G}_{\mu\nu} = \frac{i}{g}[\mathcal{D}_{\mu}, \mathcal{D}_{\nu}] = \partial_{\mu}\hat{W}_{\nu} - \partial_{\nu}\hat{W}_{\mu} - ig[\hat{W}_{\mu}, \hat{W}_{\nu}], \tag{12.22}$$

as in (11.12).

The scalar field acquires a vacuum average

$$\langle\phi\rangle = U\begin{pmatrix} 0 \\ v \end{pmatrix}, \tag{12.23}$$

where U is an $SU(2)$ matrix. If there were no gauge fields, this average would signal the spontaneous breaking of the global $SU(2)$ symmetry (in contrast to the magnon model, discussed in Sect. 11.5.2, where the $O(3)$ rotational symmetry was not completely broken, but had an $O(2)$ remnant, the $SU(2)$ symmetry is broken completely in this case). In accordance with Goldstone's theorem, we would obtain three massless particles in the spectrum, the corresponding fields representing the three parameters of the unitary matrix U.

But the gauge field is present. It swallows the massless modes and acquires mass. Indeed, we can choose a gauge where the matrix U entering (12.23) is equal to $\mathbb{1}$. Then the contributions quadratic in $\hat{W}_{\mu}$, coming from the middle term in (12.21), are

$$g^2 v^2 (0, 1)(\hat{W}_{\mu})^2 \begin{pmatrix} 0 \\ 1 \end{pmatrix} = \frac{g^2 v^2}{4}(W_{\mu}^a)^2, \tag{12.24}$$

which gives

$$m_W^2 = g^2 v^2/2. \tag{12.25}$$

We can now forget about the Goldstone modes (they have become unphysical gauge degrees of freedom) and choose a gauge where they are absent,

$$\phi(x) = \begin{pmatrix} 0 \\ v + \frac{H(x)}{\sqrt{2}} \end{pmatrix}. \tag{12.26}$$

The real field $H(x)$ gives a Higgs particle with the mass

$$m_H^2 = \lambda v^2. \tag{12.27}$$

The counting of degrees of freedom is:

$$unbroken : n_{\text{d.o.f.}} = (3 \times 2)_{\text{gauge bosons}} + 4_{\text{scalar doublet}} = 10 \,,$$
$$broken : n_{\text{d.o.f.}} = (3 \times 3)_{\text{massive vectors}} + 1_{\text{Higgs}} = 10 \,. \qquad (12.28)$$

The arithmetic works.

12.3 Standard Model: Gauge and Higgs Sectors

We are finally prepared to go over from toy models to the real world.

As we mentioned, we need altogether four gauge bosons: $W^{\pm}$, Z^0 and γ. Thus, we introduce a non-Abelian $SU(2)$ gauge field $\hat{W}_\mu$ and an Abelian field B_μ.[8] The bosonic part of the electroweak sector of the Standard Model (as the reader understands, the latter is actually not *a* model, but *the* theory describing Nature) reads

$$\mathcal{L}_{\text{bos}}^{\text{SM}} = -\frac{1}{2}\text{Tr}\{\hat{G}_{\mu\nu}\hat{G}^{\mu\nu}\} - \frac{1}{4}B_{\mu\nu}B^{\mu\nu} + (\mathcal{D}^\mu \phi)^* \mathcal{D}_\mu \phi$$
$$-\frac{\lambda}{4}(\phi^*\phi - v^2)^2 \,. \qquad (12.29)$$

The fundamental fields in this Lagrangian are $\hat{W}_\mu$, B_μ and the $SU(2)$ doublet ϕ. Now, $B_{\mu\nu} = \partial_\mu B_\nu - \partial_\nu B_\mu$ is the Abelian field strength tensor and the non-Abelian field strength $\hat{G}_{\mu\nu}$ is expressed via $\hat{W}_\mu$ as in (12.22). The field ϕ interacts with $\hat{W}_\mu$ in the same way as in the non-Abelian Higgs model discussed in the previous section, but in addition, it carries a charge with respect to B_μ. The latter fact displays itself in the modification of the covariant derivative. We will choose it in the form[9]

$$\mathcal{D}_\mu \phi = \left(\partial_\mu - ig\hat{W}_\mu + \frac{i}{2}g'B_\mu\right)\phi \,. \qquad (12.30)$$

The non-Abelian charge g and the Abelian charge $g'/2$ are *different* parameters of the Lagrangian, having little to do with each other.

The scalar vacuum average is characterized by $\phi^*\phi = v^2$. As earlier, we can choose a gauge where

$$\phi_{\text{vac}} = v\begin{pmatrix} 0 \\ 1 \end{pmatrix} \,. \qquad (12.31)$$

Substituting that into the kinetic scalar term $(\mathcal{D}^\mu \phi)^* \mathcal{D}_\mu \phi$ in Eq. (12.29), we derive the following mass term for the vector fields:

---

[8] We have called it B_μ rather than A_μ because it is *not* an electromagnetic field, as we will see very soon.

[9] Attention! Different textbooks use different sign conventions for g and g'. Our convention matches the definitions of covariant derivatives in (11.2) and (11.6).

$$v^2(0, 1) \left(g\hat{W}_\mu - \frac{g'}{2} B_\mu \right)^2 \binom{0}{1}$$

$$= \frac{g^2 v^2}{4} \left[(W_\mu^1)^2 + (W_\mu^2)^2 \right] + \frac{v^2}{4} (g' B_\mu + g W_\mu^3)^2 . \tag{12.32}$$

The first term is the mass term for the charged boson fields

$$W_\mu^\pm = \frac{W_\mu^1 \mp i W_\mu^2}{\sqrt{2}} . \tag{12.33}$$

(The opposite signs in the L.H.S. and R.H.S. are not a mistake, see below.) Their mass is given by (12.25). And the second term is the mass term for a certain field representing the linear combination

$$Z_\mu = \frac{g W_\mu^3 + g' B_\mu}{\sqrt{g^2 + g'^2}} . \tag{12.34}$$

Its mass is

$$m_Z^2 = \frac{v^2(g^2 + g'^2)}{2} . \tag{12.35}$$

The field combination orthogonal to (12.34),

$$A_\mu = \frac{-g' W_\mu^3 + g B_\mu}{\sqrt{g^2 + g'^2}} , \tag{12.36}$$

stays massless.[10] It *is* the electromagnetic field!

As was the case in the non-Abelian model of the previous section, we may choose the gauge where ϕ is real, as in (12.26). Then the field $H(x)$ describes a real Higgs particle with the mass (12.27).

A group theoretical interpretation of what we have just observed is the following:

• Suppose the gauge fields were absent. In that case, the scalar Lagrangian would enjoy the global symmetry $U(2) = SU(2) \times U(1)$. The vacuum average (12.31) would break it spontaneously down to $U(1)$ [indeed, (12.31) remains invariant under $\phi \to e^{i\alpha(1+\sigma_3)}\phi$], and three massless Goldstone bosons would appear in the spectrum.

---

[10]The normalization factor $1/\sqrt{g^2 + g'^2}$ was introduced in (12.34) and (12.36) to ensure the same standard form of the kinetic term for the fields Z_μ and A_μ:

$$\mathcal{L}_{A,Z}^{\text{kin}} = -\frac{1}{4} (\partial_\mu A_\nu - \partial_\nu A_\mu)^2 - \frac{1}{4} (\partial_\mu Z_\nu - \partial_\nu Z_\mu)^2 . \tag{12.37}$$

Fig. 12.3 Weinberg angle.

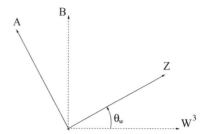

- But the gauge fields are present. They serve themselves with the Goldstone modes. Unfortunately, the latter are not in a sufficient abundance to satisfy everybody — only 3 modes for 4 gauge fields. Thus, only three out of the four gauge fields acquire mass, the fourth (the electromagnetic field) stays massless and hungry.

The counting of degrees of freedom is:

$$
\begin{aligned}
unbroken : n_{\text{d.o.f.}} &= & (3 \times 2)_W + 2_B + 4_\phi &= 12 , \\
broken : n_{\text{d.o.f.}} &= & (3 \times 3)_{W^\pm, Z} + 2_A + 1_H &= 12 .
\end{aligned} \tag{12.38}
$$

The parameter

$$
\theta_W = \arctan \frac{g'}{g} \tag{12.39}
$$

is called the *Weinberg angle*. It is nothing but the angle of rotation from the basis (W^3, B) to the basis (Z, A) of the physical fields (see Fig. 12.3). The masses of $W^\pm$ and Z can be expressed as

$$
m_W = \frac{g v}{\sqrt{2}}, \qquad m_Z = \frac{g v}{\sqrt{2} \cos \theta_W} = \frac{m_W}{\cos \theta_W} . \tag{12.40}
$$

12.4 Standard Model: Fermions

12.4.1 Gauge Interactions

After a little meditative repose and a cup of coffee,[11] we continue.

To make contact with the Fermi theory and with ordinary QED, we have to describe how the gauge bosons interact with the fermions. It is not so difficult to write down the interaction Lagrangian; it is simply $i \bar{\psi} \gamma^\mu \mathcal{D}_\mu \psi$. But a nontrivial question is what

---

[11] Oops, we are not yet done with our entrées. After few sips of wine then.

Table 12.1 Hypercharges.

Fermion	ν_L, e_L	ν_R	e_R	u_L, d_L	u_R	d_R
Hypercharge	-1	0	-2	1/3	4/3	$-2/3$

the ψ are and how to write the covariant derivative $\mathcal{D}_\mu$. One thing is clear: in contrast to ordinary QED [Eq. (11.1)] and to QCD [Eq. (11.7)], where left-handed and right-handed fermions interact with the photons (resp. gluons) in the same way, the interaction with $W^\pm$ and Z should be left-right asymmetric, to provide for parity nonconservation.

Having tried different schemas, drinking coffee, or whatever they preferred, and meditating, Weinberg and Salam understood that everything fits perfectly well, if one assumes that the left-handed fermions represent $SU(2)$ doublets interacting with the non-Abelian field $\hat{W}_\mu$ and also with B_μ, whereas the right-handed fermions are *singlets* — they are not coupled to $\hat{W}_\mu$, but only to B_μ.

Let us dwell on the fermions of the first generation. The claim is that they are grouped in the following multiplets:[12]

$$\begin{pmatrix} \nu \\ e \end{pmatrix}_L, \; \nu_R, \; e_R; \qquad \begin{pmatrix} u \\ d \end{pmatrix}_L, \; u_R, \; d_R. \tag{12.41}$$

The fermions (12.41) have the following charges Y with respect to the fields B_μ, expressed in the units of $g'/2$ (these numbers are called *hypercharges*) (Table 12.1).

The numbers in this table coincide with double the *average* electric charge in the multiplet.[13] For example, for the left-handed quark doublet,

$$Y_{u_L} = Y_{d_L} = 2 \times \frac{1}{2} \left[\frac{2}{3} + \left(-\frac{1}{3} \right) \right] = \frac{1}{3}. \tag{12.42}$$

This observation helps us to memorize the values of the supercharges in Table 12.1, but it is more correct to think in the opposite direction: we *postulate* the values of the hypercharges as in Table 12.1 and then we will *derive* the values for the electric charges of all particles. We also recall that the quarks belong to the fundamental triplet representation of strong $SU(3)$ — they exist in three colour modifications.

The reader who sees the numbers in Table 12.1 for the first time might find them rather weird and arbitrary. To some extent, they are (at least, at our present level of confident knowledge), but one is still not allowed to replace them by no matter what. The reader can verify that the sum of the *cubes* of the hypercharges over all left-handed fermions is the same as the similar sum over the right-handed fermions

---

[12]To be precise, we should have written ν_e rather than just ν in (12.41) and in all the formulas on the several subsequent pages. But your author is afraid to intoxicate his reader and get intoxicated himself by a plethora of indices. When we bring into consideration other generations, we will restore the accurate notation ν_e, ν_μ, ν_τ.

[13]Incidentally, the hypercharge of the scalar doublet is $Y_\phi = 1$, and that means that the upper component of ϕ is positively charged while the lower one is neutral.

(taking into account three quark colours):

$$\sum_L Y_L^3 = \sum_R Y_R^3 = -\frac{16}{9}. \tag{12.43}$$

This mysterious equality has a quite precise esoteric meaning — it is a condition for the internal axial anomaly (a sacred word that we occasionally dare to pronounce) to be cancelled. This makes the theory gauge-invariant and self-consistent.

Bearing in mind (12.41), Table 12.1 and the expression for the covariant derivative $\mathcal{D}_\mu = \partial_\mu - igW_\mu^a t^a + \frac{ig'}{2}Y_L B_\mu$ for the left-handed doublet fermions and $\mathcal{D}_\mu = \partial_\mu + \frac{ig'}{2}Y_R B_\mu$ for the right-handed singlet fermions, we derive the following form of the fermion Lagrangian:

$$\begin{aligned}
\mathcal{L}_f = \ & i(\bar{\nu}_L, \bar{e}_L)\gamma^\mu \left(\partial_\mu - igW_\mu^a t^a - \frac{ig'}{2}B_\mu\right)\begin{pmatrix}\nu_L \\ e_L\end{pmatrix} \\
& + i\bar{\nu}_R\gamma^\mu\partial_\mu\nu_R + i\bar{e}_R\gamma_\mu(\partial_\mu - ig'B_\mu)e_R \\
& + i(\bar{u}_L, \bar{d}_L)\gamma^\mu \left(\partial_\mu - igW_\mu^a t^a + \frac{ig'}{6}B_\mu\right)\begin{pmatrix}u_L \\ d_L\end{pmatrix} \\
& + i\bar{u}_R\gamma^\mu\left(\partial_\mu + \frac{2ig'}{3}B_\mu\right)u_R + i\bar{d}_R\gamma^\mu\left(\partial_\mu - \frac{ig'}{3}B_\mu\right)d_R.
\end{aligned} \tag{12.44}$$

Not the most simple and elegant expression that one could imagine, but that is life. We will now analyze and discuss its different parts in some improvised order.

We start with the right-handed neutrino. It is an $SU(2)$ singlet and has zero hypercharge. Thus, it interacts neither with W, nor with B, and definitely not with gluons. People call it *sterile*, though that is not completely true. As we know now, the neutrinos carry a small but nonzero mass, which brings about a term $\propto \bar{\nu}_L\nu_R$ in the Lagrangian, so that a left-handed neutrino can be converted to a right-handed one, and vice versa. We will talk about these conversions later and, for the time being, we only note that Weinberg and Salam did not know in 1967 and 1968 that the neutrinos carry mass. They thought that the right-handed neutrinos were *absolutely* sterile and did not include them into the Lagrangian for that reason.

Consider now the interactions with the charged bosons $W^\pm$. Only left-handed fermions participate there. The relevant part of (12.44) is[14]

$$\mathcal{L}_{\bar{f}fW} = \frac{g}{\sqrt{2}}W_\mu^+[\bar{\nu}_L\gamma^\mu e_L + \bar{u}_L\gamma^\mu d_L] + \text{complex conjugate}. \tag{12.45}$$

We can now express the effective Fermi constant G_F via the $SU(2)$ charge g and the mass of W^+. To that end, we compare the two graphs in Fig. 5.14, describing the process of νe scattering and redrawn here in Fig. 12.4 for the reader's benefit.

---

[14]We now see that, for $\mathcal{L}$ to carry zero net electric charge and thereby ensure charge conservation, one has to ascribe the positive charge to the combination $\propto W_\mu^1 - iW_\mu^2$, as was done in (12.33))!.

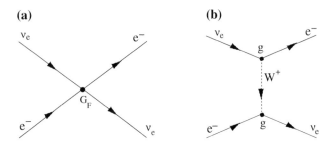

Fig. 12.4 νe scattering: (**a**) in the Fermi theory, (**b**) with W-exchange.

The left-hand graph gives the scattering amplitude in the effective Fermi theory. Bearing in mind (12.14) and (12.8), we derive

$$M^{\text{Fermi}}_{\nu_e e^- \to \nu_e e^-} = 2\sqrt{2} G_F\, \bar{u}_{eL}\gamma^\mu u_{\nu L} \cdot \bar{u}_{\nu L}\gamma_\mu u_{eL}\,, \tag{12.46}$$

where u_{eL}, etc. are the corresponding bispinors. On the other hand, the expression for the amplitude calculated by the second graph in Fig. 12.4 for *small* momentum transfer, where the massive vector propagator (10.29) reduces to $-\eta_{\mu\nu}/m_W^2$, is

$$M^{\text{SM}}_{\nu_e e^- \to \nu_e e^-} = \frac{g^2}{2m_W^2}\, \bar{u}_{eL}\gamma^\mu u_{\nu L} \cdot \bar{u}_{\nu L}\gamma_\mu u_{eL}\,. \tag{12.47}$$

Comparing (12.46) and (12.47), we obtain:

$$G_F = \frac{g^2}{4\sqrt{2} m_W^2}\,. \tag{12.48}$$

Fermi's constant is experimentally known. The relation (12.48) would allow us to find the mass of W, if we knew the value of g. But we do not know it yet.

To evaluate g, we have to look at the yet untreated part of Eq. (12.44) describing the interactions of the fermions with the electrically neutral fields W_μ^3 and B_μ. When we express them via the physical fields Z_μ and A_μ according to the picture in Fig. 12.3, we obtain the following contributions in the Lagrangian:

$$\mathcal{L}_A = g\sin\theta_W A_\mu \left[\bar{e}\gamma^\mu e - \frac{2}{3}\bar{u}\gamma^\mu u + \frac{1}{3}\bar{d}\gamma^\mu d \right]\,, \tag{12.49}$$

(where $\bar{e}\gamma^\mu e = \bar{e}_L\gamma^\mu e_L + \bar{e}_R\gamma^\mu e_R$, and similarly for $\bar{u}\gamma^\mu u$ and $\bar{d}\gamma^\mu d$) and

$$\mathcal{L}_Z = \frac{1}{2} Z_\mu \sqrt{g^2 + g'^2} \left[\bar{\nu}_L \gamma^\mu \nu_L - \cos 2\theta_W \, \bar{e}_L \gamma^\mu e_L + 2 \sin^2 \theta_W \, \bar{e}_R \gamma^\mu e_R \right.$$

$$+ \left(1 - \frac{4}{3} \sin^2 \theta_W \right) \bar{u}_L \gamma^\mu u_L - \frac{4}{3} \sin^2 \theta_W \, \bar{u}_R \gamma^\mu u_R$$

$$\left. \left(\frac{2}{3} \sin^2 \theta_W - 1 \right) \bar{d}_L \gamma^\mu d_L + \frac{2}{3} \sin^2 \theta_W \, \bar{d}_R \gamma^\mu d_R \right]. \quad (12.50)$$

The first expression describes the electromagnetic fermion interactions. It is left-right symmetric, as it should be. The combination

$$e = -g \sin \theta_W = -\frac{g g'}{\sqrt{g^2 + g'^2}} \quad (12.51)$$

is nothing but the electric charge of the electron. The quarks are fractionally charged.

Equation (12.50) describes the interaction of Z with the weak neutral current. The latter involves the V–A neutrino current including only the left-handed neutrino fields and certain nontrivial combinations of the vector and axial electron and quark currents.

Bearing in mind (12.51), (12.48), (12.40) and the experimental values of Fermi's constant and the fine-structure constant, we can now derive the nice results quoted in (5.29), which we write here again:

$$m_W = \frac{1}{\sin \theta_W} \sqrt{\frac{\pi \alpha}{G_F \sqrt{2}}} \approx \frac{37.3}{\sin \theta_W} \, \text{GeV},$$

$$m_Z = \frac{1}{\sin \theta_W \cos \theta_W} \sqrt{\frac{\pi \alpha}{G_F \sqrt{2}}} \approx \frac{74.6}{\sin 2\theta_W} \, \text{GeV}. \quad (12.52)$$

The theory thus gives a prediction for the masses which involves only one free parameter — the Weinberg angle θ_W.

Equation (12.52) is a tree-level result. Loop corrections modify it only a little. By comparing it with the experimentally known boson masses (5.30), we can determine the value of the Weinberg angle. The result was quoted in Eq. (5.31).

12.4.2 Masses

We know from experiment that all fundamental fermions, including the neutrinos, are massive, but in the framework of the theory just described, it is not so easy to understand why. Indeed, the left-handed and right-handed fermions enter the Lagrangian in a completely independent manner; they have nothing to do with each other. Moreover, they belong to different representations of the $SU(2)$ weak group: the left fermions are doublets while the right fermions are singlets. A naïve mass term like the one in (11.37) simply cannot be written down — it would not be an $SU(2)$ invariant.

What one *can* do is to write down a *Yukawa* Lagrangian describing the interaction of the left fermion doublet, a right fermion singlet and the scalar doublet:[15]

$$\mathcal{L}_e^{\text{Yuk}} = -h_e \, (\phi^-, \bar{\phi}^0) \, \bar{e}_R \begin{pmatrix} \nu_L \\ e_L \end{pmatrix} + \quad \text{complex conjugate} . \tag{12.53}$$

The Lagrangian (12.53) is an $SU(2)$ singlet and carries zero hypercharge. It is thus manifestly gauge-invariant. The Yukawa coupling constant h_e carries no dimension.

We recall now that the field ϕ acquires a nonzero expectation value (12.31). Thus, the dominant contribution in (12.53) is

$$\mathcal{L}_e^{\text{mass}} = -h_e v (\bar{e}_R e_L + \bar{e}_L e_R) = -h_e v \, \bar{e} e . \tag{12.54}$$

In other words, the electron *does* acquire a mass when spontaneous breaking of gauge symmetry is taken into account! Its value is $m_e = h_e v$.

Maybe the reader has heard some people saying that the Higgs boson is the "God particle" because it gives masses to everything. Not being an expert in theology, I cannot say whether it is really God's intention to give masses to everything and, if it is true, why He failed to do so for the photon and the graviton. But one thing is clear. A nonzero Higgs expectation value is, indeed, the reason why the W and Z bosons, the leptons and quarks have mass. On the other hand, the nucleons and most other hadrons acquire mass by a completely different mechanism, which we discussed in the previous chapter devoted to QCD.

We have written formulas showing how the electron acquires mass. Let us also do so for other fermions. The d quark, like the electron, represents a lower doublet component, and the obvious generalization of (12.53),

$$\mathcal{L}_d^{\text{Yuk}} = -h_d \, (\phi^-, \bar{\phi}^0) \, \bar{d}_R \begin{pmatrix} u_L \\ d_L \end{pmatrix} + \text{c.c.} , \tag{12.55}$$

does the job. The mass of the d quark is thus $m_d = h_d v$.

For the u quark and for the neutrino, the expression (12.53) should be somewhat modified. The key observation is that, in $SU(2)$ [in contrast to $SU(N \geq 3)$], there is *no difference* between the fundamental and anti-fundamental representations [see the remark after (6.48)]. In particular, the doublet $\begin{pmatrix} \bar{d}_L \\ -\bar{u}_L \end{pmatrix}$ transforms under $SU(2)$ in the same way as $\begin{pmatrix} u_L \\ d_L \end{pmatrix}$. The Lagrangians

---

[15] As we know from Sect. 5.3, Yukawa's theory suggested in 1935 was a phenomenological theory of the strong interaction between the nucleons and scalar meson particles (identified later with the π mesons). This seems to have nothing to do with the electroweak theory. But the Lagrangian of this interaction had the form $\mathcal{L}_{\text{Yukawa}} \propto \pi \bar{N} N$. Since then, *any* interaction of the form $\phi \bar{f} f$ (ϕ, f being scalar and spinor fields of any nature) is called a Yukawa interaction.

$$\mathcal{L}_u^{\text{Yuk}} = h_u \, (\phi^-, \bar{\phi}^0) \begin{pmatrix} \bar{d}_L \\ -\bar{u}_L \end{pmatrix} u_R + \text{c.c.} \tag{12.56}$$

and

$$\mathcal{L}_\nu^{\text{Yuk}} = h_\nu \, (\phi^-, \bar{\phi}^0) \begin{pmatrix} \bar{e}_L \\ -\bar{\nu}_L \end{pmatrix} \nu_R + \text{c.c.} \tag{12.57}$$

are therefore group singlets and gauge invariants. They generate the masses $m_u = h_u v$ and $m_\nu = h_\nu v$.

An important disclaimer is in order here. We are pretty sure that the u quark acquires mass by the mechanism just described. There is simply no other way. However, for the neutrino, another possibility exists. As was mentioned, the right-handed neutrinos do not interact with gauge bosons. Hence they carry no charges, are completely neutral. And then it is possible to endow them with mass by a different, Majorana, mechanism [see the discussion after Eq. (9.39)]! If we did *only* that and postulated the Yukawa couping constant h_ν to vanish, the right-handed neutrino would be an absolutely sterile particle and the left-handed neutrino would stay massless. We know from experiment that it does not (see the oscillation section below). Hence, we *must* postulate a nonzero h_ν. However, nothing forbids the presence of an extra Majorana mass term for ν_R. It is therefore quite possible that the right-handed neutrino mass is of a mixed, both Majorana and Dirac, nature.

12.4.3 Generations and Their Mixing

Finally we discuss what happens if we include two other generations. As we have already mentioned in the "Fermi" section, the generations are mixed. The W^+ boson interacts not just with the current $\bar{\nu}_{eL}\gamma_\mu e_L + \bar{u}_L \gamma_\mu d_L$, as in (12.45) (as promised, the notation ν_{eL} is now restored), but with the current

$$j_\mu^+ = \bar{\nu}'_{eL}\gamma_\mu e_L + \bar{u}'_L \gamma_\mu d_L + \bar{\nu}'_{\mu L}\gamma_\mu \mu_L + \bar{c}'_L \gamma_\mu s_L + \bar{\nu}'_{\tau L}\gamma_\mu \tau_L + \bar{t}'_L \gamma_\mu b_L \,,$$
$$\tag{12.58}$$

where u'_L, c'_L and t'_L are related to u_L, c_L and t_L via the CKM matrix and ν'_{eL}, $\nu'_{\mu L}$ and $\nu'_{\tau L}$ are related to ν_{eL}, $\nu_{\mu L}$ and $\nu_{\tau L}$ by the leptonic analogue of the CKM matrix, the PMNS matrix (which has nothing to do with the quark one).

"But why can one not just redefine the fields and *call* the combination u'_L the left-handed u-quark field, the combination ν'_L — the left-handed electron neutrino field, etc", our clever reader may ask at this point. The reader is basically right, as s/he was on many other occasions. Indeed, one can redefine the fermion fields in this way (we will actually do so when discussing neutrino oscillations), and if the fermions were *massless*, these "prime" states would constitute as good a Hilbert space basis as the original ones.

But the fermions are not massless. As we have just learned, the masses are generated due to nontrivial Yukawa interactions, as in (12.53), etc. The point is that the fermion fields entering the gauge part of the Lagrangian and the fermion fields entering the Yukawa part of the Lagrangian are not the same — the two sets of the quark fields are rotated with respect to each other. And the same is true for the lepton fields. It is more convenient to express the Lagrangian in term of the fields participating in the Yukawa couplings. After the nonzero vacuum expectation value of ϕ is taken into account, these fields describe the particles with definite masses — the eigenstates of the Hamiltonian.

A related question that could be asked is why are we discussing only rotations of the left-handed fields, but not of the right-handed fields. The answer is that right-handed fields do not participate in flavour-changing gauge interactions. And the Lagrangian of electromagnetic interaction and the Lagrangian of weak neutral current interaction are invariant under flavour rotations of the right fields. One can do in this case what one cannot do in the left sector — redefine the fields and bring the rotation matrices (if they exist) to unity.

12.5 Back to the Zoo

Personally, I got a first glimpse of what the electroweak theory might be in 1972 or 1973 from the popular lecture of Okun in the Moscow Polytechnical Museum. He said that, at low energies, the effects of the weak interaction, including parity nonconservation, are weak, but they grow with energy and, for energies of several dozen GeV, become 100% relevant. At that time, this range of energies was still unexplored and Okun said that, when we reach it, we will enter a new fascinating world not at all similar to that we live in and are used to.

We have been there, in this new world, since at least 1983, when $W^\pm$ and Z were experimentally discovered at CERN. We will now discuss a little bit its phenomenology, what the beasts in the high-energy section of the electroweak zoo look like.

One of my first research papers was devoted to the effects of the weak interaction in e^+e^- annihilation. Consider the process $e^+e^- \to \mu^+\mu^-$. At low energies this is an electromagnetic process described by the diagram depicted in Fig. 5.4 with the formation and subsequent decay of the virtual photon. The corresponding amplitude involves the factor e^2. But there is another diagram where the virtual photon is replaced by the virtual Z boson (see Fig. 12.5). The contribution of the second diagram is $\sim g^2 E^2/(m_Z^2 - E^2)$, where E is the total centre-of-mass energy of the colliding electron and positron. Starting from $E \gtrsim 40$ GeV, these two contributions are of the same order.

The observed effects are the following:

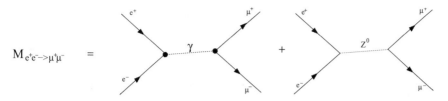

Fig. 12.5 Scattering $e^+e^- \to \mu^+\mu^-$ in electroweak theory.

- The differential cross section is no longer given by (5.7). It now involves an asymmetry: μ^- produced in the collision prefer to go in the direction of the *positron* beam.
- The final muons are now polarized even if the initial electron and positron were not.

There are similar effects in the process where, instead of $\mu^+\mu^-$ pair, a pair of quarks going over into hadron jets is produced. That was the subject of my own paper.

As was mentioned, the $W^\pm$ and Z were first observed in 1983 at CERN. This was done at the SPS proton accelerator.[16] The corresponding processes were

$$pp \longrightarrow W^\pm + \text{hadrons},$$
$$pp \longrightarrow Z + \text{hadrons}. \tag{12.59}$$

Now, Z may further decay into e^+e^- or $\mu^+\mu^-$ (these channels are especially easy to observe). W may decay into a charged lepton and neutrino. Both W and Z may also produce a pair of quark jets, the underlying processes being $Z \to \bar{u}u$, $W^- \to \bar{u}d$, etc.

Somewhat later, Z was observed in e^+e^- collisions in the process

$$e^+e^- \longrightarrow Z \longrightarrow \text{no matter what} \tag{12.60}$$

as a beautiful resonance. In contrast to the J/ψ resonance discussed in the previous chapter, the Z resonance is rather broad — its width is $\Gamma_Z \approx 2.5$ GeV (and the width of W is $\Gamma_W \approx 2.1$ GeV).

These widths are an order of magnitude larger than the characteristic widths of hadronic resonances.[17]

One of the channels of Z decay is into a couple of neutrinos, $Z \to \bar{\nu}_e\nu_e, \bar{\nu}_\mu\nu_\mu, \bar{\nu}_\tau\nu_\tau$. The neutrinos escape detection, but the partial rates of these decays,

---

[16]SPS means *Super Proton Synchrotron*. It accelerates protons up to ∼450 GeV. It is still operating, now used as the injector for the Large Hadron Collider.

[17]"How come", you may ask, "the strong interaction is stronger than the electroweak one and the strong widths (decay rates) should be larger than the electroweak ones!"

But everything is perceived in comparison. In our case, the comparison is with the resonance masses. The masses of W and Z are much larger than the characteristic hadron masses, while the *ratios* Γ_W/m_W and Γ_Z/m_Z are still small. Theoretically, these ratios are of order $G_F m^2_{W,Z}$.

$$\Gamma(Z \to \bar{\nu}_e \nu_e) = \Gamma(Z \to \bar{\nu}_\mu \nu_\mu) = \Gamma(Z \to \bar{\nu}_\tau \nu_\tau)$$

$$\approx \frac{G_F m_Z^3}{12\sqrt{2}\pi} \approx 170 \text{ MeV} , \qquad (12.61)$$

contribute to the total measurable and measured width. The experimental result coincides with the theoretical prediction, which means, in particular, that there exist exactly three light neutrinos in nature, not more. And that implies the existence of exactly three fermion generations — an experimental fact that does not have yet a theoretical explanation.

The privileged cage in our zoo is reserved for the Higgs boson. Because of its large mass,

$$m_H \approx 125 \text{ GeV} , \qquad (12.62)$$

it played hide-and-seek with the physicists for a long time. It was finally detected in experiment only in 2012, forty five years after its existence was predicted in 1967. I believe that this is the second longest time interval between the prediction and detection in the history of particle physics. (The laurel wreath in this competition indisputably belongs to the graviton. Gravitational waves were experimentally detected only in 2016, one hundred years after their existence was predicted by Einstein. We will talk about this in more detail in Chap. 15.)

The neutral Higgs field $H(x)$ describes the fluctuations of the absolute value $|\phi|$ of the scalar field around its vacuum value v. As follows from (12.53), etc., it couples to the fermions. The strength of these couplings is determined by the Yukawa constants h_e, etc. The latter are proportional to the fermion masses. Or rather the fermion masses are proportional to the h_f [cf. (12.54)]. Anyway, the Higgs boson couples stronger to heavy quarks (also to heavy leptons, but quarks are heavier!)

The first idea that comes to mind is to look for the Higgs boson by its decay into a heavy quark–antiquark pair. However, for the heaviest quark, the t quark with mass $m_t \approx 170$ GeV, this is not possible — the Higgs boson weighs less than 340 GeV (incidentally, that was known from certain indirect evidence even before it was discovered). The next best option is the decay $H \to \bar{b}b$, with the production of a pair of quark jets. Indeed, in about a half of the cases the Higgs boson follows this channel of decay. But this process is experimentally inconvenient to study due to large background — there are many processes where a pair of quark jets is produced by other mechanisms.

A much cleaner signal is provided by the decay into two photons. The amplitude of this process is described by the diagrams in Fig. 12.6. They are not tree diagrams (the Lagrangian of the Standard Model does not include the vertex $H\gamma\gamma$), but diagrams involving a W boson loop or a t quark loop. Another convenient channel, described by similar diagrams is $H \to 2Z$, with each *virtual* Z^{18} further decaying into a e^+e^- or $\mu^+\mu^-$ pair. One then searches for the peaks in the distributions over the invariant masses of two photons or of four charged leptons.

---

[18] The mass of H is below the threshold for creating two real Z bosons.

Fig. 12.6 The process
$H \to \gamma\gamma$.

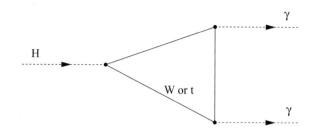

In spite of the fact that the background in the channels $H \to 2\gamma$ and $H \to 2Z \to l^+l^-l^+l^-$ is less that in the other channels, it is still a difficult experiment, much more difficult than, say, the experiment where J/ψ was discovered. The discovery of the Higgs boson was a result of the concerted efforts of many thousands of people.

I am not at all exaggerating. First of all, to be able to see the Higgs boson, the Large Hadronic Collider had to be built. The main detectors, ATLAS and CMS are also rather huge (many meters; some factories are smaller) and complicated physical devices. There are about 2900 members in the ATLAS collaboration and about 3800 members in the CMS collaboration. *All* the members of a given collaboration are the coauthors of the scientific articles published by it. Their names are all present in the print, which takes sometimes more journal space than the article itself.

Things have now changed a lot since the time when the fascinating world of elementary particles had just started to be explored and the necessary equipment could be assembled on a single laboratory table. (This was the way the electron was discovered in 1897 by Joseph Thomson...)

Going back to the background, it is also present in both "gold-plated" channels, $H \to 2\gamma$ and $H \to l^+l^-l^+l^-$, and it is still very significant. To give the reader a flavour (or rather a taste, in the reference system of this book) of how the experimentalists work today, I've reproduced in Fig. 12.7 the plot, drawn by the ATLAS collaboration, showing their original data and the way they were treated.[19]

Let me give some explanations.

- In this plot, the data, collected at total center-of-mass energies of the colliding protons $\sqrt{s} = 7$ TeV and $\sqrt{s} = 8$ TeV, were combined together. $\int L dt$ is the so-called *integrated luminosity* — the number of analyzed events is proportional to this quantity.
- The upper plot shows raw data. The dashed line is the smooth extrapolation and describes the background (*Bkg*). In the region around $m_{2\gamma} \sim 127$ GeV, the count is above the background, signaling the presence of the resonance. The solid line is a fit that takes both the signal (*Sig*) and the background (*Bkg*) into account.
- In the lower plot, the smooth background was subtracted.
- The reader is invited to compare this plot with the plot in Fig. 11.4 exhibiting the much more pronounced J/ψ resonance and feel the difference. Life is definitely

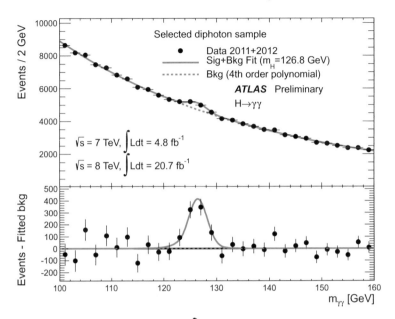

Fig. 12.7 The distribution over 2γ invariant mass, measured in ATLAS.

harder in the third millennium, but the scientists are trying hard to meet new challenges.

12.6 Neutrino Oscillations

The stars, including our Sun, burn due to the reactions of nuclear synthesis. The main fuel of a star is hydrogen. The protons, squeezed by gravity and heated by Gay–Lussac's law, collide with one another. The collisions are inelastic, and finally nuclei of ^{4}He are formed.[20]

At the intermediate stage, nuclei of deuterium are produced. The principal source of the solar energy is the process

$$p + p \longrightarrow d + e^{+} + \nu_e + 0.42\,\text{MeV}\,. \tag{12.63}$$

We see that in every such reaction an electron neutrino is liberated and, due to its very small cross section, leaves the Sun unperturbed and flies off into space. This neutrino flow meets our Earth. The vast majority of neutrinos fly through us without

---

[20]Then ^{4}He starts to burn in its turn, but for the Sun at the present stage of its evolution that is not relevant.

scattering,[21] but a small fraction of them comes into collision. This can be observed. Knowing the cross section from theory, one can measure the solar neutrino flux.

I have just described the scheme of the so-called *Homestake experiment* headed by Raymond Davis. To get rid of the "ordinary" cosmic ray background, Davis placed a tank of liquid tetrachloroethylene at a depth of $\sim$1500 m underground in the Homestake Gold Mine in South Dakota. The liquid was enriched with chlorine. The solar neutrinos induced the reaction

$$\nu_e + {}^{37}\text{Cl} \longrightarrow {}^{37}\text{Ar} + e^- . \tag{12.64}$$

Davis measured the rate of this reaction, deduced the neutrino flux and got puzzled: the flux was at least 2 times or rather even 3 times *lower* than the theoretical predictions.

The puzzle is resolved if one takes into account the effect of neutrino oscillations. There are three fermion generations and three neutrino species, but it is much easier to understand what happens assuming the existence of only two neutrinos — the electron and the muon one. For two generations, there is only one mixing angle, the analogue of the Cabbibo angle, which one can call the *Pontecorvo angle* θ_P.[22]

We now change the notation a little bit, compared to (12.58), and call the "electron neutrino" state ν_e (without "prime") the state created in the core of the Sun together with the positron in the reaction (12.63). On the other hand, the state created by a muon in processes like $\mu^- p \to \nu_\mu n$ will be called the "muon neutrino".

The key observation is that neither $|\nu_e\rangle$ nor $|\nu_\mu\rangle$ are eigenstates of the Hamiltonian. The eigenstates (call them $|\nu_1\rangle$ and $|\nu_2\rangle$) have definite masses m_1 and m_2. The states $|\nu_e\rangle$ and $\nu_\mu\rangle$ represent mixtures

$$
\begin{aligned}
|\nu_e\rangle &= \cos\theta_P |\nu_1\rangle + \sin\theta_P |\nu_2\rangle \\
|\nu_\mu\rangle &= -\sin\theta_P |\nu_1\rangle + \cos\theta_P |\nu_2\rangle .
\end{aligned}
\tag{12.65}
$$

So, the state $|\nu_e\rangle$, a superposition of the states $|\nu_1\rangle$ and $|\nu_2\rangle$, is created in the reaction (12.63) at $t = 0$. What happens next? Well, the neutrinos start to move. For an exact analysis, one should study the evolution of the localized wave packets for ν_1 and ν_2. (At the moment of creation, they have the same form and then they evolve in a different way, due to different Dirac Hamiltonians, including different masses.) To derive the correct result, it is sufficient, however, to look at the asymptotic plane-wave solutions

$$
\begin{aligned}
|\nu_1\rangle(\boldsymbol{x}, t) &= e^{i(\boldsymbol{p}_1 \cdot \boldsymbol{x} - E_1 t)} |\nu_1(0)\rangle , \\
|\nu_2\rangle(\boldsymbol{x}, t) &= e^{i(\boldsymbol{p}_2 \cdot \boldsymbol{x} - E_2 t)} |\nu_2(0)\rangle .
\end{aligned}
\tag{12.66}
$$

---

[21]The cross section of the interaction of a neutrino with energy $E \sim 1$ MeV with ordinary matter is at the level $\sigma \sim 10^{-44}$ cm^2. The corresponding mean free path of neutrinos is $\lambda \sim 1/(n\sigma)$, where n is the matter density. It is estimated at the level $\lambda \sim 10^{20}$ cm ≈ 100 light years.

[22]Bruno Pontecorvo, an Italian theorist, who spent the second part of his life in the Soviet Union, predicted this phenomenon back in 1957.

Fig. 12.8 Oscillations in the probability for an electron neutrino to keep its nature. The lower bound is $P_{min} = \cos^2 2\theta_P$.

We may here set $x = (L, 0, 0)$, $t = L$, but take into account the small differences between the energies and momenta, $E_{1,2} - |p_{1,2}| \approx m_{1,2}^2/(2E)$. (We assumed that $m_{1,2} \ll E$. For solar neutrinos, this is an experimental fact.)

This way or another, the phase factors for $|\nu_1\rangle$ and $|\nu_2\rangle$ are *different*. If the difference between the neutrino masses is small, the mismatch between the phases of two particles is also small for short times and distances. But if the latter grow, this mismatch becomes significant. As a result, the electron neutrino state created at $t = 0$ evolves at finite t into a state which is no longer a pure $|\nu_e\rangle$ state, but has a nonzero projection on $|\nu_\mu\rangle$. Muon neutrinos, which were not there at the moment of creation, pop out like rabbits from the hat of a magician!

Let us evaluate the probability for a neutrino to keep its electron nature, as a function of the distance L it travelled. We derive:

$$P_{\nu_e \to \nu_e}(L) = \left| \langle \cos\theta_P\, \nu_1(0) + \sin\theta_P\, \nu_2(0) | \cos\theta_P\, \nu_1(L) + \sin\theta_P\, \nu_2(L)\rangle \right|^2$$

$$= 1 - \sin^2(2\theta_P)\sin^2\left(\frac{\Delta m^2 L}{4E}\right), \tag{12.67}$$

where $\Delta m^2 = m_2^2 - m_1^2$. This probability oscillates with L, as shown in Fig. 12.8. If the period of these oscillations is much smaller than the annual variation of the distance between the Earth and the Sun and the Pontecorvo angle is close to 45°, so that the amplitude of the oscillations in Fig. 12.8 is close to the maximal possible one (both conditions are experimentally fulfilled), then, on the average, only about half of the neutrinos retain their electron nature and can produce an electron in the reaction (12.64) by the time they reach the Earth.[23] And that is exactly what Davis observed!

---

[23] Our rabbits, the muon neutrinos generated due to oscillations, would be eager to produce muons, but they do not have enough energy for that.

One must say that, for a long time (the Homestake Mine installation started to acquire data in the late nineteen sixties), Davis' results were accepted with scepticism by the community. People did not understand why the neutrino flux was so low and suspected a mistake. In many cases such scepticism is grounded — there were, there are and there will be wrong experiments; *errare humanum est*. But not in *this* case. The Homestake experiment was absolutely correct. In 2002, after its results were confirmed in an independent experiment, Davis was awarded the Nobel Prize.

Neutrino oscillations have now been observed in other contexts. They were seen in nuclear reactor and accelerator experiments (with the distances between the neutrino source and detector ranging from dozens of meters to hundreds of kilometers). One also has to mention the important *Super-Kamiokande* experiment, where the oscillations of atmospheric neutrinos were discovered. The atmospheric neutrinos are the muon neutrinos produced in the upper atmosphere by cosmic rays. And it was found that some of them "disappear" on their way to the terrestrial detectors. Of course, they do not really disappear, but are rather converted into ν_τ's, which escape detection. These new experiments were distinguished by the Nobel Prize in 2015.

A combined analysis of all experiments shows that the masses of the different neutrinos are very close to one another:

$$|m_2^2 - m_1^2| \approx 7.5 \cdot 10^{-5}\,\mathrm{eV}^2 \,,$$
$$|m_3^2 - m_1^2| \approx |m_3^2 - m_2^2| \approx 2.5 \cdot 10^{-3}\,\mathrm{eV}^2 \,. \tag{12.68}$$

In oscillation experiments, one does not directly measure the masses, but only the mass differences. Thus, the former are not known precisely. On the other hand, completely different astrophysical arguments provide the following bound for the sum of the masses of all neutrino species:

$$m_1 + m_2 + m_3 \lesssim 0.12\,\mathrm{eV} \,. \tag{12.69}$$

12.7 Complaints and Outlook

12.7.1 Complaining

Having read the book until this point, the reader has got a general idea of what the Standard Model of the strong, weak and electromagnetic interactions is, what its theoretical foundations are and what observed phenomena it describes. But may I ask you: *do you like it?*

You know, if you give a positive answer to this question, I will doubt your sincerity. The monumental Standard Model is *too complex* to be attractive and beautiful. It involves a lot of parameters whose origin and meaning are now obscure. At the present level of understanding, they are just God-given numbers. The total number of such parameters is at least 26. Namely, there are 12 Yukawa coupling constants $h_e, h_u,$

Fig. 12.9 A divergent
contribution in the Higgs
mass.

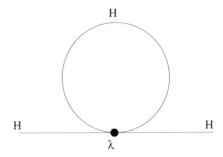

etc., four parameters in the CKM quark mixing matrix, four more parameters in the
PMNS lepton mixing matrix, the constants g and g', the scalar vacuum expectation
value v and the Higgs coupling constant λ. In the QCD sector, we have Λ_{QCD} and
an extra parameter θ_{QCD}.[24]

This makes 26. In addition, there could be Majorana mass terms for the right-
handed neutrinos, about which we know strictly nothing.

But besides being unaesthetic (all beautiful things are simple!), the Standard
Model also suffers from certain *logical* inconsistencies. The most troublesome problem is probably the mass of the Higgs boson. At the tree level, it is given by the
relation (12.27), but if one tries to calculate a 1-loop correction and draws a graph
like that in Fig. 12.9, one finds that the correction involves a quadratic divergence,

$$m_H^2[\text{tadpole}] \; = \sim \lambda \int \frac{d^4 p}{p^2} \sim \lambda \Lambda_{UV}^2 \,. \tag{12.70}$$

This divergence is not cancelled by other similar quadratically divergent 1-loop
contributions.

Our theory is renormalizable. That means that one can *absorb* this divergence in
the bare Lagrangian. In other words, one can add to the bare Lagrangian, the starting
point of all discussions, a quadratically divergent term $\propto \Lambda_{UV}^2 \phi^* \phi$ that would be
cancelled by the loop contributions. Well, it is not impossible to do that and to get
rid in this way of all divergences in perturbative calculations. But you should agree:
this procedure looks unnatural and ugly.

There is a related problem. The Lagrangian of the Standard Model involves the
Abelian gauge sector with the coupling constant g' (and also the Higgs sector with
the coupling constant λ and the Yukawa sector with the constants h_f). All these
constants grow with energy. As was discussed in Chap. 5 and at the end of Chap. 8,
such theories are not internally consistent — one can treat them only perturbatively

---

[24]This is the so-called QCD vacuum angle, the constant multiplying a possible structure
$\sim \epsilon^{\mu\nu\alpha\beta} G_{\mu\nu}^a G_{\alpha\beta}^a$ in the QCD Lagrangian. We did not discuss it before and will not do so in future.
Experimentally, θ_{QCD} is very close to zero (only the upper limit, $|\theta_{QCD}| \lesssim 10^{-10}$ is known).
Nobody understands why.

and at not too high energies, whereas their nonperturbative definition is absent: a continuum limit of the lattice path integrals does not exist.

12.7.2 Looking Through the Supersymmetric Window

To handle these two setbacks, we have to go beyond the Standard Model. An effective way to cope with the quadratic divergences in the Higgs mass is to suppose that the true fundamental theory is supersymmetric and each field has a superpartner of the opposite statistics. For example, besides the photon, a supersymmetric theory involves the *photino* (which is a fermion). Besides the quarks there are the scalar quarks (or *squarks*), which are bosons, etc.

The world we see around us does not seem to be supersymmetric — there is no symmetry between the bosons and fermions. There are no massless photinos; and scalar quarks, if they exist, have masses that are much larger than the masses of their quark partners. That means that supersymmetry, if it is there, must be *broken*, so that the masses of all superpartners are at least several hundred GeV or maybe more.

But if supersymmetry is not seen experimentally, why on Earth should one discuss it? Would it not contradict Occam's razor principle — do not introduce new notions unless you really need them?

There are three answers for that.

- Occam's razor notwithstanding, supersymmetry is a really *beautiful* concept. I will try to convince you about that in Chap. 14, while serving one of the desserts.
- It kills the quadratic divergence in the Higgs mass mentioned above. The divergent contribution of the graph in Fig. 12.9 is cancelled by the graph involving a *higgsino* loop.[25] The divergence in the graphs involving quark loops is cancelled by the corresponding graphs with the squark loops.
- Supersymmetry also helps us to handle the second problem, the problem of the growth of the Abelian gauge coupling.

Let us discuss the latter point in more detail. A usually prescribed remedy for this growth is the assumption that the strong, weak and electromagnetic interactions all take their origin in a *unified* interaction, which shows up at very high energies but is not seen at the energies accessible to accelerators, due to the fact that the large gauge symmetry on which the unified theory is based is broken. We assume that it is spontaneously broken by the Higgs mechanism, in roughly the same way that the underlying electroweak symmetry $SU(2) \times U(1)$ is broken down to $U_{EM}(1)$ in the Standard Model.

This conjecture (of *Grand Unification*) is supported by the following observation. As was mentioned, g' grows with energy. But the non-Abelian couplings $g_{su(2)}$ and

---

[25]The higgsino is a superpartner of the Higgs particle.

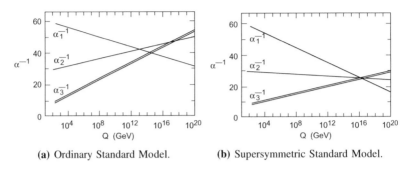

(a) Ordinary Standard Model.   (b) Supersymmetric Standard Model.

Fig. 12.10 Running couplings.

g_{strong} fall off with energy due to asymptotic freedom. We show how the three different gauge couplings *run* in Fig. 12.10.[26]

The reader is invited first to close his right eye and look at the left-hand plot. It shows how the effective couplings

$$\alpha_1^{-1} = \frac{4\pi}{g'^2}, \qquad \alpha_2^{-1} = \frac{4\pi}{g^2}, \qquad \alpha_3^{-1} = \frac{4\pi}{g_{strong}^2}, \tag{12.71}$$

depend on the logarithm of the characteristic energy Q in the Standard Model.[27]

We see that the three constants, rather different at $Q \sim 100$ GeV, approach one another as the energy grows and become *roughly* equal at $Q \sim 10^{15}$ GeV. This is the characteristic scale of Grand Unification.

We may assume that the Grand Unified Theory (GUT) is based on a large gauge group G with a subgroup $SU(3) \times SU(2) \times U(1)$. The minimal possibility is $G = SU(5)$. Such a GUT should include a large number of gauge bosons — at least 24, as would be the case for $SU(5)$. When G is broken by the Higgs mechanism down to $SU(3) \times SU(2) \times U(1)$, the bosons associated with the generators of the unbroken group (the gluons, **W** and **B**) stay massless, while the other gauge bosons (call them X bosons[28]) acquire large masses of order 10^{15} GeV.

X bosons convert quarks into leptons and the other way round (see Fig. 12.11). That means, in particular, that the proton is no longer stable. It can decay into $\pi^0 e^+$ or $\pi^+ \bar{\nu}_e$. However, as the mass of X bosons is very large, the probability of the proton decay is very small.

---

[26]From [*M. Peskin, hep-ph/9705479, published in the Proceedings of the 1996 European School of Particle Physics.*].

[27]We have plotted the inverse couplings rather than the couplings themselves simply because the dependence of α^{-1} on $\ln Q$ is very simple — at one-loop order, it is just a linear function [see (11.22) and (5.13)].

[28]Besides the vector X bosons, the GUT counterparts of W and Z, there are scalar X bosons, the counterparts of the ordinary Higgs boson. We will meet the latter monsters again at the end of Chap. 14.

Fig. 12.11 Quark-lepton
transitions.

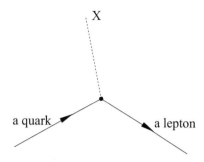

But everything is perceived in comparison. One can evaluate this probability and derive, in particular, that the minimal nonsupersymmetric GUT with $G = SU(5)$ contradicts experiment. The proton lifetime would be equal in this case to $\tau_p \sim 10^{28}$ years, whereas the current experimental bound is $\tau_p > 10^{34}$ years. Other unification groups do not save the situation. They all predict a very large, but not large enough, proton lifetime. In addition, the three lines in the left-hand plot do not intersect in *exactly* the same point, as the GUT predicts. We should conclude that, in spite of their attractive features, the simple-minded nonsupersymmetric GUT models do not describe Nature.

And now the reader is allowed to open his second eye and look at the right-hand plot. S/he will see that, in the minimal supersymmetric extension of the Standard Model, the runnings of the couplings are modified such that the lines cross at the same point! The scale of unification is somewhat larger than in the nonsupersymmetric models,

$$M_{\text{SUSY GUT}} \approx 10^{16} \, \text{GeV}. \tag{12.72}$$

The larger unification scale means larger masses of the X bosons. Because of this and for some other reasons, the probability of proton decay is further suppressed, and there is no longer any contradiction with experiment!

People rightly consider these three circumstances:

(i) the fact that supersymmetry kills quadratic divergences in the Higgs mass,

(ii) exact matching of the three constants at the unification scale,

(iii) a possibility for the proton to live long enough to please experimentalists,

as a strong indication that supersymmetry is really there.

However, it does not allow one to diminish the number of the free parameters and make the model aesthetically attractive. On the contrary. Supersymmetric models include additional fields and particles, we do not know now at what scale supersymmetry is broken and, as a result, one has to deal with at least a hundred *new* parameters. Evidently, this is not a satisfactory situation.

But we have now reached the frontier of our knowledge, and can only make guesses on what happens beyond.

Part V
Trou Normand

Chapter 13
The Human Dimension

When I had just started to write this book, I told one of my French colleagues about my project. I listed its different future parts: *Hors d'Oeuvres, Entrées, Dessert, …,* and he interrupted me, "What about the *Trou Normand*?"[1] I immediately realized that that was a good idea. The preceding chapters were full of rather nontrivial, high-calory scientific information. The reader needs a rest.

Thus, I will stop writing formulas for a while and tell you instead a couple of stories about the *people* who wrote them for the first time, about some physicists of the last century who formed our present knowledge of the theories of fundamental interactions, which I have tried to share with you. Of course, it is impossible to write about the hundreds and thousands of scientists who made important contributions in this quest. My selection is necessarily subjective and personal.

13.1 Richard Feynman

If you ask me which physicist contributed the most in the development of quantum field theory during the last century, I would hesitate — there were many. But if you insist, I will probably mention Feynman.

His best-known achievement is the invention of a very convenient language, the language of Feynman diagrams, in which physicists have mostly spoken since that time. This allowed them to perform their calculations (which were mayhem before) in a fast, efficient and visual way. I now invite you to reread a long quotation from his Nobel lecture on pp. 63–64.

But Feynman also did many other things. I will mention only three.

---

[1] As I've already mentioned in the Introduction, it is a restive digestive pause between the main dish and the dessert.

© Springer International Publishing AG 2017   245
A. Smilga, *Digestible Quantum Field Theory*,
DOI 10.1007/978-3-319-59922-9_13

1. He was one of the people who established the V–A structure of the charged weak current (see Chap. 12 for explanations).
2. He introduced the notion of path integrals in quantum mechanics and quantum field theory. We discussed that in Sect. 8.3.
3. He performed the first attempt to calculate loop graphs in non-Abelian gauge theories and in gravity. He anticipated the appearance of the ghosts, whose accurate description was later given by Faddeev and Popov (see the discussion on p. 182).

In addition, Feynman is broadly known in the United States as the person who first understood the cause of the Challenger disaster.[2]

But Feynman was not only a brilliant physicist. He was also a brilliant personality. He has written two books of memoirs: *Surely You're Joking, Mr. Feynman!* and *What Do You Care what Other People Think?* Both are very interesting, but I especially recommend reading the first one. Here are some stories from this book.

13.1.1 Safecracker

During the war, young Feynman worked in Los Alamos and participated in developing nuclear weapons. He was the head of a computational group (at that time there were no real computers yet — all calculations were done by humans equipped by overgrown mechanical adding machines — and Feynman's job was to organize their work efficiently). The Los Alamos activities were obviously top secret and the documents were stored in locked safes. Feynman developed a hobby — to *pick* the locks in the offices of his colleagues. He achieved a rather high proficiency in this art. Standard tumbler locks were just peanuts for him. He could pick any of them with a paper clip and a screwdriver.

... To demonstrate that the locks meant nothing, whenever I wanted somebody's report and they weren't around, I'd just go in their office, open the filing cabinet and take it out. When I was finished, I would give it back to the guy: "Thanks for your report."

"Where'd you get it?"

"Out of your filing cabinet."

"But I locked it!"

"I know that you locked it. The locks are not good."

When more sophisticated filing cabinets with combination locks arrived, Feynman considered it as a challenge. To understand how they work, he took apart the one in his own office. These locks resisted paperclips and screwdrivers. To open them, one needed to know three 2-digit numbers of a secret code and dial them by turning wheels. At first, Feynman did not know how to handle them. But then he found out that he could determine the two last numbers of the combination by turning the

---

[2]A rubber joint which became less resilient due to unusually low temperatures in the morning of the launch. As a result, the joint failed.

wheels of an *open* safe back and forth and listening to weak clicks.[3] So whenever Feynman went into a colleague's office, he unobtrusively fiddled with the lock of an open safe, learned the two last numbers and after coming back to his office, noted them in his own secret notebook.

Feynman acquired a reputation as a skilled safecracker. Thus, if somebody was absent, but people urgently needed some particular secret document locked in his safe, they asked Feynman to open it. If Feynman knew the two last numbers of the code, he agreed, took a large box of tools (which he actually did not need), went to the office, locked the entrance door, so that nobody could see what he was doing, quickly opened the safe by simply trying all possible first numbers that he did not know (actually, he needed only about twenty tries — that safe also opened if one dialled the number that differed by one or two from the correct one) and then spent another a quarter of an hour inspecting the wallpaper design and shaking his toolbox from time to time to give the impression that he was working hard and that the assigned task was not so easy...

I skip other related amusing stories, addressing the reader to the special "safe-cracker" chapter in Feynman's book.

13.1.2 Drum Player

Feynman had a lifelong hobby — he was a good amateur percussionist. His preferred instrument was the *bongo* — a special small drum popular in Cuba, but he also played other instruments. He started to play drum frequently in Los Alamos, where there were not many other distractions — no theatres, no concert halls... I am now giving him the floor:

... Sometimes I would take the drums with me into the woods at some distance, so I wouldn't disturb anybody, and would beat them with a stick, and sing. I remember one night walking around a tree, looking at the moon, and beating the drum, making believe I was Indian.

One day a guy came up to me and said, "Around Thanksgiving you weren't out in the woods beating a drum, were you?"

"Yes, I was," I said.

"Oh! Then my wife was right!" Then he told me this story:

One night he heard some drum music in the distance, and went upstairs to the other guy in the duplex house that they lived in, and the other guy heard it too. Remember, all these guys were from the East. They didn't know anything about Indians, and they were very interested: the Indians must have been having some kind of ceremony, or something exciting, and the two men decided to go out to see what it was.

---

[3]In his book Feynman explained how he did it, but frankly I did not understand much. Anyway, there is not much practical point — I guess that the particular safes that Feynman learned to crack are not in use any more.

As they walked along, the music got louder as they came nearer, and they began to get nervous. They realized that the Indians probably have scouts out watching so that nobody could disturb their ceremony. So they got down on their bellies and crawled along the trail until the sound was just over the next hill, apparently. They crawled up over the hill and discovered to their surprise that it was only one Indian, doing the ceremony all by himself — dancing around a tree, beating the drum with a stick, chanting. The two guys backed away from him slowly, because they did not want to disturb him: He was probably setting up some spell, or something.

They told their wives what they saw, and the wives said, "Oh, it must have been Feynman — he likes to beat drums."...

After leaving Los Alamos after the war, Feynman seemed to lose interest in safes — he did not encounter any more really exciting safes where important things were stored. But he continued to play drums. During his trip to Brazil in 1951, where he taught physics at the University of Rio de Janeiro, he joined a samba club in the vicinity of Copacabana beach and learned to play a special Brazilian instrument called the *frigideira*, "... a toy frying pan made of metal, about six inches in diameter." Feynman was proud to acquire enough skills to become probably the best "frigideiraist" in this particular club. He participated with the band of his club in the competition between different Copacabana samba clubs, and they won this competition.

In Caltech (California Technological Institute), where Feynman worked since 1951 until the end of his life, he often played bongos by himself and together with his friend Ralph Leighton.[4]

Gradually, he acquired sufficient skills in this art to play together with professionals. He and Leighton were asked to accompany some modern Carribean-style ballet in San Francisco, where there was no other music but percussion. Feynman had a problem, however; he had never learned to read music, always played by ear, but this time he had to learn his percussion part *just so*, to avoid a mismatch with the dancers. He managed to do it, and the performance was finally a success!

Later, the choreographer of this ballet entered it in a contest for choreographers first in the United States (where they finished first or second) and then in an international competition in Paris. The winner was a Latvian group who danced a classical ballet. And the Caribbean ballet was rated the second.

... She went to the judges afterwards to find out the weakness in her ballet.

"Well, Madame, the music was not really satisfactory. It was not subtle enough. Controlled crescendoes were missing..."

---

[4]His former student who followed Feynman's course in Caltech, made notes and is a co-author of the *Feynman lectures* together with Feynman and Matthew Sands.

13.1.3   Artist

Another hobby of Feynman's was art. He learned drawing in his mid-forties, when he worked at Caltech. Feynman tells the following story. He had a friend, an artist called Jerry Zorthian. They had a philosophical discussion about the comparative value of art and science for humanity, but soon agreed that the discussion was difficult, because at that time Feynman knew nothing about art and Zorthian knew nothing about science. Each of them was sorry about that and expressed a sincere and honest desire to learn more. They agreed to teach each other.

Feynman accepted these two tasks — of a student and of a teacher — quite seriously. To explain Faraday's law to his friend, he made a little coil of wire and attached to it an iron nail on a piece of string. When the current went through the coil, the nail swung into it. The artist looked and said, "Wow, it's just like making love!" But he managed to learn neither Faraday's law, nor other laws of physics in their mathematical form...

But Feynman's studies were more successful and efficient. His friend turned out to be a pretty good art teacher, better than a physics student. Feynman made progress. Soon he took other courses in drawing: first, by correspondence and then he joined a drawing class in the Pasadena Art Museum. The students of this class drew different things, including the human body — nude models helped them to learn this. At first, it was difficult for Feynman.

... I started to draw the model, and by the time I'd done one leg, the ten minutes (that was the time for which the model usually posed) *were up, I looked around me and saw that everyone else had already drawn a complete picture, with shading in the back — the whole business.*

But finally he *learned* drawing. Not understanding much about art myself, I cannot judge whether he became a good artist or just a mediocre one. Personally, I first saw one of the Feynman's pictures (a carefully drawn beautiful woman's head) in Princeton, in the office of one of my colleagues, and liked it. Many sketches (and photos!) of Feynman can be found in the book *The Art of Richard P. Feynman: Images by a Curious Character* written by his daughter Michelle Feynman (the reader can also easily find many images on Internet). Please take a look and decide yourself!

Feynman took the pseudonym *Ofey*, the name blacks use for "whitey". In addition, the three first letters of his name made their way into the pseudonym.[5] He did so to ensure that the people would not know that he was a famous physicist, a Nobel Prize winner and would not think, "Wow, a physics professor can draw too. How wonderful!", but judged the quality of his pictures as they were.

Feynman (or rather *Ofey*) managed to sell, I guess, a couple of dozen of his pictures. It goes without saying that the money he got for them was not the main source of his income, but he rightly writes that the fact that people were prepared to

---

[5]Feynman also writes that it originally came from the French expression "au fait" suggested by one of his friends. But then he writes that "au fait" means "it is done" in English, whereas it is rather "in fact" or "by the way". Actually, neither "it is done", nor "by the way" seems to be a sensible choice for a pseudonym, and this part of the story is still not completely clear to me.

pay for his pictures meant that they genuinely liked them. And naturally, Feynman was happy about it. Once Ofey was even honoured by a personal artistic exhibition — a "one-man show".

I skip an interesting story how Ofey drew a picture for a private massage parlour — see Feynman's book for that, but I want to mention another of his pictures that he called *Madame Curie observing the radiations from radium*. It was a picture of a naked woman illuminated by a pointlike source of light, giving an interesting shadow pattern. This picture aroused the anger of activists of the feminist movement, "How come. Professor Feynman insults the famous woman scientist by representing her nude. This is unacceptable!.." Feynman even had some troubles because of that, but not so serious ones...

13.2   Lev Landau

One of the lesser-known (and not quite scientific) ideas of Landau was the logarithmic "genius scale". Each theorist (in physics) was attributed a class number or a rank, equal up to an additive constant to the decimal logarithm of his impact in science, taken with a negative sign. Thus, a physicist in the first class had ten times the impact of someone in the second class and so on.

This classification was never published, and one has to rely in this case on the recollections of his disciples, which do not completely coincide with one another. But all agree that Landau attributed rank 0 to Newton and rank 0.5 to Einstein. Several theorists of the 20th century were placed in the first class. According to one of the variants, they were: Bohr, Bose, de Broglie, Dirac, Heisenberg, Schrödinger and Wigner[6] — the creators of quantum mechanics. By the end of his life, Landau had added Feynman to this list, but nobody else. According to Landau's logic, the impact of a physicist of the 5th class was $\sim 10^{-4}$ in Bohr units, which Landau considered being undistinguishable from zero (but one can, of course, argue this point). Landau called such rank-5 physicists "pathologists".

In the nineteen thirties and forties, he modestly attributed to himself the rank 2.5. By the end of his life, he agreed that he was a second-class physicist. But if you ask my personal opinion, Landau was probably inferior by *half* of a class to Bohr or Dirac or Feynman, but had essentially higher rank than Bose or de Broglie.

A distinguishing feature of Landau as a scientist was his universality. He made fundamental contributions in *all* branches of theoretical physics, from hydrodynamics to the theory of the weak interactions. He received his Nobel Prize in 1962 *for his pioneering theories for condensed matter, especially liquid helium.* Indeed, the

---

[6]Some of these names may be new for the reader. Sathyendranath Bose established simultaneously with Einstein the Bose–Einstein statistics for the photon and other particles with integer spin, see p. 143. Louis de Broglie was the first to understand that not only the photon, but also the electron and other elementary particles have both particle and wave nature; see p. 10. Eugene Wigner was the first to understand the role of symmetry in atomic spectra and introduced the methods of group theory into physics.

Fig. 13.1 Landau.

theories of superfluid helium, both ^{4}He and ^{3}He[7] are probably his most spectacular achievements.

The effective quantum field theories of helium that Landau constructed were nonrelativistic. But Landau also made important contributions to relativistic quantum field theories, the main subject of our book.

1. As we mentioned, the Ginzburg–Landau model that describes superconductivity gave the first example of the Higgs mechanism when a gauge field acquires mass due to interaction with a scalar vacuum condensate.
2. Landau, Abrikosov and Khalatnikov derived the asymptotic formula (5.13) for the electron charge and were the first to understand that QED was not internally consistent.
3. Landau introduced the notion of CP symmetry in particle physics. He conjectured that this symmetry is exact, but that was found later not to be the case, though the effects of its breaking are weak.

But Landau was not only a scientist. He was also a Teacher. He created in the Soviet Union a brilliant *school* of theoretical physics. Landau died in 1968, and I had no chance to know him personally. But I know (and, alas, *knew*) many of his disciples, heard their recollections about Landau, which were not all published. I would like to share with my reader a few interesting things concerning his biography and personality.

---

[7]The physics of ^{3}He is richer and more tricky. In contrast to the atoms of ^{4}He, ^{3}He atoms carry a half-integer spin and are fermions. They cannot form a Bose condensate bringing about superfluidity right away. It is the *pairs* of ^{3}He atoms that condense, much as the Cooper pairs of electrons condense in a superconductor.

13.2.1   Youth

Everybody knows the Chinese proverb about interesting times. Well, the 20th century was definitely quite interesting. And the Russian Empire and then the Soviet Union was an interesting place, more interesting than the United States. As a result, Landau's biography was more interesting than the biography of Feynman — that is why we are spending more time talking about it. Fortunately, it was not interesting enough to prevent him from doing theoretical physics, which he did with great passion for the greater part of his life. (In contrast to Feynman, Landau had no marked hobbies.)

Landau was born in 1908 in Baku (now Azerbaijan) in the family of an oil engineer. He displayed his exceptional mathematical abilities already in early childhood, was a real whizz kid and graduated from high school at the age of 13. That was already in 1921, after the Bolshevik revolution and civil war. During the war, Baku changed hands several times; at some point Landau's father was arrested and soon released. But it seems that all these dramatic events did not affect young Landau so much.

He spent 3 more years in Baku studying at the University there. In 1924 he moved to Leningrad, which was at that time the main scientific centre of the country, and joined the Leningrad State University. He finished his undergraduate studies there and entered the graduate school at the Leningrad Physical Technical Institute (LPTI) in 1927. Landau was a brilliant student and then a brilliant young researcher (his first scientific paper was published in 1926), but in Leningrad he was not the only one.

Among the members of the theorical group that Landau joined were Vladimir Fock, Matvei Bronstein (an expert in gravity whose fate was tragic: he was arrested and executed in 1937 during Stalin's purges), Dmitry Ivanenko (when the neutron was discovered in 1932, Ivanenko was the first to understand that atomic nuclei are made of protons and neutrons and not of protons and electrons as people thought in the twenties) and George Gamov (he later emigrated to the United States and is known for his prediction of the existence of the cosmic microwave background and for decrypting the genetic code). It was a good, tight-knit and merry company of young, very gifted people.

In 1929, Landau won a grant from the People's Commissariat for Education and left Leningrad for Western Europe. At that time, the USSR was not yet a closed society and such trips abroad were not uncommon. The situation changed in the middle of the thirties. In 1934 Pyotr Kapitsa, who worked in the Cavendish Laboratory with Rutherford and spent his summer vacations in the dacha of his family near Moscow, was suddenly denied permission to return to Great Britain and had to stay in the USSR, where he worked for fifty more years in the Institute of Physical Problems in Moscow, which he founded.

We will talk about Kapitsa and his institute later, but let us now go back to 1929. Landau made a grand European tour, including Berlin, Göttingen, Leipzig, Copenhagen, Cambridge, Zürich... He met Born, Heisenberg, Dirac, Pauli... Especially fruitful for Landau was the visit to Copenhagen. The Institute of Theoretical Physics on Blegdamsvej 17, founded and headed by Niels Bohr, was the main spiritual and intellectual centre for physics in Europe at that time. Bohr had a seminar there, which

Landau enthusiastically joined and became an active participant, meaning that, besides giving his own talks, he asked more questions at the talks of others than anybody else, except probably Bohr himself.

Landau worked a lot during his trip, even more than at home. His best-known paper of this period is the paper on the spectrum of an electron in a uniform magnetic field — he discovered the phenomenon that is now called *Landau diamagnetism*. This is important to understand the magnetic properties of materials, but still more important was the unravelled double degeneracy of the excited energy levels of the associated Hamiltonian. We now understand that this double degeneracy is nothing but *supersymmetry*; we will discuss this in detail in the next chapter.[8]

Landau made a strong impression on the greats of European physics, and it would have been easy for him to find a position in Europe, if he had wanted. He even got some offers, which he did not take up. He wanted to return back home, and did so in the spring of 1931. One should say that at that time Landau had genuine communist revolutionary views and was a good Soviet citizen. Well, indeed, by 1931 he had not seen anything but good from the Soviet regime and from the authorities. Troubles would come later...

13.2.2 Kharkov □ Landau's Minimum

Landau came back to the graduate school at LPTI and was soon offered a senior research position (the head of the theory department), though not in Leningrad, but in Kharkov, at the Ukrainian Physical Technical Institute, which had just been founded (that was in 1930, while Landau was travelling in Europe). It was established as a daughter institute of LPTI, and many physicists from Leningrad moved to Kharkov. Landau did research at UPTI and also taught physics at Kharkov University.

At that time, Kharkov rather than Kiev was the administrative capital of the Ukraine, and UPTI had become a strong scientific centre. Conferences and symposia were constantly being arranged there. Bohr, Pauli, Dirac, Fowler and many other famous physicists attended them. Not only Russians worked at UPTI, but also some western scholars. One can mention Fritz Houtermans, who emigrated to the USSR in 1935 and got a permanent position at UPTI.[9] Victor Weisskopf, a good theorist, who might have missed the Nobel Prize because he did not publish the results of a certain important and quite correct calculation, not being sure that it was correct, fled from the Nazis to Kharkov and seriously considered staying there for good, but finally he found a position in the US.

---

[8]Of course, Landau did not call it supersymmetry — the word did not exist yet. Landau can be compared in this respect to Mr. Jourdain, the *bourgeois gentilhomme*, who spoke prose without knowing what prose is.

[9]That was not a fortunate choice for him. He was arrested during the Great Purge in 1937 and spent a couple of years in the jail of the NKVD. Then, after the Soviet-German pact of 1939, he was turned over to Germany and was immediately arrested by the Gestapo there... But I am afraid that we are now losing our own narrative thread.

In Kharkov Landau acquired his first disciples. The school of Landau was born. The first four disciples were Kompaneets, Lifshitz, Akhiezer and Pomeranchuk. I write "disciple" rather than "student" or "pupil" all the time, because this word better characterizes the status of a member of Landau's school. Indeed, the latter carried some features of a medieval guild or a monastic order. To join it, a candidate had to pass a probation period of several years in the status of an apprentice.

In practice this meant the following. Landau demanded that people who wanted to work with him should first pass 8 to 11 difficult exams (*theoretical minimum*) on all branches of theoretical physics and the necessary maths. In its mature version, the list included:

1. Mathematics-I (integration, differential equations, vector and tensor analysis).
2. Classical analytic mechanics.
3. Classical field theory.
4. Mathematics-II (complex analysis, special functions).
5. Quantum mechanics.
6. Quantum electrodynamics (evidently, this exam was added to the list in the fifties, after QED was constructed).
7. Classical statistical physics.
8. Fluid mechanics and elasticity.
9. Electrodynamics of continuous media.
10. Quantum statistical physics.
11. Physical kinetics.

In the fifties and sixties the syllabus of each exam roughly corresponded to the contents of volumes in the Course of Theoretical Physics by Landau and Lifshitz — there is an obvious relationship between this fundamental work and Landau's minimum. But in the thirties this course was not yet written, there were no other good textbooks, and often an apprentice should have studied the original papers by Bethe, Pauli and others written in German or in English. And if you did not command these languages well enough, that was your problem.

Shortly before the dreadful car accident in 1962 (I will talk about it later), Landau made a list of all his disciples who passed all the exams. There were 43 names. The first half of this list involved a Nobel Prize winner (Abrikosov) and many other brilliant theorists who in the course of years acquired a high "genius rank". Since ~ 1955, Landau decided to examine personally only the maths skill of a candidate and delegated the task of examining other skills to his disciples. And for the second half of the list, the average genius rank went down... Ioffe (number 13 in the disciple list) writes in his memoirs:

... The examination proceeded as follows.

An aspiring student would call Landau and say: I would like to take an exam on such and such a course (the order was more or less arbitrary).

— OK, please come on a certain day and time.

When students arrived at Landau's appartement, Landau would ask them to leave all their books, notes, etc. in the hall and invite them into a small room with a round table, with a few pages of blank paper on it, and nothing else. Then Landau would formulate a problem and leave, but every 15 to 20 min he would reappear and look over one's shoulder.

If he was silent, then this was a good sign, but sometimes he would say "hmm" — this was a bad sign. I have no failed examination experience of my own. However, once, when I was sitting statistical physics, I started solving a problem in a way that Landau did not expect. Landau came, looked and said, "hmm". Then he left. In 20 minutes he came back, looked again and said "hmm" in an even more dissatisfied tone. At that moment Evgeny Lifshitz appeared, who also looked at my notes and shouted, "Dau[10], do not waste time, throw him out!" But Dau replied, "Let us give him another 20 minutes." During this time I got the answer and it was correct! Dau looked at the answer, looked again at my calculations and agreed that I was right. After that, he and Lifshitz asked a few easy questions, and the exam was over...

Since the death of Landau, the tradition of *theoretical minimum* has been kept alive by his disciples and the disciples of disciples. It is alive today. The students of MIPT (the Moscow Institute of Physics and Technology), who want to become theorists, are examined by the physicists working at Landau's Institute for Theoretical Physics.

13.2.3 Arrest □ The Bomb

In February 1937, Landau left Kharkov and moved to Moscow, to the new Institute of Physical Problems headed by Kapitsa. That happened at the beginning of the infamous Great Purge, which ravaged in the country for two years in 1937–1938. About 1.4 million people were condemned, mostly by false accusations of high treason and espionage. Half of them were executed. Scientists in general and physicists in particular did not escape the common fate. I have already mentioned the names of Bronstein and Houtermans. There were others... Looking at all this, Landau changed his political views. He still believed in socialism and its superiority compared to the "inhuman capitalist system", but thought now that "fair Lenin's principles" had been badly distorted and betrayed by Stalin.

In April 1938, K., a gifted romantic poet, a student of the Literary Institute in Moscow, met Moisei Korets, a friend and colleague of Landau, who moved to Moscow from Kharkov at about the same time. He told Korets that a group of revolutionary-minded students, who wanted to overthrow the anti-popular Stalin regime, existed in his institute and proposed to Korets that he should join. Korets went to Landau and they together wrote a leaflet, which Korets transmitted to K. who was supposed to reproduce it by hectograph (Xerox machines were not invented yet) and send it around to some addresses. Four days after that, Korets and Landau were arrested. However, K. was not, and a natural suspicion is that he was an agent provocateur of the NKVD. But it was never proved, and that is why I preferred not to put his full name here, bearing also in mind that K. volunteered to join the army in 1941 and died as a hero in September 1942.

---

[10]That was how his disciples called him.

The leaflet was written in strong terms.

... Comrades!

The great cause of the October Revolution is now scoundrelly betrayed... In his wild hate of genuine socialism, Stalin can be compared with Hitler and Mussolini... The proletariat who overthrew the rotten power of the Tsar and the capitalists will be able to do the same with the fascist dictator and his clique!...

By the standards of that time, such a leaflet meant an immediate execution — hundreds of thousands were executed without any reason, for doing nothing. But Landau escaped that fate.

One should say now that, all his evil deeds notwithstanding, Stalin had a certain respect for science and for the scientific elite. Kapitsa was deprived of the right to work in Cambridge, but the government bought his expensive equipment and financed the organisation and construction of a new scientific institute, tailored for Kapitsa.

Kapitsa wrote several letters to Stalin, and these letters were written with dignity, the letters of a free man, not of a "trembling subject". In particular, he addressed Stalin in February 1937 when Fock was arrested in Leningrad. He wrote that this arrest was harmful for Soviet science and for its image abroad. Such a letter in the midnight of the Great Purge was a courageous and risky deed. Kapitsa could well have been arrested himself. But no, not only was Kapitsa not arrested, but Fock was released!

A year later, having learned about Landau's arrest, Kapitsa wrote to Stalin again. This time Landau was not immediately released, but it seems that the letter had some effect. Landau was neither executed, nor condemned, but stayed in jail under investigation.

In November 1938, Ezhov (the Commissar of Internal Affairs, the main perpetrator of the Great Purge) was released from his post, to be later arrested and executed in his turn. The Great Purge was over. Kapitsa waited some time, and in April 1939 he wrote the second letter on Landau's behalf, this time to Molotov, the second person in the government. And it worked; Landau was released!

He resumed his work at Kapitsa's institute, and this work was efficient. In 1940–1941 Landau developed the theory of liquid helium, which explained the phenomenon of superfluidity discovered by Kapitsa in 1938 and for which he was later awarded the Nobel Prize.

In 1941 the Institute of Physical Problems was evacuated to Kazan, where Landau spent the war years.

In 1945–1953, Landau participated in the Soviet nuclear project. It is interesting that his role was in a certain sense similar to Feynman's role in the American project; it was rather the role of a mathematician.[11] Landau invented efficient "cheap" numerical algorithms for ponderous calculations that had to be performed. In particular, Landau calculated the dynamics of the first Soviet thermonuclear bomb, Sakharov's *sloyka* (a layer cake). For his contributions to the atomic and hydrogen bomb, Landau

---

[11]Of course, the scope of Landau's responsibilities was much larger than Feynman's — after all, Landau was 10 years older.

received two Stalin Prizes: in 1949 and 1953. In 1954 he was awarded the title "Hero of Socialist Labour".

Though quite important, his contributions to the project were less decisive than those of Sakharov or Kurchatov or Zeldovich. According to the evidence of his disciples, Landau worked for the bomb not really with reluctance, but, better to say, not with the same enthusiasm as those three people. There is no wonder; the latter had had a different life experience than Landau: they had not spent a year in jail...

13.2.4 Accident

On Sunday, January 7, 1962, the roads in Moscow and its vicinity were transformed into a skating ring. The previous day it had rained, and then the frost came. Around 10 a.m. a car stopped at Landau's door — he had to go to the Nuclear Institute in Dubna (about 100 km from Moscow) to give a talk at the seminar and discuss physics with colleagues. At some point the driver wanted to pass a bus, but then saw a truck going from the opposite direction. He braked, but a bit too hard, the car went out of control and started to turn around on the ice like a hockey puck. The truck hit the car right at the place where Landau was sitting.

The first entry in the medical record was "multiple brain contusions, a laceration in the forehead-temple region, fractures of the skull, fractures in seven ribs, a lung damaged, fracture of pelvis, shock". Normally, such injuries are not compatible with life. But Landau survived. He did so due to his healthy constitution (Landau was a healthy man, though he never did much sport), due to the efforts of physicians and due to the devotion of his disciples.

On January 12, five days after the disaster, Landau almost ceased to breathe. The only thing that could save him was an artificial lung — a machine that oxygenizes the blood and removes from it carbon dioxide. Such devices were very rare in Moscow at that time and were very bulky. The hospital where Landau was placed did not have it, but this machine was available in another hospital. So as not to waste time, physicists carried out the heavy machine by hand, stopped a truck that passed by on the street and finally brought it to Landau's ward. He was saved.

But a week later Landau had a brain oedema and was again on the brink of death. The doctors told the disciples that a certain medicine had recently been developed that might save him, but it was so new and rare that one could not find it in Moscow; the only places in the world where it was available were London and Prague. Russian physicists called their British and Czech colleagues, the medicine was found and sent to Moscow. The package from London came first. When Sir John Douglas Cockroft (Nobel Prize in physics in 1951) had got the required medicine, he realized that could not deliver it to the airport in time. He called the air company, explained the situation and the flight was delayed by an hour. Landau was saved again.

It took three more months before Landau recovered to the point that he could talk. And then walk. Bearing in mind the gravity of his injuries, this was a small, or maybe not so small miracle. In November 1962, he was awarded the Nobel Prize, but he

Fig. 13.2 Ter-Martirosyan.

was not yet in good enough shape to go to Stockholm, and the prize was delivered by the Swedish ambassador, who came to his hospital in Moscow.

Landau lived 6 more years. He could talk, slowly walk and socialize to some extent. But unfortunately, he could not do science any more. He died in April 1968.

13.3 Karen Ter-Martirosyan □ ITEP

Karen Avetovich Ter-Martirosyan[12] was a disciple of Landau (No. 11 in the list) and one of my teachers. His genius rank was probably lower than that of Landau or of Feynman, but he did a lot of solid and interesting scientific work. His best-known results refer to Regge theory.

That is something we have not yet talked about in our book. Regge theory is a phenomenological theory of the strong interactions describing high-energy hadron scattering at small angles (this kinematical region gives the dominant contribution to the total cross section). It was very popular and was actively developed (the especially important developments being due to Tullio Regge, Gabriele Veneziano, Ter-Martirosyan's former student Vladimir Gribov and to Ter-Martirosyan himself) during the sixties and the seventies, when QCD did not yet exist. It keeps its significance today: hadron scattering is a complicated process, and we are still not able to describe it in the framework of the basic QCD principles in a model-independent way.

Ter-Martirosyan worked in the Institute of Theoretical and Experimental Physics in Moscow. In the second half of the last century, ITEP was probably the strongest Soviet research centre for particle physics, where many bright people worked. But none

---

[12]This section represents an adaptation of my notes written for the memoirs volume *Under the Spell of Landau*, edited by Mikhail Shifman.

of them (and nobody whom I had a chance to meet personally in other institutions) could match K.A. in the quality of a Teacher.

The explanation is simple: K.A. really *cared* about his students. After they got their master's, he tried to obtain an ITEP PhD scholarship and then (or maybe directly after graduation; in those days that was sometimes possible) a permanent ITEP position for as many of them as he could. When this was not possible, he called all his colleagues in other research institutes, asking whether they had open positions, telling them what brilliant student he happened to have right now,[13] etc. Like a father trying to help his children as much as he can...

I first met K.A. in July 1970, while taking the admission exams to MIPT. To be more precise, by that date I had already passed the regular exams: written and oral maths, written and oral physics and the Russian composition. But, following the special procedure set up in MIPT, I had also to pass a kind of interview where the professors looked at their future students, asked them all kind of questions about physics and life, and the impression they formed played a role in the final selection and in distributing the students over different specialisations (in MIPT this was done already at the freshman level, though it could be changed later). I was asked which branch of physics I liked the best. I answered — *elementary particles* — and was sent to K.A. I saw a man with an expressive, mobile face, who briskly asked few questions (I remember, it was something about an electron going simultaneously through two holes in the barrier and thus producing an interference pattern), and I was accepted.

MIPT was a special university. The best in physics in Soviet Union and maybe the best in the world at that time. During the first two years, all students studied together on the university campus in Dolgoprudny (a small town very close to Moscow). But starting from the third (sometimes, from the second) year, they spent more and more time at the "base" — a research Institute with this or that specialisation. In our case, it was ITEP. There we followed many special courses (the strong interactions, the weak interactions, theory of accelerators, nuclear electronics...) and also spent some time in one of the experimental labs. By default, we were assumed to write a master thesis on some experimental subject and then continue to work as high energy experimentalists.

Those who wanted to become theorists had to pass some extra special exams — the same idea as Landau's minimum, but in an alleviated form. I had to pass not eleven, but only three exams — on quantum mechanics, quantum electrodynamics and on modern field theory (whatever was modern back in 1975). The first exam was on quantum mechanics. Its programme was basically the Landau & Lifshitz textbook, but without some sections. Thus, the instructions for the exam consisted simply of the list of those sections of L & L that could be omitted in preparation. I had got such a list from some senior student, but it involved a *bug*. Instead of "113,

---

[13]One must say that many of his disciples became brilliant theorists, indeed. The most known of them are Gribov and three Alexanders: Alexander Migdal, Alexander Polyakov and Alexander Zamolodchikov.

147", it read "113–147", and I understood that there was no need to study the sections from 113 to 147 — the whole of scattering theory!

Happily I called K.A., we met, he quickly understood my audacity and ignorance and sent me home. I came to him again several months later, having already learned the Sects. 113, 114, etc. This time, I was more fortunate. After posing some more simple questions, he suggested to me a problem: to find the shift of the Coulomb levels in the $p\bar{p}$ atom due to strong interactions. This problem is not so simple. According to the standard procedure set up by K.A., I first spent several hours in his office trying to solve it, then he listened to what I was able to babble after that and sent me home: *come back when you solve it!* It took me about a week. To find the shift, one had to write a solution of the radial Schrödinger equation as a generic hypergeometric function (*of course, a hypergeometric function* — commented K.A. — *this is known from the high school curriculum*) and to impose proper boundary conditions. Later I learned that this problem was based on some unpublished paper of K.A. He solved it, found it too simple to publish, but suggested it instead to students.

While taking two other theoretical exams, my examiner was not K.A., but Boris Lazarevich Ioffe who became the adviser of my master thesis and then PhD thesis. And meanwhile I was following the regular course of K.A. on the theory of the strong interactions. That was the spring of 1975. QCD had just been invented and had not yet established itself as the true indisputable fundamental theory of the strong interactions. The course of K.A. was not about the quarks and gluons, but about the Regge theory mentioned above. K.A. wanted to write a book on the basis of his course (it did not finally work out for reasons unknown to me). So he asked us to take notes of his lectures, write them up in a more or less decent literary form and give them back to him. I remember that, when it was my turn, I tried to convey exactly what he had said during the lecture and wrote: *as is well known from the high school curriculum* and then it was, if I remember correctly, an integral representation for the Bessel function $J_0(x)$.

At that time K.A. was excited about the bootstrap philosophy (we mentioned that on p. 69; bootstrap adepts claimed that there are no fundamental particles and no fundamental fields whatsoever, but everything depends on itself and everything else in some self-consistent way), rather than about quarks. I've already recalled on p. 190 an experimental seminar in ITEP (or was it a thesis defence?) when K.A. (it was him!) expressed dissatisfaction when the speaker compared his data with the predictions of the "speculative and suspicious" quark model.

After the November Revolution of 1974 when charmonium was discovered, he changed his mind, of course. His scientific interests changed too. In the late seventies and eighties, he worked not on Regge theory any more, but on the physics of Grand Unification.

K.A. was one of the pillars of the ITEP seminar. Similar to Landau's seminar at the Kapitsa Institute and Bohr's seminar in Copenhagen, it was a genuine "Russian seminar", where people in the audience have the habit of asking as many questions as is necessary to clarify what the speaker says — whether it is a major scientific discovery, pure nonsense or something in between. Discussions at the ITEP seminar were sometimes rather heated, and K.A. (together with some other members of the

ITEP theoretical department) never missed an opportunity to express his opinion. That *was*, I believe, the most efficient way to do science and to learn science. It is difficult (at least, for me personally) to learn a new subject by listening to a formal talk without questions being asked. I am missing the ITEP seminar now...

Feynman, Landau and Ter-Martirosyan were rather different people, but they shared in common two important qualities:

1. A passionate love of science and absolute devotion to it.
2. Absolute scientific integrity. All three of them never hesitated to say honestly what they thought, but if they realized that they were wrong, they never hesitated to admit it.

Going back to K.A., his absolute integrity was his distinguishing feature not only in physics, but also in human relationships. In some cases, people were offended by his too frank conduct. But I think that no-one could bear a grudge against K.A. for long. People understood that K.A. might have been careless about their vanity, but he was equally careless about *his own* vanity. Never did I meet another scientist for whom the words of Boris Pasternak: "It is unseemly to be famous, celebrity does not exalt..." would so adequately express what he *really* felt.

Part VI
Dessert

Chapter 14
Supersymmetry

We now resume physics discussions. Let us first recall the characteristic energy and distance scales relevant for the theories of the fundamental interactions (Table 14.1).

The region with energies $E \lesssim 300\,\text{GeV}$ is now pretty well charted. This is the realm where the Standard Model and all low-energy effective theories that are derived from it are valid. We see, however, that this charted region is very tiny, compared with the uncharted one. The Planck mass exceeds the maximal energy accessible at existing accelerators by ~ 16 orders of magnitude!

And there is no hope of building much larger accelerators in future. The length of the accelerator ring is determined by the energy that we want to achieve and the magnetic field that bends the beam and makes it run along a circular orbit,

$$L \sim \frac{2\pi E}{|e|B}.\tag{14.1}$$

The magnetic field at the best existing superconducting magnets is of the order of 10 Tesla = 100,000 gauss, and it cannot be made essentially stronger: too strong a magnetic field destroys superconductivity — this limit is brought about by the atomic structure of ordinary matter, which also puts limits on the density of matter and its rigidity.

The length of the LHC ring (where the protons are now accelerated up to 7 TeV) is 27 km, and the cost of its construction (not counting digging the tunnel, which was done at CERN earlier for another accelerator and not counting the cost of the huge detectors) was about 5 billion Swiss Francs. The total operating budget of the LHC is about a billion CHF per year. To achieve an energy 10 times higher, one would need an accelerator 10 times larger and (at least!) 10 times more expensive. Maybe this will be done one day, I do not know, but that would already be the absolute limit of mankind's technological possibilities.

And even assuming the construction of this gigantic machine, the advance into the ocean of unknown (on whose shore we are still in some sense sitting, as Newton

© Springer International Publishing AG 2017
A. Smilga, *Digestible Quantum Field Theory*,
DOI 10.1007/978-3-319-59922-9_14

Table 14.1 Characteristic physical scales.

	Atom	Nucleus	Electroweak	Grand unification	Planck
Energy	1 eV	1 MeV	100 GeV	$10^{15}-$ 10^{16} GeV	10^{19} GeV
Distance	10^{-10} m	10^{-15} m	10^{-18} m	$10^{-31}-$ 10^{-32} m	10^{-35} m

did) would not be so fantastic: we would still explore only less than a 10^{-14}th part of it.

Without any hope of studying the underlying structure of the world experimentally, we have no choice but to try to perceive it, using only white paper, multicolour pens, maybe some computers and, the most important tool, our brains, in the same way as Einstein used his brains to construct general relativity.

We are trying hard, but we cannot at the moment boast of spectacular successes. The main conceptual problem is gravity, which currently resists all attempts to quantize it — we will discuss it at length in Chap. 16. Leaving gravity aside, one can ask what theory describes physical phenomena at energies much larger than the electroweak scale, but much less than the Planck scale. Many theorists now believe that this is a supersymmetric field theory.

The word "supersymmetry" has already appeared in the pages of our book. The last time was in Sect. 12.7, right before the break. There we gave certain arguments why supersymmetry is considered an interesting, solid and promising idea. And now it is time to explain what supersymmetry really *is*. We will do that in this chapter, presenting and discussing some examples of supersymmetric systems.

Quantum field theory is not a simple branch of physics. One needs to apply considerable effort to learn and really understand it. But it happens very often that a field theory phenomenon, seeming to be esoteric and complicated, has a very close analogue in quantum mechanics, for systems with a finite number of degrees of freedom. And there it looks much more simple and transparent.[1]

We have already profited from this fortunate circumstance when discussing the diagram technique for scattering amplitudes. We derived the Feynman rules quite rigorously for quantum mechanics and then asked the reader to believe that they look similar for relativistic field theories. Now we use this philosophy (*quantum field theory is a complicated form of quantum mechanics, and the latter is a simple version of field theory*) to study supersymmetry.

Actually, one can *define* supersymmetry as *double degeneracy of all excited states in a quantum Hamiltonian*. The word "excited" is important here. The ground state may have this degeneracy or it may not.

---

[1] Feynman would appreciate this approach. He complained in his "*Surely you're joking*" book that he could not understand a new concept without first visualizing it, constructing a simple model example in his mind. And if he was able to do that, he more often than not understood things better than the others.

14.1   Electrons in a Magnetic Field

As was mentioned in the previous chapter, one of the simplest supersymmetric quantum problems is the motion of an electron in a homogeneous magnetic field — the problem solved by Landau in 1930. The generic Hamiltonian describing the motion in a magnetic field (the Pauli Hamiltonian) reads

$$H = \frac{\left[\sigma \cdot \left(P - \frac{e}{c}A\right)\right]^2}{2m}, \tag{14.2}$$

where A is the vector potential, $B = \nabla \times A$, and σ are the Pauli matrices. [The explicit form of $t = \sigma/2$ was given in Eq. (6.40).] The Pauli matrices satisfy the algebra

$$\sigma_j \sigma_k = \delta_{jk} \mathbb{1} + i\epsilon_{jkl}\sigma_l. \tag{14.3}$$

Bearing this and the definition $P_j = -i\hbar \partial/\partial x^j$ in mind, we may rewrite (14.2) as

$$H = \frac{\left(P - \frac{e}{c}A\right)^2}{2m} - \frac{e\hbar}{2mc}B \cdot \sigma. \tag{14.4}$$

The Hamiltonians (14.2) and (14.4) act on spinor wave functions

$$\Psi(x) = \begin{pmatrix} a_+(x) \\ a_-(x) \end{pmatrix}. \tag{14.5}$$

The second term in (14.4) describes the interaction of the Dirac magnetic moment (10.30) of an electron with an external magnetic field.

Generically, the Hamiltonian (14.2) is not supersymmetric. Let us assume, however, that the *direction* of the magnetic field is fixed. We may choose it to lie along the z axis:

$$B = [0, 0, B(x, y)]. \tag{14.6}$$

Landau assumed that the field is constant. But that is not necessary for supersymmetry; one can allow it to depend on x and y.[2] If the magnetic field has the form (14.6), one can choose the vector potential A to have only x and y components. In that case, the Hamiltonian represents a sum of two terms. The first term describes a nontrivial motion in the (x, y) plane, while the second term has the form $p_z^2/2m$ and describes free motion along the direction of the field (indeed, when $v \| B$, the Lorentz force is zero). Consider only the first nontrivial part. It coincides with (14.2), only now we assume that the vectors P and A lie in the (x, y) plane.

---

[2] But not on z, otherwise the divergence $\nabla \cdot B$ would be different from zero, which is not allowed by Maxwell's equations.

Consider the operators

$$Q = \frac{1}{\sqrt{2m}} \, \boldsymbol{\sigma} \cdot \left(\boldsymbol{P} - \frac{e}{c} \boldsymbol{A} \right) \frac{1 + \sigma_3}{2} \tag{14.7}$$

and

$$Q^\dagger = \frac{1}{\sqrt{2m}} \, \boldsymbol{\sigma} \cdot \left(\boldsymbol{P} - \frac{e}{c} \boldsymbol{A} \right) \frac{1 - \sigma_3}{2}. \tag{14.8}$$

Using the identities (14.3) and bearing in mind that the vector $\boldsymbol{\sigma}$ in the scalar product has only the components $\sigma_{1,2}$, which anticommute with σ_3, it is easy to show that

- The operator (14.8) is the Hermitian conjugate of (14.7),
- The operator Q is nilpotent, $Q^2 = 0$. The same is true for $Q^\dagger$.
- The anticommutator $\{Q, Q^\dagger\}_+$ coincides with the Hamiltonian.

It follows that the commutators $[Q, H]$ and $[Q^\dagger, H]$ vanish. In other words, the operators Q and $Q^\dagger$ (called *supercharges*) are new nontrivial integrals of motion. By Noether's theorem, this implies the existence of a new nontrivial symmetry — supersymmetry. The algebra

$$Q^2 = (Q^\dagger)^2 = 0, \quad \{Q, Q^\dagger\}_+ = H \tag{14.9}$$

is the simplest possible *supersymmetry algebra*.

The algebra (14.9) implies that the Hamiltonian has *two different* anticommuting Hermitian square roots:

$$S_1 = \frac{Q + Q^\dagger}{\sqrt{2}} \quad \text{and} \quad S_2 = \frac{i(Q^\dagger - Q)}{\sqrt{2}}. \tag{14.10}$$

That is why the algebra (14.9) is usually called in the literature the algebra of $\mathcal{N} = 2$ supersymmetric quantum mechanics (SQM). I must warn the reader, however, that this universally adopted terminology is confusing. If the Hamiltonian had only *one* square root, as for a generic Pauli Hamiltonian (14.2), this would not bring about any new symmetry: there would be no double-degenerate levels. The only assertion that could be made in this case is that the spectrum of the Hamiltonian (14.2) is positive definite, but this can always be achieved for a ghost-free Hamiltonian with a spectrum bounded from below simply by adding a constant. Thus, $\mathcal{N} = 1$ SQM is an oxymoron; the counting starts with $\mathcal{N} = 2$.

Let us now discuss the properties of a genuine supersymmetric Hamiltonian satisfying the algebra (14.9). Before doing that, it is instructive to recall the solution of a familiar problem — the structure of the spectrum of the hydrogen atom. This problem has rotational symmetry; the operator of the orbital angular momentum $\boldsymbol{L}$ commutes with the Hamiltonian. The ground state has zero angular momentum. This means that its wave function does not depend on the angles, but only on the distance between the electron and the proton. The ground state is thus *annihilated* by the action of the symmetry generator L:

$$L|\text{ground state}\rangle = 0.$$

The full spectrum involves the eigenstates of L^2 and L_3 with different angular momenta. If the latter is nonzero, the state is not annihilated by L. The action of $L_{1\pm i2}$ on such a state gives a state with the same eigenvalue of L^2, but the eigenvalue of L_3 is shifted by ± 1. The states are grouped into multiplets — the multiplet of 3 degenerate states with $L = 1$, the multiplet of 5 degenerate states with $L = 2$, etc.

We go back to supersymmetry. The statement is that all the states in the spectrum of a supersymmetric Hamiltonian with nonvanishing energy are paired: the spectrum is thus split into a set of degenerate doublets. In addition, the spectrum *might* involve unpaired ground states annihilated by the action of both Q and $Q^\dagger$ and having zero energy. In fact, such a structure of the spectrum can be derived from the algebra (14.9) as a simple mathematical theorem. We even contemplated giving a rigorous proof of it here, but then decided that proving theorems would not fit well with the style and spirit of our book.

Thus, we continue to follow our inductive *down* ↗ *up* approach and study the Landau Hamiltonian — the Hamiltonian (14.4) with a *homogeneous* magnetic field, $\boldsymbol{B} = (0, 0, B)$. In the calculations below we make a simplification, getting rid of all dimensional constants $\hbar, c, m, |e|$ (so that e goes over to -1). We wrote them before to make it clear that the problem under consideration is not just a mathematical construction, but a real physical problem. Its solution gives an explanation of a real physical phenomenon, the so-called *Landau diamagnetism*.[3] But it is the mathematical structure of the problem that we are now mostly interested in, and keeping $\hbar$, etc. would obscure it.

Now the vector potential can be chosen as $\boldsymbol{A} = (-By/2, Bx/2, 0)$. The Hamiltonian and the supercharges acquire the form

$$H = \frac{1}{2}\left[\left(-i\frac{\partial}{\partial x} - \frac{By}{2}\right)^2 + \left(-i\frac{\partial}{\partial y} + \frac{Bx}{2}\right)^2\right] + \frac{B}{2}\sigma_3. \qquad (14.11)$$

$$Q = -i\sqrt{2}\begin{pmatrix} 0 & 0 \\ \frac{\partial}{\partial w^*} - \frac{Bw}{4} & 0 \end{pmatrix},$$

$$Q^\dagger = -i\sqrt{2}\begin{pmatrix} 0 & \frac{\partial}{\partial w} + \frac{Bw^*}{4} \\ 0 & 0 \end{pmatrix}, \qquad (14.12)$$

where $w = x + iy$.

---

[3]Electrons rotating in an external magnetic field $\boldsymbol{B}$ induce their own small magnetic field in the direction opposite to $\boldsymbol{B}$.

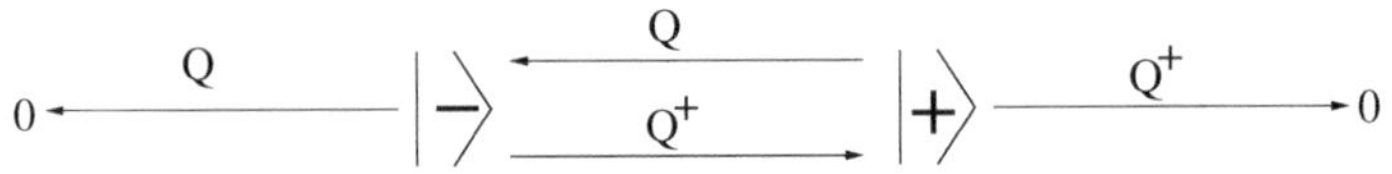

Fig. 14.1 A supersymmetric doublet.

It is clearly seen that the whole Hilbert space (14.5) is split in this case into two subspaces: subspace $|+\rangle$ with positive electron spin projection and subspace $|-\rangle$ with negative spin projection. The transition matrix elements $\langle+|H|-\rangle$ all vanish.

The supercharge (14.7) involves a projector onto spin-up states. In other words, $Q|-\rangle = 0$. By the same token, $Q^\dagger|+\rangle = 0$. The matrices $\sigma_{1,2}$ are off-diagonal, and that means that when Q acts on a spin-up state, one obtains as a result a spin-down state, $Q|+\rangle = |-\rangle$. Also, $Q^\dagger|-\rangle = |+\rangle$. Now note that for any spin-up eigenstate of the Hamiltonian, the result of the action of Q is a spin-down state with the same energy; this follows from $[H, Q] = 0$. Consequently, the spectrum is indeed mostly split into degenerate doublets: one of the states in the doublet is annihilated by Q, another state by $Q^\dagger$, and one state can be obtained from the other by the action of either Q or $Q^\dagger$ (see Fig. 14.1).

One can observe that the energy of the states in a supersymmetric doublet is strictly positive. Indeed, for any normalized state Ψ,

$$E_\Psi = \langle\Psi|QQ^\dagger + Q^\dagger Q|\Psi\rangle = \|Q^\dagger\Psi\|^2 + \|Q\Psi\|^2, \qquad (14.13)$$

where $\|X\| = \sqrt{\langle X|X\rangle}$ is the norm.

One also sees that the energy is zero for the states annihilated both by Q and $Q^\dagger$. Whether or not such supersymmetric ground states exist, how many such states are there and what they are is a dynamical question, which requires a special analysis. For the Landau Hamiltonian, this analysis is not complicated, however. The results depend on the sign of B. Let B be positive. Then the normalized zero-energy states are absent in the spin-up sector. This can be easily seen by inspecting Eq. (14.11). The first term is positive definite, and the second term shifts the energy up by $B/2$. One cannot have zero. On the other hand, there are infinitely many spin-down zero-energy states. The condition $Q^\dagger\Psi = 0$ ($Q\Psi = 0$ is satisfied automatically) gives the first-order equation

$$\left(\frac{\partial}{\partial w} + \frac{Bw^*}{4}\right) a_- = 0 \qquad (14.14)$$

with the solutions

$$a_-(w, w^*) = \exp\left\{-\frac{Bw^*(w + c)}{4}\right\}. \qquad (14.15)$$

Different complex c correspond in the classical limit to different positions of the centre of the circular electron orbits in the (x, y) plane.

Fig. 14.2 Spectrum of the
Landau Hamiltonian
($B > 0$). $n = 0, 1, \ldots$ mark
the oscillator levels.

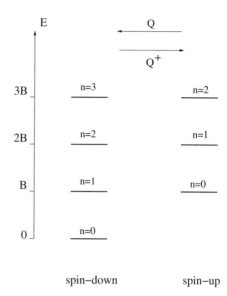

If B is negative, the picture is opposite. There are no zero-energy states in the spin-down sector and infinitely many of them in the spin-up sector.

We obtained the ground-state wave functions (14.15) by solving a simple first-order equation (14.14). This situation is typical for supersymmetry. The supersymmetric ground states satisfy not only the Schrödinger equation, but also the equations $Q\Psi = Q^\dagger\Psi = 0$, which are much simpler.

To find the whole spectrum, one still has to solve the Schrödinger equation. For the Landau problem, the latter is also rather simple. In both sectors the Hamiltonian reduces to the Hamiltonian of the harmonic oscillator. The spectrum is thus equally spaced. All levels are infinitely degenerate due to the symmetry under translations in the (x, y) plane. (And for the real physical problem, one also should add $p_z^2/2m$; do not *completely* forget about it.) The spacing between the levels is equal to $|B|$.[4] The difference between the two sectors is that the spin-down states are shifted down with respect to the spin-up states by B (we assume, for definiteness, that $B > 0$), which exactly coincides with the spacing between the oscillator levels (see Fig. 14.2). This is a *physical* reason for the double degeneracy of all excited levels. [And we remember: the mathematical reason unsuspected by Landau is the supersymmetric algebra (14.9)!]

If the magnetic field in (14.6) is not homogeneous, but depends nontrivially on x, y, the problem is still supersymmetric, and even though one cannot solve it

---

[4]If the constants $\hbar$, etc. are restored, the spacing is

$$\Delta = \frac{\hbar|eB|}{mc}. \qquad (14.16)$$

analytically any more, the generic pattern of the spectrum is the same as for the homogeneous field — some number of supersymmetric zero-energy states and a set of double-degenerate supersymmetric doublets. One can show (we will not do so) that the number of zero energy states is proportional to the magnetic flux:

$$n_{E=0} \; = \; \frac{|\Phi|}{2\pi} \; = \; \frac{1}{2\pi} \left| \int B(x, y) dx dy \right| . \tag{14.17}$$

For a homogeneous field, the flux is infinite, and the number of supersymmetric states is infinite too. If the flux is finite, so is the number of supersymmetric states. The latter are spin-down states if the flux is positive and spin-up states if the flux is negative.[5]

"And what if the flux is zero", our inquisitive reader may ask. Well, in that case there are no supersymmetric ground states annihilated by both supercharges whatsoever, and all the states are paired into supersymmetry doublets. The absence of supersymmetric ground states (for field theories — the absence of a supersymmetric vacuum) means that the supersymmetry is *spontaneously broken*. Indeed, there is no conceptional difference with the models involving spontaneous breaking of an ordinary global symmetry, which we discussed in Sect. 11.5.2. In all these cases, the Lagrangian is invariant under symmetry transformations,[6] while the vacuum is not.

14.2   Grassmannian Description

The Pauli Hamiltonian has a matrix form. That is OK if you only want to solve the Schrödinger equation and find the spectrum. But it is not convenient if you want to establish the connection between the Hamiltonian and Lagrangian formulation, if you want to understand, following Noether's ideas, to what kind of supersymmetry transformations of dynamic variables, leaving the action invariant, are the new integrals of motion, the supercharges, related.

All that is achieved if we use Grassmann dynamic variables. We introduced Grassmann numbers in Chap. 6 and used them to describe the dynamics of fermion fields in Chap. 9. We mentioned there that Grassmann variables are tailor-made to describe supersymmetric systems and promised to elucidate this in Chap. 14, i.e. *now*. The time has come to fulfill our promise.

In Chap. 9 we explained how the standard oscillator Hamiltonian can be expressed via holomorphic variables [see Eq. (9.9)]. We also presented the Lagrangian and the classical and quantum Hamiltonians for a "Grassmann oscillator", the system involving a single Grassmann variable ξ [Eqs. (9.18), (9.23), (9.28)].

---

[5]Incidentally, $\Phi/2\pi$ cannot be fractional — the flux is *quantized*. If $\Phi/2\pi$ were not an integer, one could not consistently pose the quantum problem: the electron wave function could not be uniquely defined.

[6]We have not learned yet what is the supersymmetric Lagrangian and what are supersymmetry transformations, but we will do so soon.

The Hamiltonian of a supersymmetric oscillator is simply the *sum* of the ordinary oscillator Hamiltonian and the Grassmann oscillator Hamiltonian. Adding (9.12) and (9.28), we obtain

$$H_{\text{sup-osc}}^{\text{qu}} = \frac{\omega}{2}\left(a^*\frac{\partial}{\partial a^*} + \frac{\partial}{\partial a^*}a^*\right) - \frac{\omega}{2}\left(\xi\frac{\partial}{\partial\xi} - \frac{\partial}{\partial\xi}\xi\right)$$

$$= \omega\left(\frac{\partial}{\partial a^*}a^* - \xi\frac{\partial}{\partial\xi}\right), \tag{14.18}$$

where a^* is an ordinary complex number and ξ is a Grassmann complex number.

The expression (14.18) is the quantum Hamiltonian. In classical theory, the differential operators $\partial/\partial a^*$ and $\partial/\partial\xi$ go over to the conjugated holomorphic variables a and $\xi^\dagger$ playing the role of the canonical momenta. As follows from the definitions (9.10) and (9.24), the canonical Poisson brackets are

$$\{a^*, a\} = \{\xi^\dagger, \xi\} = -i. \tag{14.19}$$

The classical counterpart of (14.18) is

$$H_{\text{sup-osc}}^{\text{cl}} = \omega(aa^* - \xi\xi^\dagger). \tag{14.20}$$

The normalized eigenfunctions and eigenvalues of the quantum Hamiltonian (14.18) are

$$\Psi_n(a^*, \xi) = \frac{(a^*)^n}{\sqrt{n!}}, \qquad E = \omega(n+1),$$

$$\tilde\Psi_n(a^*, \xi) = \frac{(a^*)^n}{\sqrt{n!}}\xi, \qquad E = \omega n. \tag{14.21}$$

Now it is natural to call the states without the Grassmann factor ξ in the wave function *bosonic* and the states with such a factor *fermionic*.

There is one unpaired ground state with zero energy, $\Psi = \xi$, while all the excited states are paired: the states $\Psi = 1$ and $\Psi = a^*\xi$ have the energy ω, the states $\Psi = a^*$ and $\Psi \sim (a^*)^2\xi$ have the energy 2ω, etc.

The ground-state wave function involves the factor ξ and is thus fermionic. This can be interpreted in the logic of Chap. 9: the vacuum state is the state with filled Dirac sea. But people also often use an alternative convention where it is $\xi^\dagger$ rather than ξ that plays the role of a holomorphic dynamic variable on which the wave functions depend. In this representation,

$$H^{\text{qu}} = \omega\left(a^*\frac{\partial}{\partial a^*} + \xi^\dagger\frac{\partial}{\partial\xi^\dagger}\right) \tag{14.22}$$

and the vacuum is bosonic, $\Psi_{\text{vac}}(a^*, \xi^\dagger) = 1$. Using one or the other convention is a matter of taste.

According to our definition, the system (14.18) is supersymmetric. It is not difficult to write down the operators of the quantum supercharges:

$$Q = \sqrt{\omega}\, a^* \xi, \qquad Q^\dagger = \sqrt{\omega}\, \frac{\partial^2}{\partial a^* \partial \xi}. \tag{14.23}$$

The supercharges (14.23) are nilpotent, and their anticommutator gives the Hamiltonian (14.18).

The Lagrangian can be obtained from (14.20) by a Legendre transformation. We derive:

$$L = i(\dot{a}a^* - \dot{\xi}\xi^\dagger) + \omega(\xi\xi^\dagger - aa^*). \tag{14.24}$$

A diligent reader can now easily verify that:

- The equations of motion following from (14.24) coincide with the Hamilton equations of motion (9.22) following from (14.20).
- The Lagrangian (14.24) is invariant under the transformations

$$\begin{aligned} \delta a &= i\epsilon^\dagger \xi, & \delta a^* &= i\epsilon\xi^\dagger, \\ \delta\xi &= ia\epsilon, & \delta\xi^\dagger &= -ia^*\epsilon^\dagger, \end{aligned} \tag{14.25}$$

where ϵ is a complex anticommuting transformation parameter. These are the supersymmetry transformations.

We have just described the simplest possible supersymmetric system, but all other such systems can also be conveniently described in the Grassmannian language. In particular, the matrix supercharges (14.12) and the Hamiltonian (14.11) can be translated into this language by substituting

$$\sigma_- = \frac{\sigma_1 - i\sigma_2}{2} \rightarrow \psi, \qquad \sigma_+ = \frac{\sigma_1 + i\sigma_2}{2} \rightarrow \hat{\psi}^\dagger \equiv \frac{\partial}{\partial\psi}$$

$$\sigma_3 = [\sigma_+, \sigma_-] \rightarrow \hat{\psi}^\dagger \psi - \psi\hat{\psi}^\dagger. \tag{14.26}$$

[cf. the comment after (9.30)]. Indeed, ψ and $\hat{\psi}^\dagger$ have the same properties as the quantum operators ξ and $\hat{\xi}^\dagger$ in the model above: they are nilpotent, Hermitially conjugate to each other and their anticommutator is equal to unity.

The supercharges (14.12) acquire the form[7]

---

[7]Not wishing to overcharge our formulas, we do not display the hats for Q and $Q^\dagger$ and do not do that for π and other bosonic operators, neither for ξ and ψ. But we have to do so for $\xi^\dagger$ and $\psi^\dagger$ in order not to confuse quantum operators with classical variables.

$$Q = \sqrt{2}\psi \left(\pi^\dagger + \frac{i B w}{4} \right),$$

$$Q^\dagger = \sqrt{2}\hat{\psi}^\dagger \left(\pi - \frac{i B w^*}{4} \right), \tag{14.27}$$

where $\pi = -i\partial/\partial w$ and $\pi^\dagger = -i\partial/\partial w^*$ are the bosonic quantum canonical momenta. The operators (14.27) act on the wave functions $\Psi(w, w^*; \psi)$.

The states annihilated by $Q^\dagger$, which we earlier called spin-up states, do not involve the factor ψ in the wave function and are now interpreted as bosonic. The spin-down states annihilated by Q are now interpreted as fermionic.

The algebra (14.9) is the simplest supersymmetry algebra with only one pair of complex supercharges. An obvious way to extend it consists in including another such pair and postulating the anticommutators

$$\{Q_j, Q_k\}_+ = \{Q_j^\dagger, Q_k^\dagger\}_+ = 0,$$

$$\{Q_j, Q_k^\dagger\}_+ = \delta_{jk} H, \qquad j, k = 1, 2. \tag{14.28}$$

In this case, the Hamiltonian has *four* different Hermitian square roots, and we are dealing with the $\mathcal{N} = 4$ extended supersymmetry algebra.

There are many quantum mechanical systems for which this algebra is realized. We give one example:

$$Q_1 = \psi_1 \pi + i(w^*)^2 \hat{\psi}_2^\dagger, \qquad Q_2 = \psi_2 \pi - i(w^*)^2 \hat{\psi}_1^\dagger,$$

$$Q_1^\dagger = \hat{\psi}_1^\dagger \pi^\dagger - i w^2 \psi_2, \qquad Q_2^\dagger = \hat{\psi}_2^\dagger \pi^\dagger + i w^2 \psi_1, \tag{14.29}$$

where $\hat{\psi}_j^\dagger = \partial/\partial\psi_j$.

The fermion operators in (14.29) can be traded for 4×4 matrices. For example,

$$\psi_1 \equiv \sigma_- \otimes \mathbb{1}, \quad \psi_2 \equiv \sigma_3 \otimes \sigma_-, \quad \hat{\psi}_1^\dagger \equiv \sigma_+ \otimes \mathbb{1}, \quad \hat{\psi}_2^\dagger \equiv \sigma_3 \otimes \sigma_+. \tag{14.30}$$

The Clifford algebra

$$\{\psi_j, \psi_k\}_+ = \{\psi_j^\dagger, \psi_k^\dagger\}_+ = 0, \qquad \{\psi_j, \psi_k^\dagger\}_+ = \delta_{jk}$$

holds.

The spectrum of the excited states of such a system is 4-fold degenerate. Theorists also know nontrivial $\mathcal{N} = 8$ SQM systems with 8-fold degeneracy and still higher extended SQM systems up to $\mathcal{N} = 32$.

14.3 Field Theories

For supersymmetric field theories, the Grassmann variables, which are a convenient but optional tool in quantum mechanics, become indispensible. An SQM Hamiltonian involving n Grassmann degrees of freedom can be alternatively represented as a

$2^n \times 2^n$ matrix. But in field theories the number of the variables is infinite. If we do not use Grassmann variables, the Hamiltonian represents an infinite-dimensional matrix, which is difficult to handle.[8]

As we know, anticommuting field variables describe fermions. And the double degeneracy of the excited eigenstates of the Hamiltonian, characteristic of all supersymmetric theories, is the degeneracy between the boson and fermion states. Supersymmetry is not (yet?) seen in experiment, but many different supersymmetric field Lagrangians can be written down as a mathematical exercise.

One has to admit that theoretical physics has changed its appearance rather fundamentally during the last 30–40 years. It has become more and more difficult to build new accelerator machines and, as a result, the influx of experimental data has decreased dramatically. This influx cannot keep all the theorists busy, and many of us have no other choice but to study different imaginary, fantasy worlds instead of the real physical world. This actually means that what most of us are now doing is not physics (the science studying Nature), but *mathematics*.

Indeed, geometry originated in Babylon as an applied discipline to serve the needs of farmers and tax officials. At that time it was not yet mathematics in the proper meaning of the word. But geometry became mathematics when the Greeks discovered its beauty and got interested in purely *abstract* geometric constructions. According to a legend, on being asked by a visitor what was the practical use of geometric theorems, Euclid called for his slave and ordered him to give the visitor an *obol*[9] and see him off, because *this man is looking for profit, not for truth*.

Euclid was doing planimetry and stereometry, which still described nature and could bring about some profits in agriculture and architecture. In some sense, that was theoretical, but physics. But later people started to study with enthusiasm 4-dimensional and higher-dimensional geometries — the geometries of imaginary worlds. And that was already hardcore maths.

Observing Nature and formulating it in economic mathematical terms provides one with the language, the tools, which one can further use for abstract games and studies that are not immediately profitable. But you understand, of course: if something is not profitable *immediately*, it does not mean that it will *never* be. For example, due to Einstein we now know that the geometry of our quite real and quite physical world is in fact 4-dimensional, and that the fabric of space-time is not flat, but curved (see more discussion in the next chapter). To describe it, Einstein had to use the formalism of Riemannian geometry worked out several decades previously by pure mathematicians, who sought not for profit, but for beauty and pleasure.

In other words, playing abstract mathematical games is not in all cases a completely asocial behaviour. Quite often these games allow one to develop an appropriate

---

[8]Well, it can somehow be handled in numerical path integral calculations (there is no other way; the path integral over fermion variables is *defined* as the determinant of an infinite-dimensional matrix going over into a matrix of large finite dimension in practical calculations), but not analytically, if a theory involves nontrivial interactions.

[9]A Greek silver coin.

language and tools to describe physical phenomena. The dialog between pure maths and theoretical physics is mutually benefitial.[10]

Going back to supersymmetry, there are serious reasons to believe that it *is* relevant for our world and does describe certain physical phenomena. We talked about that in Chap. 12 and will add several more words on this issue at the end of this chapter. But we do not know yet in what particular guise supersymmetry will turn up one day. While waiting for its triumphal arrival, it makes sense to explore all theoretic possibilities.

The first example of a supersymmetric field theory Lagrangian was given by Yuri Golfand and Evgeny Likhtman in their seminal paper of 1971. That was the Lagrangian of *supersymmetric electrodynamics*. The spectrum of this theory includes a massless "photon" (we have put the quotation marks because it is not the physical photon that we know), its massless fermionic superpartner, the photino, the massive "electron" and "positron" and two charged "scalar electrons" of the same mass. You may ask: why do we need two charged scalars, would not one be enough? No, not enough. We have four physical fermion charged massive states: two helicity states for the electron and two helicity states for the positron. These should be matched by four massive scalar states, which only can be achieved if we include two different complex scalar fields in the Lagrangian.

We will not write down the full expression for the latter here, as it is a little long, but restrict ourselves to its photon and photino sector,

$$\mathcal{L}_{\text{photon/ino}} = -\frac{1}{4} F_{\mu\nu} F^{\mu\nu} + i\psi\sigma^\mu\partial_\mu\psi^\dagger, \tag{14.31}$$

where the photino ψ_α is a neutral (Majorana) 2-component spinor. There are two helicity states for the photino and two helicity states for the photon. They are paired by supersymmetry.

We now want to write the full Lagrangian of the simplest nontrivial 4-dimensional supersymmetric model, the *Wess–Zumino model*. This Lagrangian involves a 2-component chiral fermion field ψ_α, descrining a Majorana fermion, and a complex scalar field ϕ. It reads

$$\mathcal{L}_{WZ} = \partial^\mu\phi\,\partial_\mu\phi^* + i\psi\sigma^\mu\partial_\mu\psi^\dagger$$
$$-\frac{1}{2}\left[\mathcal{W}''(\phi)\psi^\alpha\psi_\alpha + \text{c.c.}\right] - \mathcal{W}'(\phi)\mathcal{W}'(\phi^*), \tag{14.32}$$

where the function

---

[10]Unfortunately, it is now often hampered by a difference of scientific languages used by the two communities; it is difficult for me to understand a paper written by a pure mathematician, even if it is written on a subject that I know fairly well, because it is expressed in a different form, to which I am unaccustomed. On the other hand, I have heard complaints from mathematician colleagues that it is difficult for them to understand papers written by physicists. This rift between the two communities is a comparatively recent development. It did not exist 200 years ago, while 100 years ago it was already there, but not so deep. I cannot say that I like this rift, but such are the facts of life and one has to accept them.

$$\mathcal{W}(\phi) = \frac{m}{2}\phi^2 + \frac{\lambda}{6}\phi^3 \tag{14.33}$$

is called the *superpotential*. $\mathcal{W}'(\phi)$ and $\mathcal{W}''(\phi)$ are single and double derivatives of the superpotential with respect to its argument.

If $\lambda \neq 0$, the model involves rich dynamics. The scalar potential includes, besides the mass term $m^2\phi\phi^*$, cubic and quartic interactions. Fermions have the same mass m[11] and participate together with the bosons in the Yukawa interaction. It is important that the *same* function (14.33) determines the fermion interactions and the scalar potential. Otherwise the system would not be supersymmetric.

And the system (14.32) *is*. In particular, loop corrections to the fermion and scalar masses are exactly the same in all orders of perturbation theory. The degeneracy between the boson and fermion states is *exact*.

The latter statement follows from the invariance of the Wess–Zumino action under the supertransformations

$$\begin{aligned}
\delta\phi &= \psi^\alpha \epsilon_\alpha\,, \\
\delta\phi^* &= \epsilon^\dagger_{\dot\alpha} \psi^{\dagger\dot\alpha}\,, \\
\delta\psi^\alpha &= i\partial_\mu\phi\,\epsilon^\dagger_{\dot\beta}(\bar\sigma^\mu)^{\dot\beta\alpha} - \epsilon^\alpha \mathcal{W}'(\phi^*)\,, \\
\delta\psi^{\dagger\dot\alpha} &= -i\partial_\mu\phi^*\,(\bar\sigma^\mu)^{\dot\alpha\beta}\epsilon_\beta - \epsilon^{\dagger\dot\alpha}\mathcal{W}'(\phi)\,,
\end{aligned} \tag{14.34}$$

with $\bar\sigma^\mu$ defined in (2.11). You may check it explicitly, but probably this exercise would not bring you satisfaction. You still would not understand *why* it is invariant, how Julius Wess and Bruno Zumino actually *derived* the Lagrangian (14.32) in 1974, and how the pioneers, Golfand and Likhtman, derived the Lagrangian of supersymmetric QED three years earlier. The question is legitimate, but the only answer to which I will restrict myself here is, "they were the pioneers and they were clever". The reasoning of the pioneers is more often than not somewhat awkward. More direct and transparent derivations and explanations come later.

For example, one could ask how Archimedes derived his famous formula $V = 4\pi R^3/3$ for the volume of a ball. One should say that Archimedes was very proud of his result and even asked for the appropriate beautiful and complicated geometric construction to be engraved on his tombstone. However, since the time people learned how to do the integral $\int_0^\pi \sin\theta\,d\theta$ analytically, the Archimedes method of derivation has lost its relevance and is interesting now only to historians of science.

Going back to supersymmetry, an adequate formalism, the *superspace* and *superfield* formalism, which allows one to derive the Lagrangians discussed above and many other supersymmetric Lagrangians in a rather simple, one may say industrial, way, was developed by Abdus Salam and John Strathdee in 1974. We will not describe it here, but will try to explain its essence by analogy.

The volume of the parallelepiped formed by the vectors $\mathbf{a}$, $\mathbf{b}$, $\mathbf{c}$ is

$$V = |[\mathbf{a} \times \mathbf{b}] \cdot \mathbf{c}|\,. \tag{14.35}$$

---

[11]This mass is of Majorana nature, see p. 154.

Obviously, the volume is invariant under rotations. This invariance would not be immediately seen, however, if we expressed the volume in components:

$$V = |a_1b_2c_3 + a_2b_3c_1 + a_3b_1c_2 - a_1c_2b_3 - a_2c_3b_1 - a_3c_1b_2|. \quad (14.36)$$

Now we want to draw an analogy between the expressions (14.36) and (14.32). Both are *component* expressions. The point is that the Salam and Strathdee formalism allows one to express supersymmetric actions in a short superfield form in the same way as the formula (14.35) expresses the volume of the parallelepiped in a short vector form. In particular, the bosonic field $\phi(x)$ and the fermionic field $\psi_\alpha(x)$ in (14.32) represent different components of the appropriate (so-called *chiral*) superfield — in the same sense as the numbers a_1, a_2, a_3 represent different components of the vector **a**.

The underlying mathematical structure of the supersymmetric quantum mechanical models of the preceding section was the algebra (14.9). What is the analogue of this algebra in four dimensions?

Well, we want to keep Lorentz invariance. This means that the role of the Hamiltonian should be taken by the operator of 4-momentum P_μ. The supercharges should also belong to a representation of the Lorentz group. Bearing in mind that the generators of supersymmetry make fermions out of bosons and that fermion fields are the spinors ψ_α and $\psi_{\dot\alpha}^\dagger$, it is easy to conclude that the supercharges should also have the spinor nature. It is then almost trivial to guess the form of the algebra:

$$\{Q_\alpha, Q_\beta\}_+ = \{Q_{\dot\alpha}^\dagger, Q_{\dot\beta}^\dagger\}_+ = 0$$

$$\{Q_\alpha, Q_{\dot\beta}^\dagger\}_+ = (\sigma^\mu)_{\alpha\dot\beta} P_\mu. \quad (14.37)$$

Equation (14.32) is the simplest supersymmetric model. There are many others. An interesting one is *supersymmetric Yang–Mills theory* (SYM). Its Lagrangian represents a non-Abelian generalization of the Lagrangian of supersymmetric photodynamics (14.31) and reads

$$\mathcal{L}_{\text{SYM}} = -\frac{1}{4}G_{\mu\nu}^a G_{\mu\nu}^a + i\psi^a \sigma^\mu \mathcal{D}_\mu \psi^{\dagger a}, \quad (14.38)$$

where a is the adjoint colour index and $\mathcal{D}_\mu$ is the covariant derivative, $\mathcal{D}_\mu X^a = \partial_\mu X^a + g f^{abc} A_\mu^b X^c$.

At tree level, the spectrum of the theory involves the massless gluons, which are bosons, and the massless fermions (the quantum states for the field ψ_α^a), called the *gluinos*. Let the gauge froup be $SU(N)$. Taking into account two helicity polarizations, $h = \pm 1$, for the gluons and two helicity polarizations, $h = \pm 1/2$, for the gluinos, we obtain $2(N^2 - 1)$ quantum gluon states and the same number of quantum gluino states of a given momentum **p**. Each gluon state has a gluino superpartner with the helicity differing by $1/2$ from the gluon helicity. They form together a separate

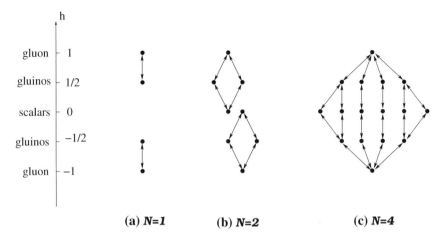

Fig. 14.3 Supermultiplets in the perturbative spectrum of different supersymmetric Yang–Mills theories. The arrows pointing down describe the action of Q's while the arrows pointing up — the action of $Q^\dagger$'s. In the case of the $\mathcal{N} = 4$ multiplet, not all the supercharge actions are shown.

supermultiplet. Thus, the properly defined supercharge[12] Q acting on a gluon state of helicity $h = 1$ gives a gluino state of the same momentum and colour carrying helicity $h = 1/2$, whereas $Q|h = 1/2\rangle = 0$. On the other hand, $Q^\dagger|h = 1/2\rangle = |h = 1\rangle$, whereas $Q^\dagger|h = 1\rangle = 0$. There are also supermultiplets formed by the gluon states with $h = -1$ and the gluino states with $h = -1/2$ (see Fig. 14.3a).

However, the theory (14.38) is asymptotically free, and that probably means that it is confining. Then the physical spectrum involves only massive hadron-like states made of the fundamental gluon and gluino fields: *glueballs*, which are bosons, and *glueballinos*, their fermionic superpartners.

As was also the case in quantum mechanics, the algebra (14.37) can be extended to include extra supercharges. Forget for a moment about confinement. The extra supercharges mean the presence of extra gluino states in the *perturbative* spectrum. In the extended $\mathcal{N} = 2$ supersymmetric Yang–Mills theory, there are two different gluino fields and twice as many gluino states as the gluon ones. To provide for the equal number of bosonic and fermionic states, as is required by supersymmetry, some extra bosonic states should be found. And indeed, the full $\mathcal{N} = 2$ SYM Lagrangian also includes, besides the fields A_μ^a, $\psi_\alpha^{a\,(1)}$ and $\psi_\alpha^{a\,(2)}$, two real scalars in the adjoint representation of the group. For each value of the colour index a and momentum **p**, we have in this case two supermultiplets, each of them including four degenerate states, as shown in Fig. 14.3b.

---

[12]For example, if the 3-momentum is directed along the positive z axis, so that $p^\mu = (E; 0, 0, E)$ and $p_\mu = (E; 0, 0, -E)$, it is the component $Q_{\alpha=2}$ of the supercharge and its Hermitian conjugate which are relevant — out of all the anticommutators $\{Q_\alpha, Q_{\dot\beta}^\dagger\}_+$ in (14.37) only $\{Q_2, Q_{\dot 2}^\dagger\}_+$ has in this case a nonzero projection on the particle states, while the supercharges Q_1 and $Q_{\dot 1}^\dagger$ decouple.

The $\mathcal{N} = 4$ SYM theory is especially interesting. Besides the gauge field it involves four different gluino fields and six different real scalars. At the tree level, it involves for each momentum **p** and colour a a *single* supermultiplet (rather than two different ones as was the case for $\mathcal{N} = 1$ and $\mathcal{N} = 2$). This multiplet has sixteen degenerate states: $1 + 1 = 2$ gluon states with the helicity $h = \pm 1$, $4 + 4 = 8$ fermion states with $h = \pm 1/2$ and 6 scalar states of zero helicity. The wonderful equality

$$2 + 6 = 8$$

holds, so that the numbers of the bosonic and fermionic states are equal.

In $\mathcal{N} = 4$ theory, the coupling constant does not "run" — the different perturbative contributions in its renormalisation exactly cancel out in all orders. That means that the theory is not confining and the tree-level perturbative states are the true physical eigenstates of the Hamiltonian.[13]

Another interesting theory is supergravity. We defer our discussion of this topic to the final chapter.

14.4 The Case for Supersymmetry

Some of the reasons why most people now believe that the supersymmetry is really there somewhere, were given in Sect. 12.7. We will rediscuss them here, while making some extra remarks.

You have probably heard the story how Lord Kelvin (to whom we owe the laws of thermodynamics in their modern form together with the Kelvin scale of absolute temperature) gave a speech back in 1900 in the Royal Society, where he said that the building of theoretical physics (with a design very different, of course, from what we talked about in Chap. 4) had been mainly constructed, though its beautiful harmony was obscured by two small clouds:

1. The strange result of the experiment of Michelson and Morley, who sought to observe the luminiferous aether, but failed.
2. The ultraviolet catastrophe for black body radiation (according to classical theory, the luminocity of a black body was infinite).

Kelvin called attention to these two imperfections hoping that they would be eliminated soon, after which physicists would be able to rest on their laurels. On a different occasion, Albert Michelson expressed a similar idea in an even stronger form: ... *It seems probable that most of the grand underlying principles have been firmly established....future truths of physical science are to be looked for in the sixth place of decimals.*

---

[13] Well, I am simplifying here. Just as the charged asymptotic states in QED are not simply electrons, but electrons accompanied by the electromagnetic field that they create, this field representing a coherent mixture of many soft photons, the physical states in the $\mathcal{N} = 4$ SYM theory represent similar coherent mixtures. But if the coupling is small, this effect is not so relevant.

We now know how wrong Kelvin and Michelson were. The "clouds" mentioned by Kelvin refused to dissolve. Instead, the first cloud induced Lorentz, Poincaré and Einstein to formulate special relativity, and the second cloud was the germ from which the whole of quantum theory emerged.

I recalled this story because nowadays we are in a similar situation. The building of the Standard Model is constructed. But its beauty is obscured by two clouds:

1. As was discussed in Sect. 12.7, the Standard Model is not internally consistent due to the absence of asymptotic freedom in the Abelian gauge and Yukawa sectors and quadratic divergences in the scalar field sector.
2. It is absolutely unclear at the moment why the vacuum energy in our Universe is so close to zero [cf. the discussion in Sect. 4.7]. The observed cosmological constant (dark energy) is less than the natural scale $\sim m_P^4$ by ~ 122 orders of magnitude!

Let us concentrate on the second cloud. You have just learned that the energy of a supersymmetric vacuum state that is annihilated by the action of all supercharges must be zero. Bearing this in mind, the unbelievably small value of the vacuum energy in our World represents a definite smoking gun indicating the presence of supersymmetry.

On the other hand, we do not see much supersymmetry around. Hence, if supersymmetry exists at the fundamental level, it must be *broken*. But broken in such a way that the vacuum energy is still pretty close to zero, while the (low-energy) particle spectrum carries no trace of supersymmetry.

A symmetry can be broken explicitly when a term not invariant under the corresponding symmetry is added to the Lagrangian. For example, the quark mass term explicitly breaks the chiral symmetry $SU_L(N_f) \times SU_R(N_f)$ of massless QCD (see Sect. 11.5.3). Supersymmetry might also be broken in this way.

But supersymmetry can also be broken spontaneously when the action is still exactly symmetric, but there is no vacuum supersymmetric state. This scenario is much more aesthetically appealing and has in my opinion a better chance of turning out to be true. Following our habitual approach — to explain complicated things by studying simple toy models — we present a very simple SQM model involving spontaneous breaking of supersymmetry, where the vacuum energy is nonzero, but is exponentially small and can be made as close to zero as one wishes.

As we mentioned at the end of Sect. 14.1, supersymmetry is broken spontaneously for the electron moving in a magnetic field (14.6) of zero flux. But we would like now to talk about a model that is even simpler. This model (suggested by Edward Witten) has the following quantum supercharges and the Hamiltonian:

$$Q = \frac{1}{\sqrt{2}} \psi [p - i W'(x)],$$

$$Q^\dagger = \frac{1}{\sqrt{2}} \hat{\psi}^\dagger [p + i W'(x)]; \qquad (14.39)$$

$$H = \frac{p^2 + [W'(x)]^2}{2} + \frac{1}{2}W''(x)(\hat{\psi}\hat{\psi}^\dagger - \hat{\psi}^\dagger\hat{\psi}). \tag{14.40}$$

where $p = -i\partial/\partial x$ and ψ and $\hat{\psi}^\dagger$ can be identified in the matrix formulation with σ_- and σ_+, so that the Hamiltonian (14.40) describes the one-dimensional motion of a spin $\frac{1}{2}$ particle in the external potential $V = \frac{1}{2}[W'(x)]^2$ and the magnetic field $W''(x)$.

Let $W(x)$ be a polynomial. If the order of this polynomial is even, a supersymmetric vacuum state exists:

$$\Psi_{\text{vac}} = e^{-W(x)} \quad \text{or} \quad \tilde{\Psi}_{\text{vac}} = \psi e^{W(x)}, \tag{14.41}$$

depending on whether $\lim_{x\to\pm\infty} W(x)$ is plus or minus infinity. But if the order of $W(x)$ is odd, the wave functions (14.41) are not normalized, and a supersymmetric vacuum is absent. Let

$$W(x) = \frac{x^2}{2} + \frac{gx^3}{3} \tag{14.42}$$

and let g be small. In the bosonic spin-up sector, the effective potential is

$$V_{\text{bos}}(x) = \frac{[W'(x)]^2 - W''(x)}{2}. \tag{14.43}$$

It represents a skewed double-well potential, the right well being a little deeper than the left one. In the fermionic sector,

$$V_{\text{ferm}}(x) = \frac{[W'(x)]^2 + W''(x)}{2}, \tag{14.44}$$

and the left well is deeper (Fig. 14.4).

If $g \ll 1$, the barrier between the two wells is high, of order $1/g^2$. There are two degenerate ground states — bosonic and fermionic. The bosonic ground state is concentrated in the right well and its fermionic superpartner in the left well. Their energy is positive, but small:

$$E_0 \sim \frac{1}{g^2} \exp\left\{-\frac{1}{3g^2}\right\}. \tag{14.45}$$

It would be nice if a similar simple model described Nature. Unfortunately, life is more complicated. *Simple* models fail for different reasons. One of them is that we do not see around a massless *goldstino* particle.

The latter is a neutral massless fermion that necessarily arises in any theory involving spontaneous breaking of global supersymmetry by the same token that a massless Goldstone boson necessarily arises in a theory involving spontaneous breaking of an

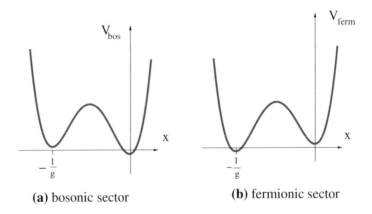

(a) bosonic sector   **(b)** fermionic sector

Fig. 14.4 Skewed double-well potentials in Witten's model.

ordinary global symmetry.[14] People can get rid of the goldstino, introducing more tricky models that include supergravity. Then the goldstino may be eaten up by the *gravitino*,[15] which becomes massive. This is the so-called super-Higgs effect, discovered by Dmitry Volkov and Vyacheslav Soroka in 1973, which is quite analogous to the ordinary Higgs effect when gauge bosons acquire mass by consuming ordinary Goldstone bosons.

Well, one can get rid of the goldstino this way, but then the dark energy strikes back with a vengeance. A natural scale for the vacuum energy density is in this case of order $m_{\text{gravitino}}^4$. As we do not see the gravitino in experiment, its mass should be at least $\sim 100\,\text{GeV}$, which gives $\epsilon_{\text{vac}} \gtrsim (100\,\text{GeV})^4$. And this is still larger than the observed value of the cosmological constant, maybe not by 122, but at least by 60 orders of magnitude. One has to admit that the second cloud mentioned on p. 282 is still in the sky…

We now examine the first cloud. The situation is better there. It is true that supersymmetry may mend certain ugly features and inconsistencies of the Standard Model. The main point is that supersymmetry may engender interesting cancellations and thus eliminate certain disturbing divergences. We already talked about the cancellation of different divergent contributions to the vacuum energy. Another such important cancellation is the cancellation of quadratic divergences in the Higgs mass: in supersymmetric versions of the Standard Model they are absent.

A related fact is the nice resolution of the so-called *hierarchy problem*, provided by supersymmetry. It is a problem characteristic of the Grand Unified Theories. As was discussed in Sect. 12.7.2, Grand Unified Theories assume the existence of a large underlying gauge group G that is broken down to $SU(3) \times SU(2) \times U(1)$ at the scale $M_{\text{GUT}} \sim 10^{15}\,\text{GeV}$ by the Higgs mechanism. The corresponding Higgs particles (the

---

[14]This effect was first pointed out by Dmitry Volkov and Vladimir Akulov in their pioneer paper in 1972.

[15]A hypothetical superpartner of the graviton of spin 3/2 — see the last chapter.

scalar X bosons in Fig. 12.11) also should have mass of order M_{GUT}, which is much, much larger than the mass of the Higgs boson we are familiar with. In an ordinary non-supersymmetric model, it is not clear why this large GUT scale does not show up in the loop graphs describing perturbative corrections to the Higgs mass.

And in supersymmetric GUT models, the two different scales M_{GUT} and M_{EW} do not talk to each other: a mouse (the ordinary Higgs boson) is not disturbed by supersymmetric elephants (the scalar X bosons and their superpartners) and its mass stays small.

Another nice feature of supersymmetry was mentioned before. If we assume underlying supersymmetry, the three different gauge couplings (12.71) of the Standard Model meet after "running" at the same point $\sim 10^{16}$ GeV, whereas that is not quite so in the non-supersymmetric Standard Model (see Fig. 12.10).

The fact that the unification scale in supersymmetric GUT models is somewhat greater than in nonsupersymmetric ones means that the proton decay probability is further suppressed there, and there is no contradiction with experiment any more!

An additional point (an additional cloud) where supersymmetry might prove helpful is the problem of *dark matter*. It is an experimental fact that, in addition to the stars and interstellar gas that we can observe with our optical and radio telescopes, our Universe contains another kind of matter, the "dark matter", which shows up only through its gravitational interaction (affecting the orbital velocities of the stars in galaxies) and not in any other way.[16] According to modern estimates, the total mass of dark matter in the Universe is about five times larger than the visible mass.

We are now pretty sure that dark matter is there, but we do not know what it really *is*. It should represent some neutral massive particles participating neither in the electromagnetic, nor in the strong interactions, relics from the epoch when the Universe was hot. The reader is now probably thinking about the neutrinos. Indeed, they are the first natural candidates for the role of dark matter. It turns out, however, that their masses are too small to provide for the total observed dark matter mass.

Thus, dark matter consists not of the neutrinos, but of a different kind of WIMPs (*Weakly Interacting Massive Particles*). Supersymmetric models *do* provide appropriate candidates. These WIMPs might be the gravitinos that acquired mass by the super-Higgs mechanism (we briefly mentioned this possibility above). Or they might be massive photinos, or something else…

It is my duty to warn the reader, however. Not all the theorists now believe that supersymmetry is there in our physical world. An argument against it is the fact that no trace of supersymmetric partners of known particles listed in Table 3.1 has been seen in the accelerator experiments at LHC at CERN. The search for supersymmetry was among the main motivations (possibly *the* main motivation) for building this accelerator. People really hoped to see something supersymmetric there, but alas, these hopes have not been realized. Only lower bounds on the masses of supersymmetric particles were established.

---

[16]I mean that it *has* not been observed in any other way, not that it *cannot* be observed. Experiments on the direct dark matter search are now running and more sensitive experiments are planned.

Let us concentrate on one particular such bound. We now know that if the scalar superpartner of the top quark exists, its mass should be large enough,

$$m_{\text{stop}} \gtrsim 650 \, \text{GeV} \, . \tag{14.46}$$

This brings about the following problem. As was mentioned, in supersymmetric theory quadratic ultraviolet divergences in the Higgs mass cancel out. But if super-symmetry is broken at low energy, the cancellation is not in some sense complete — *finite* contributions to m_H^2 survive. These contributions are proportional to the scale at which supersymmetry is broken and which determines the masses of the superpartners. The most significant contributions come from the loops involving the heaviest quark (the top quark) and its scalar superpartner. Roughly speaking,

$$\Delta m_H^2 \approx 4 m_{\text{stop}}^2 \, . \tag{14.47}$$

Bearing in mind the bound (14.46), that gives a value that is at least ~ 100 times larger than the experimental value for m_H^2!

No, this is not a disaster. One can always assume that the original bare Lagrangian involves a negative mass term for the Higgs field[17] that cancels the large loop contri-bution. But you probably agree that such *fine tuning*, when 99% of a large quantity is cancelled out and only 1% is left, is a little unnatural...But not impossible...

At present, we cannot say much more.

---

[17]Or rather for the fundamental complex scalar doublet ϕ.

Chapter 15
General Relativity

It took me some time to decide on the proper place for this chapter — in the *Entrées* or in the *Dessert* part of the book. On one hand, Einstein's classical theory of gravity describes Nature (we know that from experiment) and we now understand it fairly well. It would be natural to consume this dish among the *Entrées*, leaving conjectures and unanswered questions for the *Dessert*. But on the other hand, this book is mostly about *quantum* rather than classical field theory. My main intention was to give you an idea about what the Standard Model is. That was achieved by the end of Chap. 12, providing a good opportunity to take a break. And thirdly, the general relativity chapter is closely related to the last chapter in our book, devoted to attempts to construct quantum gravity (and full of conjectures and unanswered questions). I even wanted first to write up one large gravity chapter, but decided later to separate the classical and quantum stuff.

I have to say that my second and third considerations prevailed, and we are going to discuss classical gravity in detail *now*. In Chaps. 3 and 4 we already said few words about it. The time has come to write some formulas.[1]

Theoretical physics is a team venture. Many brilliant scientists worked on it before and are working now. And as you noticed, for the vast majority of parts (like quantum mechanics, QED, QCD) or chapters (like e.g. the constituent quark model) in the book of theoretical physics, it is impossible to name a single author. This is also true of special relativity. The contribution of Einstein was the most important one, but he had predecessors (Lorentz, Michelson, Poincaré), without whom he would not have been able to write his own seminal paper.

---

[1]Do not be afraid, however. We will not write so many of them. As is also true for the other parts of the book, our goal is not to learn the subject at the professional level. For those who want to do that, I recommend either the five last chapters in the *Field Theory* volume of the Landau course, or *Gravitation and Cosmology* by Weinberg. (Be careful, however. The sign conventions in these two books are different. For example, Weinberg defines the flat Minkowski metric as $\eta_{\mu\nu} = (-+++)$. We follow the Landau–Lifshitz conventions.).

© Springer International Publishing AG 2017
A. Smilga, *Digestible Quantum Field Theory*,
DOI 10.1007/978-3-319-59922-9_15

But general relativity is almost an exception: it is largely the creation of one man, of Einstein. I said, "almost", because there was also an important contribution from Hilbert, who figured out the form of the action from which Einstein's equations can be derived. One can also mention Karl Schwarzschild, who obtained an important and beautiful solution to Einstein's equations; we are going to discuss that later. But the contribution of Einstein was definitely the decisive one. I do not know for how long the discovery of general relativity would have been delayed without him. For 10 years? Or maybe more?

Einstein started thinking about the problem of constructing a relativistic theory of gravity very soon after publishing his papers on special relativity in 1905. It took him 10 years to solve it. The theory was finally formulated in November 1915.

15.1 Curved Space-Time

15.1.1 Mathematics

In ordinary field theories, the dynamical variables are functions of the time and spatial coordinates, but space and time themselves are fixed and flat. The Lorentz-invariant interval between two events is $s^2 = c^2 t^2 - \boldsymbol{x}^2$.

But gravity is different. Its dynamical variables are the components of the metric tensor $g_{\mu\nu}$, which depend on $\boldsymbol{x}$ and t. This tensor defines the invariant interval between two infinitesimally close events:

$$ds^2 = g_{\mu\nu}dx^\mu dx^\nu . \tag{15.1}$$

Thus, space-time is not flat any more, and its fabric changes from point to point.

We have just written the word "tensor". But it is not a tensor in the same sense as we used it before. In Sect. 6.1 we defined tensors in flat space or in flat space-time as sets of components that transformed in a particular way under rotations and Lorentz boosts. In curved space a tensor is a set of components that transform in a particular way under *general coordinate transformations*, when the coordinates x^μ go over into the new coordinates $x'^\mu(x^\nu)$, representing arbitrary smooth functions of the old ones.

Similar to what is the case for flat Minkowski space tensors, there are indices of two types — contravariant and covariant. A contravariant vector transforms in the same way as the differential dx^μ:

$$V'^\mu = \frac{\partial x'^\mu}{\partial x^\nu} V^\nu . \tag{15.2}$$

A covariant vector transforms in the same way as the gradient operator ∂_μ:

$$V'_\mu = \frac{\partial x^\nu}{\partial x'^\mu} V_\nu . \tag{15.3}$$

For two vectors of different types, V^μ and W_μ, one can define a scalar product $V^\mu W_\mu$, which is invariant; it does not depend on coordinate choice.

Further, one can define tensors with several contravariant and covariant indices. Their transformation law is an obvious generalization of (6.13). For example, the metric tensor entering (15.1) is a covariant second-rank symmetric tensor, and it transforms as

$$g'_{\mu\nu}(x') = \frac{\partial x^\sigma}{\partial x'^\mu} \frac{\partial x^\rho}{\partial x'^\nu} g_{\sigma\rho}(x). \qquad (15.4)$$

For any contravariant vector V^μ, the vector $V_\mu = g_{\mu\nu} V^\nu$ is covariant. To make a contravariant vector out of the covariant one, we define the inverse metric tensor $g^{\mu\nu}$ by $g^{\mu\nu} g_{\nu\sigma} = \delta^\mu_\sigma$. Then $V^\mu = g^{\mu\nu} V_\nu$.

We have said "curved space-time". But what *is* curvature? Everybody has an intuitive notion about it, but one must be careful. In the technical sense that mathematicians and physicists use the term, a sphere is curved, but the surface of a cylinder or of a cone is not.

Consider three geographical points: N — the North Pole, Q — Quito in Ecuador and L — Libreville in Gabon. Both cities are very close to the equator, but Quito is located at 78°35′W, whereas Libreville is located at 9°27′E. They are separated by ≈88° of longitude. We will take an approximate value 90°, a quarter of the equator.

We now would like to ask Sophie (who is still with us, I am sure) to take a plane to Quito and after doing proper sightseeing and shopping to move from there to Libreville by land and sea along the equator (it is the shortest way, a geodesic!), carrying with her a vector tangent to the Earth surface, represented by an umbrella (Fig. 15.1). I would specially ask her to keep the direction of this tangent vector constant, to perform its *parallel transport*. Mathematically, this means that the angle between the vector and the geodesic path is kept constant. In the first part of her trip, Sophie can choose this angle to be zero.

On arriving in Libreville, Sophie should change the direction of her motion and go to the North Pole along the meridian (which is also a geodesic). She still has to carry her umbrella, keeping its angle with the meridian equal to $\pi/2$. At the North Pole she should turn again, now heading to Quito, still taking care not to change the value π of the angle that the tangent vector makes with the Quito meridian. It is clear that, when she arrives at Quito and finally puts the umbrella down, she will find out that it has turned by $\pi/2$, compared to its direction in the beginning of her quest (Fig. 15.2).

And this means that the surface of the Earth is curved! One can *define* the curvature as the ratio[2]

$$\mathbb{R} = \frac{\Delta\phi}{\mathcal{A}}, \qquad (15.5)$$

where $\Delta\phi$ is the angle by which a tangent vector is rotated after a parallel transport around a closed circuit and $\mathcal{A}$ is the area that our circuit encloses.

---

[2]To avoid confusion, this is the so-called *Gaussian curvature*, to be distingushed from the *Riemannian scalar curvature*. We will define the latter very soon.

Fig. 15.1 Eastbound.

The area of the curvilinear triangle QLN is $\mathcal{A} = \pi r_E^2/2$, r_E being the radius of the Earth, and hence the Gaussian curvature of the Earth surface is

$$\mathbb{R}_E = \frac{1}{r_E^2}\,. \tag{15.6}$$

Fig. 15.2 Rotation of a
tangent vector after going
around a closed circuit.

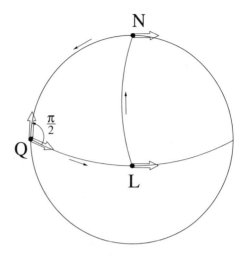

Note that the tangent vector that Sophie carried turned *counterclockwise*, the same
way as her head, while she moved along the circuit $Q \to L \to N \to Q$. This means
that the curvature is *positive*. There are also surfaces with negative curvature where
the tangent vector is rotated in the direction opposite to the direction of rotation of
traveller's head (one example is a horse's saddle, another one is the Lobachevsky
plane). To make the definition (15.5) universal, one should take care of the sign of
$\Delta\phi$ and also include there the sign factor labelling the direction of the circuit.

The sphere has a constant curvature, and to determine it this way, one can consider
a large circuit (as we did). But generically the curvature is different at different points
and should be locally defined as the angle of rotation of a tangent vector due to a
parallel transport around an infinitesimal circuit divided by the infinitesimal area
enclosed by this circuit.

Note that the Gaussian curvature (15.5) of the lateral surface of a cylinder is zero
— an umbrella does not change its direction during parallel transport there. The
reason is simple: a cylinder cap can be made by folding a flat sheet of paper, and the
Gaussian curvature is *invariant* under such foldings, one should only take care that
there are no contractions or dilatations.

For manifolds of dimension 3 and higher, one can define many different curvatures.
The rotation angles now depend on the orientation of an infinitesimal circuit in the
manifold and on the initial direction of the tangent vector. The infinitesimal area
element is now an antisymmetric tensor $\delta\mathcal{A}^{\alpha\beta}$ (for example, if this tensor has only
one component $\delta\mathcal{A}^{12} = -\delta\mathcal{A}^{21}$, the circuit lies in the (12) plane). The different
curvatures are the components of the *Riemann curvature tensor* $R_{\mu\nu\alpha\beta}$. After going
around a small closed circuit, a vector V^μ is shifted by

$$\delta V^\mu = -\frac{1}{2} R^\mu{}_{\nu\alpha\beta} V^\nu \delta\mathcal{A}^{\alpha\beta}. \tag{15.7}$$

The Riemann tensor has the dimension $[1/l^2]$. It can be expressed via second spatial derivatives and the squares of first derivatives of the metric tensor. This expression is not so simple:

$$R_{\mu\nu\alpha\beta} = \frac{1}{2}\partial_\nu\partial_\alpha g_{\mu\beta} + \text{many other terms},\qquad(15.8)$$

and it is one of the facts that makes calculations in general relativity technically difficult.

A generic 4-dimensional rank-4 tensor has 256 different components, but the Riemann tensor satisfies certain constraints:

$$R_{\mu\nu\alpha\beta} = -R_{\nu\mu\alpha\beta} = -R_{\mu\nu\beta\alpha} = R_{\alpha\beta\mu\nu}\,,$$
$$\epsilon^{\mu\nu\alpha\beta}R_{\mu\nu\alpha\beta} = 0\,.\qquad(15.9)$$

As a result, it has only 20 independent components.[3]

We will also define the second-rank symmetric *Ricci tensor*

$$R_{\mu\nu} = g^{\alpha\beta}R_{\mu\alpha\nu\beta}\qquad(15.10)$$

and the Riemannian scalar curvature

$$R = g^{\mu\nu}R_{\mu\nu} = g^{\mu\nu}g^{\alpha\beta}R_{\mu\alpha\nu\beta}\,.\qquad(15.11)$$

For two-dimensional surfaces, the latter coincides with the Gaussian curvature up to a factor of 2: $R = 2\mathbb{R}$.

15.1.2 Physics

Following Einstein, we are going to apply this formalism to the geometry of our (3+1)-dimensional world. The idea that our space-time is curved is counterintuitive. It may be met with protests by a person who hears it for the first and even for the second time. These protests are quite legitimate. I explained the notion of curvature by sending Sophie to travel along the Earth surface. "But OK", you might say, "the surface of the Earth is curved, but the Earth itself is a 3-dimensional object, and we seem not to see any traces of curvature in our 3-dimensional space."

This question is so important that it deserves giving not one, but several answers. First, you may question your intuition:

• It is very good to have one. It helps us to learn and understand complicated things, to make them simple and handy. But the intuition of any person depends on his/her previous experience. I think our stone age ancestors (who were not much more

---

[3]In 3 dimensions, $R_{\mu\nu\alpha\beta}$ would have 6 independent components and in two dimensions only one independent component, R_{1212}, related to the Gaussian curvature.

stupid than we are) would vigorously protest against the idea that the *Earth* (leave alone space) is curved, the presence of the horizon notwithstanding.

- Suppose that the Earth is perfectly smooth and round, and that gravity is so strong that you cannot even think about lifting your head and the only thing you can do is to *crawl* along the surface. In this case, your Universe would be 2-dimensional. The claim that it is curved could at first seem intuitively crazy to you, but you could verify it by performing the Sophie experiment on parallel transport of tangent vectors.

To give a more satisfactory and scientific answer, we go back to three spatial dimensions. It is true that the geometry of the world that we know is very close to the Euclidean one, but it makes sense to ask — *how do we know that?* Well, one of the theorems of Euclidean geometry says that the sum of the angles in a triangle is equal to π. One can check it experimentally. Draw a triangle, measure the angles by a protractor and add them. You get π, OK. But *how* did you draw your triangle? Presumably, on a sheet of paper using a ruler. The size of your triangle was thus not larger than 30 cm, the size of the A4 sheet. To check the validity of your theorem at an essentially larger scale, you need to modify the procedure.

One of the possibilities is the following. Pick three high mountains at a distance of several dozen kilometers between one another. Climb up each mountain and measure by a theodolite the angles between the directions by which you see the summits of two other mountains. Then go down to your desk and sum the angles.

In fact, this is not a *gedankenexperiment*. It was actually performed by the king of mathematicians of the first half of the 19th century Carl Friedrich Gauss, when he was charged to map the Kingdom of Hanover. He misused the allocated public funds and the geodesics data he had collected[4] for the highly theoretical purpose irrelevant to the taxpayers of finding out whether space was Euclidean or not — Gauss understood the problem and had suspicions that it might not be. But the sum of the angles coincided with π within the experimental uncertainties.

A similar experiment was contemplated by one of the creators of non-Euclidean geometry, Nikolai Lobachevsky. In 1826 he proposed going to the sky and studying the curvature of space by measuring *stellar parallaxes*.[5] One can contemplate two sources for a nonzero parallax: *(i)* a finite distance from the Earth to the star and *(ii)* a finite spatial curvature. One could in principle compare the parallaxes of different stars and extract the value of the curvature (or put an upper limit on it). At that time the parallaxes and the distances to stars had not yet been measured, and Lobachevsky could only conclude that the radius of curvature of our space is larger than ~5 light years. When the parallaxes were measured in the middle of the 19th century, they were all explained by finite stellar distances, and no traces of curvature were detected.

When describing the settings of the Gauss experiment, I wrote "measure the angles between the directions by which you see..." This actually means that the sides of our

---

[4]Probably, not directly him. The Kingdom of Hanover was not so small, and there would have been a whole team of topographers.

[5]The shifts of stellar positions on the celestial sphere between summer and winter, when the Earth is on the opposite points of its orbit.

triangle are represented not by the straight lines drawn by a ruler, but by rays of light — there is nothing straighter in our physical World!

The point is that even the rays of light are not *quite* straight. They are deflected in a gravitational field. This is one of the predictions of general relativity, but the effect is very natural; Newton would not be surprised by it. Indeed, photons are material particles. They carry energy and they are sensitive to gravity. In particular, if they fly by the Sun, the latter attracts them, so that the photons slightly change their direction of motion. As a result, the position of a distant star (the source of the photons) on the celestial sphere is shifted a little, if the Sun happens to pass nearby. This effect was first observed by the expedition of Arthur Eddington during a total solar eclipse in 1919. The observed value of the deflection triumphally agreed with the predictions of general relativity.

Thus, saying that space-time is curved in the presence of gravitating matter is tantamount to saying that the rays of light are deflected by gravitational fields; the rays are *bent*, they are not straight any more. Of course, this concerns not only rays of light, but everything else; for example, neutrinos with energy much larger than their mass behave in exactly the same way as photons. A distinguishing feature of the gravitational interaction is its universality, and that is why it is convenient to describe the gravitational effects not as forces applied to material bodies dwelling in Newtonian absolute flat space, but as a distortion in the fabric of space-time itself.

One had to wait a century and a half before the curvature of space was actually detected following the Gauss-Lobachevsky scheme. I am talking about *gravitational lenses*.

Suppose that somewhere in deep space there exists a very massive gravitating object X (it can be a cluster of galaxies). Suppose that we are observing a bright quasar or a galaxy behind this massive object. Then the photons flying by the object X are deflected. If they pass X from above, they are deflected down and if they pass X from below, they are deflected up.[6] As a result, two different rays can meet in the eye of an observer. The latter will see two twin quasars, instead of one!

We thus obtained a "diangle", the sum of whose angles is not zero, as Euclid would insist, but nonzero and positive. This signals a positive spatial curvature. The first such gravitational lens was discovered in 1979. One sees a twin quasar at a distance ∼7.8 billion light years from us. The angle of this "anomalous diangle" (the angular separation between the twin images on the celestial sphere) is about 6 arcseconds (Fig. 15.3).

---

[6]It is like a free kick performed by a football professional. Due to the Magnus effect, it can go around the wall right into the goal!

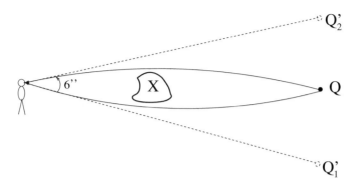

Fig. 15.3 A gravitational lens. X — a cluster of galaxies, Q — a quasar, $Q'_{1,2}$ — its images.

We explained how the fact that our space-time is curved can be established by doing experiments *in* it.[7] Curvature is an *intrinsic* property of a manifold,[8] and to explain it, one does not need to claim that our Universe is embedded somewhere.

But maybe it still *is* embedded in flat space of some higher dimension and its curvature has in fact an external origin, similar to the curvature of the Earth surface or of a soap bubble? We do not know. But I can say right now that many theorists, including your author, believe that this might be indeed the case. We will discuss this speculative idea in the last chapter.

15.2 Actions □ Equations of Motion □ Their Solutions

Einstein's theory of gravity is generally considered to be a complicated theory. The complications are there, indeed: one has to learn the unusual formalism of Riemannian geometry and, in addition, there are some complications of a fundamental nature that we will talk about in the last chapter. However, once the formalism is set, it becomes a very simple, even trivial exercise to write down the action of the theory.

The invariant action describing the motion of a particle (or any point mass, a planet) in an external gravitational field reads

---

[7] As you understand, this is the essence of the scientific method. To explain what any notion in physics means, one should specify the way it can be measured. This applies to curvature in general relativity, as well as the notions of time and distance in special relativity. The only difference is that one does not need fancy maths to explain how the light watch works and why time measured in a moving frame is not the same as time measured at rest. Whereas for curvature, for general relativity, one still has to develop a certain mathematical formalism not present either in the high school or in the standard university curriculum.

[8] This concerns the Riemann curvature tensor and its contractions. There exists also the so-called *extrinsic* curvature, which depends on how the manifold is folded (or *embedded* in a higher-dimensional Euclidean space). It is not zero for a cylinder. But we will not talk more about this extrinsic curvature, as it seems to be irrelevant for physics.

$$S = -mc \int ds \,. \tag{15.12}$$

In fact, it was already written down in (7.16). The only difference is that now space-time is curved and the interval ds is given by (15.1) rather than (7.17).

It is also not so difficult to write down the action of the gravitational field. We want it not to depend on the choice of coordinates. The simplest invariant of general coordinate transformations is the invariant volume

$$V = \int \sqrt{-g}\, d^4x \,, \tag{15.13}$$

where $g = \det(g_{\mu\nu})$ is the determinant of the metric tensor.[9]

To obtain the dimensionless action, we should multiply (15.13) by a constant Λ carrying the dimension m^4. This is nothing but the cosmological constant giving the vacuum energy density. As we have already mentioned on p. 50 and discussed again in the previous chapter, experiment says that it is nonzero, but very small, $\Lambda \approx 10^{-122}$ in Planck units.

A more complicated invariant involves the scalar curvature R in the integrand. It gives the *Einstein–Hilbert action* of general relativity:

$$S_{\text{grav}} = -\frac{m_P^2}{16\pi} \int R\sqrt{-g}\, d^4x \,. \tag{15.14}$$

In this expression, m_P is the Planck mass. The coefficient $1/16\pi$ is chosen such that at the end of the day the combination $1/m_P^2$ acquires the dynamical meaning of Newton's constant G_N. The negative sign provides for the universal gravitational *attraction*, rather than repulsion [see Eq. (15.27) below].

Consider now the action

$$S = -\frac{m_P^2}{16\pi} \int R\sqrt{-g}\, d^4x + \int \mathcal{L}\sqrt{-g}\, d^4x \,. \tag{15.15}$$

The Lagrangian $\mathcal{L}$ in the second term describes matter — all the ordinary (non-gravitational) fields filling out our space-time. To derive the gravitational equations of motion, we have to take the variation of the action over the metric $g_{\mu\nu}$ and equate it to zero.

The variation of the second term gives the object called the *energy-momentum tensor* of the matter field. By definition,

---

[9] As an example, you can take the metric on the surface of a unit sphere, $ds^2 = d\theta^2 + \sin^2\theta d\phi^2$. Then $\sqrt{g} = \sin\theta$ and the total area of the sphere is $\int_0^{2\pi}\int_0^{\pi} \sin\theta d\theta = 4\pi$. The determinant of a metric with Minkowski signature is negative, and that is why the extra minus sign was plugged in under the root.

$$T_{\mu\nu} = \frac{2}{\sqrt{-g}} \frac{\delta(\sqrt{-g}\,\mathcal{L})}{\delta g^{\mu\nu}}. \tag{15.16}$$

One can show that the component T_{00} is the energy density carried by the field.[10] The components $T_{0j} = T_{j0}$ give the momentum density and the spatial components T_{jk} are the stress tensor. If $T_{jk} = P\delta_{jk}$, P is the pressure.

Taking also into account the variation of the first term in (15.15) (we leave this calculation to the reader), we derive *Einstein's equations*:

$$R_{\mu\nu} - \frac{1}{2}g_{\mu\nu}R = \frac{8\pi}{m_P^2}T_{\mu\nu}. \tag{15.18}$$

In usual units:

$$R_{\mu\nu} - \frac{1}{2}g_{\mu\nu}R = \frac{8\pi G_N}{c^4}T_{\mu\nu}. \tag{15.19}$$

15.2.1 Gravitational Waves

Let space be empty, with no matter. Then $T_{\mu\nu} = 0$ and Einstein's equations imply the vanishing of the Ricci tensor:

$$R_{\mu\nu} = 0. \tag{15.20}$$

Let the metric be almost flat,

$$g_{\mu\nu} = \eta_{\mu\nu} + h_{\mu\nu}, \qquad h_{\mu\nu} \ll 1. \tag{15.21}$$

Substituting it into the exact expression for $R_{\mu\nu}$ via the metric [we have presented here only the stub (15.8)], expanding it in $h_{\mu\nu}$, keeping only the linear term and choosing the coordinates appropriately, the Eq. (15.20) reduces to

$$\Box h_{\mu\nu} = 0, \tag{15.22}$$

where $h_{\mu\nu}$ can be chosen such that the components h_{00} and h_{0i} vanish and $\partial_i h_{ij} = h_{ii} = 0$; only two independent parameters are left.

This is the wave equation that describes the propagation of gravitational waves. In a gravitational wave the distances between neighbouring spatial points oscillate. This oscillation has a quadrupole nature: if the wave propagates along the x-axis, then

---

[10]For example, for a complex scalar field in flat space, $\mathcal{L} = \partial_\mu\phi^*\partial^\mu\phi - m^2\phi^*\phi$, one can derive

$$T_{00} = 2|\partial_0\phi|^2 - \mathcal{L} = |\partial_0\phi|^2 + |\partial_j\phi|^2 + m^2|\phi|^2, \tag{15.17}$$

cf. footnote 11 in Chap. 7.

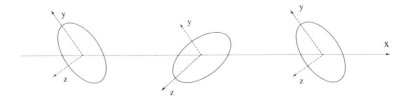

Fig. 15.4 Propagation of a linearly polarized gravitational wave.

a small increase of distance in the y-direction is accompanied by a small decrease in the z-direction, and vice versa. The maximal relative variation of the transverse distances is the amplitude of a gravitational wave called the *strain* (Fig. 15.4).

The direct experimental observation of gravitational waves is a very recent achievement. It was announced in February 2016. I will not describe this beautiful experiment in detail here. However, I will mention that the astrophysical source of the observed waves was the process of fusion of two black holes into a single one. The masses of the merging black holes were about 29 and 36 solar masses and their sizes and their relative distance were of the order of $d \sim 100$ km. The mass of the newly-merged black hole was ~ 62 solar masses, which means that energy of order $3M_\odot$ was radiated in the form of gravitational waves.

This catastrophe occurred far away, at a distance $L \sim 1.3$ billion light years from us. When the gravitational wave arrived at terrestrial detectors, it was very weak. A rough estimate for its strain is

$$h \sim \frac{d}{L} \sim 10^{-20}, \tag{15.23}$$

but the actual value was somewhat smaller, $h \sim 10^{-21}$. At a distance of 4 km (this was the arm-length of the Michelson interferometer used in the experiment), this corresponded to displacements of order 10^{-18} m, one thousandth of a proton size! It is really remarkable that it is possible to detect such tiny effects.

In 2017 this discovery was distinguished by the Nobel Prize. It was awarded to Rainer Weiss, Barry Barish and Kip Thorne *for decisive contributions to the LIGO detector and the observation of gravitational waves.*

Today, in November 2017, when I am writing this sentence, we are aware of five more absolutely credible observations of gravitational waves. In four such cases, the gravitational waves were generated by fusion of black holes and the characteristics of these events were similar to what was discussed above. But in one of the cases, the released energy was much less (only about 0.03 $M_\odot$) and the nature of this event (which occured much closer to us, only 100 million light years away) was different: it was a fusion of two neutron stars, accompanied by a burst of gamma-radiation. The possibility of such an event was first predicted by Sergei Blinnikov, Igor Novikov, Tatiana Perevodchikova and Alexander Polnarev in 1984.

15.2.2 *Schwarzschild Solution and Black Holes*

We now discuss another very important solution to Einstein's equations, which describes the static gravitational field of a poinlike massive source — the gravitational analogue of the Coulomb potential. In electrodynamics, the Coulomb static potential keeps its form also in Maxwell's relativistic theory — relativistic effects come into play there only if charges move and/or fields have nontrivial time dependence. It is not so in general relativity.

The expression for the static metric in the presence of a pointlike mass was found by Karl Schwarzschild in 1916, immediately after he read the famous Einstein paper on general relativity.[11] It has the form

$$ds^2 = \left(1 - \frac{r_g}{r}\right)c^2dt^2 - r^2(\sin^2\theta d\phi^2 + d\theta^2) - \frac{dr^2}{1 - r_g/r}, \tag{15.24}$$

where the gravitating mass M is placed at the origin and

$$r_g = \frac{2G_N M}{c^2} \tag{15.25}$$

is a constant called the *gravitational radius* of the gravitating mass. We have already met it before in Eq. (3.2).

The Ricci tensor calculated by this metric vanishes everywhere except at the origin, where it is infinite.

Let us first look at what is going on at large distances, $r \gg r_g$. In that case, we can neglect the ratio r_g/r in the last term, but not in the first term, where it is multiplied by the large factor c^2. We obtain:

$$ds \approx \sqrt{dt^2\left(c^2 - \frac{2G_N M}{r}\right) - dx^2}. \tag{15.26}$$

Expanding the root, plugging it into (15.12) and adding to S/t the irrelevant constant mc^2, we derive the familiar nonrelativistic Lagrangian

$$L = \frac{mv^2}{2} + \frac{G_N m M}{r}. \tag{15.27}$$

It is amusing how the Newton gravitational potential appears in this calculation — as a contribution in the temporal component g_{00} of the metric tensor!

---

[11] As we know and had another chance to recall during our *Trou Normand* break, human stories are often tragic. Schwarzschild served as an artillery officer on the Eastern front, but died not of shells or bullets but of a rare skin decease. This happened in May 1916, several months after he derived his brilliant result, for which he will always be remembered.

The next term in the expansion in r_g/r gives the so-called *post-Newtonian approximation*. In 1915 Einstein, not yet knowing the exact solution (15.24), calculated the post-Newtonian corrections to the Lagrangian (15.27) perturbatively. These corrections bring about distortions in the elliptical orbits of the planets. The major axes of the ellipses begin to turn a little. The effect is maximal for Mercury, the planet closest to the Sun. Einstein calculated the rate of this rotation[12] and the result (43 arcseconds in a hundred years) perfectly coincided with what was known from astronomic measurements.[13]

Interesting things happen at $r = r_g$. The temporal component of the metric vanishes there, and the radial component g_{rr} becomes infinite. This singularity seems to tell us that the solution loses its meaning at $r \leq r_g$, but that is not so. The *components* of the metric tensor and of the other tensors depend on the coordinate choice and as such do not have much of physical meaning. It is the scalar curvature *invariants*, like $R_{\mu\nu\alpha\beta}R^{\mu\nu\alpha\beta}$ or $R_{\mu\nu\alpha\beta}R^{\alpha\beta\rho\sigma}R^{\mu\nu}_{\rho\sigma}$ (the scalar curvature R or the invariant $R_{\mu\nu}R^{\mu\nu}$ vanish due to the equations of motion $R_{\mu\nu} = 0$) that have physical meaning. And they are not at all singular at $r = r_g$.

The sphere $r = r_g$ (called the *event horizon*) has, however, quite a distinct and important meaning. One can show that a ray of light emitted at some point below the horizon, $r < r_g$, has no chance to go outside; the only way it can go is *towards* the true singularity at $r = 0$. In other words, for $r \sim r_g$, when the gravitational field is strong (the deviations from the flat Minkowski metric are essential), the metric (15.24) describes a *black hole*.

We discussed the physics of black holes many centuries ago, in Chap. 3, but I want to add here one remark to this early discussion. Black holes are formed during gravitational collapse of stars or of still larger clumps of matter. One may ask — how fast does the collapse occur? The answer is not trivial. If we observe this collapse from outside, it never ends. We would see how material particles *approach* the event horizon, but will never see how they *cross* it. At the technical level, this is clear from Eq. (15.24). Suppose we send a ray of light in the radial direction directly to the centre. The trajectory of light is determined by the condition $ds = 0$. And we well see that, as r tends to r_g, the apparent velocity dr/dt becomes smaller and smaller. Our photon seems to be "frozen" there, it approaches the horizon, but is never able to reach it.

---

[12] The calculation is difficult, but the result is simple. It is especially simple when the ellipse has a small eccentricity, being close to a circle. The angular velocity Ω of rotation of such a nearly circular orbit is given by

$$\Omega = \frac{3v^2}{c^2}\omega, \qquad (15.28)$$

where v and ω are the linear and angular velocities of the planet.

[13] The effect of general relativity is not the only and not the most important cause urging the perihelion of Mercury's orbit to precess. This precession is mostly due to the gravitational tugs of other planets, giving 531" per century. However, precise measurements give the value 574", and it was the mismatch of 43" that general relativity explained.

The fact that the crossing of the horizon is not observable from outside is also clear heuristically: if you can see nothing from the region inside the horizon, how on Earth are you going to observe the particles crossing it?

But if, instead of dropping matter into the hole, you decide to go there *yourself*, carrying your watch with you, you will find out that travelling all the way to the horizon takes a finite time. If the black hole is large, like a quasar, which can be as large as the Solar System, you may not even notice the moment of crossing — the physical gravitational fields, the curvature invariants $R_{\mu\nu\alpha\beta}R^{\mu\nu\alpha\beta}$, etc. are still rather small at $r = r_g$. There is, of course, hot matter that falls in the hole together with you and makes the journey not nice,[14] but nothing particularly drastic happens right at the horizon.

But soon after that, as you continue your fall, the gravitational fields and the associated tidal forces rapidly grow. At the true singularity at the origin, they are infinite.

15.2.3 Friedmann Solution and Cosmology

The Friedmann solution is the *cosmological* solution describing not the "details" (a gravitational wave here, a massive gravitating body there, etc.), but the structure of the Universe as a whole. We described this solution in words and told the story about it in the several paragraphs on pp. 48–50, which you may now wish to reread.

But if you are too lazy to flip back through the whole book, I will briefly repeat here the essentials.

- Einstein looked for a stationary cosmological solution and was able to find it only by introducing the cosmological constant, giving a constant negative vacuum pressure.
- Friedmann found non-stationary solutions where the Universe either first expanded and then contracted (the *closed* model) or kept expanding forever (the *open* model).
- The expansion of the Universe was later confirmed experimentally.

We now have the tools to understand how these solutions were actually obtained and to describe them quantitatively.

The Universe is not homogeneous. It is made of clumps of matter of different size. There are stars, galaxies, clusters of galaxies and their superclusters. However, all those are *local* inhomogeneities. The largest known superclusters have the size of the order of half a billion light years, 20 times less than the total size of the Universe (its observable part). Thus, we can assume that the Universe is roughly homogeneous at the *large* cosmological scale.

This fact is also confirmed by the observations of the cosmic microwave radiation background. This is the relic light, the oldest light in our Universe left from the epoch

---

[14]Actually, I do not recommend this particular charter flight. There are more pleasant ways to spend one's vacation.

when the Universe was very young, no galaxies or stars had yet been formed, and space was filled with plasma, including protons (with the admixture of deuterons and the nuclei of helium), electrons and photons.[15]

The plasma expanded and cooled down. At some point (about 400 000 years after the Big Bang) the temperature became so low (only $\sim$3000 K) that the protons and electrons *recombined* and formed neutral hydrogen atoms. The plasma was converted into a gas. In contrast to the plasma before the recombination, this gas was transparent to photons. The latter could now travel through the whole Universe without being scattered. Thus, they survived until the present day, though they changed their dress. Since the moment of recombination the size of the Universe increased roughly 1000 times and so did the wavelength of the photons.[16] Correspondingly, their frequency decreased. At the moment of recombination they were the photons of red and near-infrared light. And now they are in the microwave range.

The cosmic microwave background was discovered by Arno Penzias and Robert Wilson in 1964. This is thermal radiation with the temperature $\sim$2.7 K and is highly homogeneous, indicating high homogeneity of the Universe at the moment of re-combination. The Universe became clumpy later.

How may the metric of the uniform and isotropic Universe look?

- First of all, it cannot have mixed components g_{0j}. The latter could be interpreted as a 3-dimensional vector that would single out a certain distinguished direction in space, which is not there by isotropy.
- By an appropriate coordinate choice, the component g_{00} of the metric tensor can be brought to unity (we revert to the convention $c = 1$). This gives

$$ds^2_{\text{Universe}} = dt^2 - g_{jk}dx^j dx^k . \tag{15.29}$$

- The spatial part of the metric describes a certain 3-dimensional manifold. By homogeneity, its curvature should be the same at all points. There are only two possibilities: *(i)* A constant positive curvature gives a 3-sphere (not an ordinary 2-sphere, but a 3-sphere, the border of a 4-dimensional ball). *(ii)* A constant negative curvature gives a 3-dimensional *pseudosphere* (or Lobachevsky space).

Consider first the case of positive curvature. The metric acquires the form

$$ds^2 = dt^2 - a^2(t)[d\chi^2 + \sin^2 \chi(d\theta^2 + \sin^2 \theta d\phi^2)] , \tag{15.30}$$

where χ, θ, ϕ are three angles parametrizing the 3-sphere and $a(t)$ is its radius. The latter determines the total spatial volume of the Universe,

$$V = 2\pi^2 a^3(t) . \tag{15.31}$$

---

[15] Neutrinos and mysterious dark matter were also there, but let us now disregard them.

[16] Alternatively, one can explain this increase by interpreting it as a red shift — the source of the microwave photons that we observe now is at a very large distance from us, almost at the border of the observed part of the Universe. And it moves away from us almost with the speed of light.

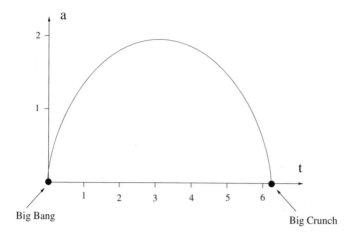

Fig. 15.5 Evolution of the Universe in the closed Friedmann model. t and $a(t)$ are expressed in units of a_0.

The only thing not fixed by symmetry is the time dependence $a(t)$, and our goal is to find it. To do this, we have to plug the ansatz (15.30) into the Einstein equations (15.18), derive an ordinary differential equation for $a(t)$ and solve it.

The right-hand side of (15.18) includes the energy-momentum tensor of matter $T_{\mu\nu}$. Following Friedmann, we accept for matter the model of homogeneously distributed *dust*. By "dust" we mean matter with no pressure.[17] Then the spatial components T_{jk} vanish, mixed components T_{0j} also vanish (matter is at rest in the chosen frame), and only T_{00}, the energy density ϵ, is nonzero. We assume it to be the same at all spatial points. Multiplying it by the total volume (15.31) of the Universe, we obtain its total mass M, which is an integral of motion.

The solution can be represented in the parametric form,

$$a = a_0(1 - \cos\eta)$$
$$t = a_0(\eta - \sin\eta)\,, \tag{15.32}$$

where

$$a_0 = \frac{2G_N M}{3\pi}\,. \tag{15.33}$$

The dependence (15.32) is represented in Fig. 15.5. We see that the Universe is created out of nothing at $t = 0$. At first, it rapidly expands,[18]

---

[17]In this sense, the stars undoubtedly represent a "dust" — high pressure in the interior of the stars is first of all not so high, being much smaller than the energy density; and secondly, it exists in a negligibly small spatial region and is irrelevant for cosmology.

[18]The law (15.34) does not apply at very small times when matter was hot and ultrarelativistic. The spatial components $T_{jk} = P\delta_{jk} = (\epsilon/3)\delta_{jk}$ (ϵ being the energy density) should be taken into

$$a \approx \sqrt[3]{\frac{9a_0t^2}{2}}. \tag{15.34}$$

Then the expansion slows down; $a(t)$ reaches the maximal value of $2a_0$ at $t = \pi a_0$.[19] After that the Universe starts to contract. At $t = 2\pi a_0$, its size is zero again. This is the moment of the Big Crunch, the Universe disappears and there is Nothing again. This is the *closed* cosmological scenario.

Consider now the case when the Universe has the geometry of a pseudosphere. The proper ansatz for the metric is now

$$ds^2 = dt^2 - a^2(t)[d\chi^2 + \sinh^2 \chi(d\theta^2 + \sin^2 \theta d\phi^2)]. \tag{15.35}$$

Following the same steps as before, we obtain the solution

$$a = a_0(\cosh \eta - 1)$$
$$t = a_0(\sinh \eta - \eta). \tag{15.36}$$

The pseudosphere is not a compact manifold, its volume is infinite, and so is the total mass of the Universe. We cannot therefore express a_0 via M_{Universe}, as we did in (15.33).[20]

One can instead do the following. *(i)* Rewrite the expression (15.33) of the closed model by trading the total mass for the mass density, $M = V\mu = 2\pi^2 a^3 \mu$. We obtain:

$$a_0 = \frac{4\pi G_N a^3 \mu}{3}. \tag{15.37}$$

In this formula, $a(t)$ and $\mu(t)$ can be chosen at any moment of time. It does not matter: a_0 is an integral of motion. *(ii)* One can now show that the *same* relationship between $a(t)$, $\mu(t)$ and a_0 holds in the open model.

The solution (15.36) is plotted in Fig. 15.6. At the beginning, the Lobachevsky scale factor $a(t)$ grows in the same way as the radius of the 3-sphere in (15.34). But then the behavior is radically different. Without hesitation $a(t)$ continues to grow. For large t, $a(t) \approx t$; the Universe keeps expanding with the speed of light. This is the *open* cosmological model.

(Footnote 18 continued)
account in this case. The Universe still expands, but its size grows in this case as $a \propto \sqrt{t}$ rather than $a \propto t^{2/3}$.

[19]The estimate $a_0 \sim G_N M$, implied by Eq. (15.33), can be understood heuristically from the condition that the gravitational potential energy of the Universe at the peak of its expansion, $U \sim G_N M^2/a_0$, is of the same order as the energy M stored in its mass. It is practically the same argument that gave us the estimate (3.8) for a characteristic stellar mass.

[20]Actually, the notion of the "total mass of the Universe" is in any case not physical, because at any given moment t, only a finite volume of the Universe at the distances smaller than ct is accessible to observations. Partially, this also applies to the previous closed scenario, where the spatial geometry was a 3-sphere. The whole Universe becomes accessible to observations there only at the contraction stage.

Fig. 15.6 Evolution of the Universe in the open Friedmann model.

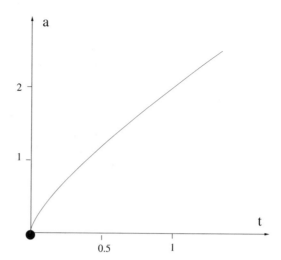

A very interesting question is, of course, our own fate. Will the Universe in a dozen billion years or so reach its maximal size and then, first slowly and then faster and faster, start contracting and heating up to disappear finally in the igneous implosion? Or will it keep expanding forever, getting cooler and cooler?

The future can be predicted from precise knowledge of the contemporary conditions. We know relatively well the present rate of expansion, the *Hubble constant*[21]

$$H = \frac{\dot{a}}{a} \approx 70 \, \frac{\text{km/sec}}{\text{Mpc}}. \tag{15.38}$$

The theoretical predictions for it in the closed and open models are

$$H_{\text{closed}} = \frac{\sin \eta}{a_0 (1 - \cos \eta)^2},$$

$$H_{\text{open}} = \frac{\sinh \eta}{a_0 (\cosh \eta - 1)^2}. \tag{15.39}$$

In both models, H decreases with η and with time. In the open model, it stays positive, while in the closed model, it reaches zero at $\eta = \pi$ and then changes sign.

One can derive the following important inequalities (we leave this exercise to the dedicated reader):

---

[21]"Mpc" means *megaparsec*, the unit of distance beloved by astronomers and cosmologists. One megaparsec is about 3.26 million light years. The experimental value of H given in (15.38) means that a galaxy at the distance ~ 100 Mpc from us runs away with a velocity of about 7000 km/sec.

$$\mu > \frac{3H^2}{8\pi G_N} \qquad \text{(closed model)},$$

$$\mu < \frac{3H^2}{8\pi G_N} \qquad \text{(open model)}. \qquad (15.40)$$

In other words, to determine our future, we need to know the average mass density μ of the Universe. If it exceeds a certain critical value, at some point the gravitational attraction overcomes the impulse of the explosion and the cloud of debris will gradually shrink back to a point. And if the density and the associated gravitational pull are less than critical, it will not. For the present epoch, the critical density is

$$\mu_{\text{crit}} \approx 10^{-9} \, \frac{\text{J}}{\text{m}^3}, \qquad (15.41)$$

or $\approx 10^{-26}$ kg/m^3 in mass units. (This is 100000 times less than the density of ordinary baryon matter in a galaxy.)

Experiment says that the *total* energy density of the Universe is rather close to the critical value; it is difficult to say whether it exceeds μ_{crit} or not. Does it mean that we cannot now say whether the final stage of the Universe evolution will be the Big Crunch of the closed model or the Big Freeze of the open model?

Actually, it does not and, according to what we know now, the Big Freeze scenario is much more probable. As was mentioned, the Hubble constant depends on time. Measuring the red shifts of very distant galaxies, one can extract information about the Hubble constant in the distant past and thereby about its time dependence. When such measurements were performed, people saw with great surprise that the Universe expands with *acceleration*. This contradicts the prediction (15.39) of the "standard" Friedmann cosmology and can be explained only if we assume that the energy-momentum tensor includes, besides the dust term $T_{00} = \mu$, also the vacuum term, $T_{\mu\nu} = \Lambda g_{\mu\nu}$ with a nonzero positive cosmological constant Λ.[22] The associated

Fig. 15.7 The components of the Universe.

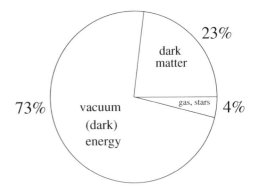

---

[22]In the original Friedmann papers, the equations with nonzero Λ were also written and solved, but until the experimental evidence for nonzero Λ was obtained, people did not need to include

negative vacuum pressure will drive the expansion forever, even in the distant future when the density of ordinary and dark matter decreases to negligible values.

We conclude this chapter with a chart for the different components in the total energy density pie, as we know it today (Fig. 15.7).

(Footnote 22 continued)

Λ in the analysis to describe the data and usually did not do so. We have chosen not to include Λ in the equations above simply to avoid extra complications. It is better first to learn and understand Friedmann's equations with $\Lambda = 0$. Adding the cosmological term presents no conceptional difficulties.

Chapter 16
Mysterious Quantum Gravity

As was mentioned before, we do not know what quantum gravity *is*. In this final chapter we will explain why it is so difficult to quantize it. Then I will say what most people now think about the directions where the solution to this hard problem should be sought and finally abuse your patience by sharing with you my own heretical and irresponsible guesses.

16.1 Attempts at Perturbative Calculations

16.1.1 Graviton Scattering

Let us try to carry through the same programme for gravity that we did for other field theories. Assume the metric to be close to the Minkowski flat metric, as in Eq. (15.21). We can then disregard the geometric nature of our theory, treat the *difference* $h_{\mu\nu} = g_{\mu\nu} - \eta_{\mu\nu}$ as a tensor field in ordinary flat space-time and quantize it. We obtain the graviton — a massless particle of spin 2, carrying helicities $h = 2$ or $h = -2$.

The gravitons are not free; they can scatter on ordinary particles (that is how gravitational waves have been observed) or on each other. Let us concentrate on the latter process. To calculate the graviton-graviton scattering amplitude we have to expand the Einstein–Hilbert action (15.14) in $h_{\mu\nu}$. Symbolically, one can write

$$\mathcal{L}_{\text{grav}} \sim m_P^2[(\partial h)^2 + ah(\partial h)^2 + bh^2(\partial h)^2 + \cdots], \tag{16.1}$$

where a, b are numerical coefficients of order 1. It is now convenient to go to "perturbative" normalization and introduce the field $\tilde{h}_{\mu\nu} = m_P h_{\mu\nu}$. The Lagrangian acquires the form

© Springer International Publishing AG 2017

A. Smilga, *Digestible Quantum Field Theory*,

DOI 10.1007/978-3-319-59922-9_16

Fig. 16.1 Tree-level graphs contributing to the graviton scattering amplitude.

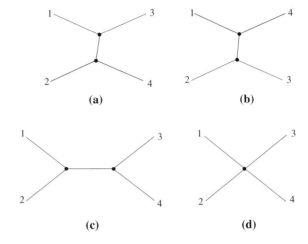

(a)   (b)

(c)   (d)

$$\mathcal{L}_{\text{grav}} \sim (\partial \tilde{h})^2 + \frac{a}{m_P} \tilde{h}(\partial \tilde{h})^2 + \frac{b}{m_P^2} \tilde{h}^2 (\partial \tilde{h})^2 + \cdots . \qquad (16.2)$$

The field $\tilde{h}$ carries the dimension of mass, like the scalar field in the model (10.12). However, in contrast to that model, the interaction terms in the Lagrangian involve extra derivatives, whose dimension is compensated by the factors of m_P in the denominator.

The diagrams describing graviton scattering are drawn in Fig. 16.1. The cubic vertices there include a momentum factor upstairs and the factor m_P downstairs. The quartic vertex includes two powers of momenta and the factor m_P^{-2}.

The amplitudes depend on the helicities of the ingoing and outgoing gravitons. Consider the amplitude where all the helicities are positive. It reads

$$M_{++|++} = 8\pi i G_N \frac{s^3}{tu}, \qquad (16.3)$$

where $s = (p_1 + p_2)^2$, $t = (p_1 - p_3)^2$, $u = (p_1 - p_4)^2$ are the conventional kinematic invariants.

This amplitude is singular at $t = 0$ or $u = 0$, corresponding to scattering angles 0 or π. This singularity is due to a massless graviton exchange in the t-channel or u-channel; the same singularity is well known for the Coulomb scattering amplitude and differential cross section. The novel feature is that the graviton cross section *grows with energy*, $d\sigma \propto G_N^2 E^2$, when the scattering angles are fixed. This growth has the same origin (a dimensionful coupling constant G) as in the Fermi theory. We discussed the latter in Chap. 5 [see Eq. (5.22)] and in Chap. 12.

By the same token as the Fermi theory, Einstein's gravity is *not renormalizable*! In particular, loop corrections to the tree-level graviton scattering amplitudes involve troublesome power ultraviolet divergences. The role of the unitary bound (5.24) in the Fermi theory, the energy beyond which perturbative calculations are meaningless, is now played by the Planck mass m_P.

16.1.2   What Is Wrong with Non-renormalizability?

I would ask the reader to mark the statement above: *The quantum counterpart of Einsten's gravity is not a renormalizable theory. It is thus not self-consistent at the perturbative level.*

But the next several pages are addressed not so much to my target reader, who may safely skip this section at first reading, but to a more practised person. I hesitated whether or not I should include this discussion in the book, but decided finally to do so. By my experience, this issue (the Wilsonian Lagrangian and its relationship to perturbative calculations in nonrenormalizable theories) is not so simple — even some experts might sometimes be confused about it. It is not completely excluded that one of them would open our book...

For renormalizable theories, the notion of the Wilsonian Lagrangian[1] was in fact heuristically discussed in Chap. 5. Please look again at Eq. (5.12). It expresses the effective physical charge in QED via the bare charge e_0^2 and the ultraviolet cutoff Λ. The meaning of this expression is the following. Suppose that we draw a loop diagram, as in Fig. 5.7, and integrate over all the momenta q in the loop. This also involves the kinematical region where q are much larger than the energy scale we are interested in. Large momenta mean small distances, and our loop effectively shrinks to a point, giving simply a Λ-dependent factor, which multiplies the tree amplitude. Such a shrinking is similar in spirit to the effective contraction of certain parts of diagrams involving propagators of heavy particles. We discussed this before: see Fig. 5.14, where the heavy W propagator was contracted to a point, and Fig. 5.15, where an electron loop was contracted to a point. (The electrons can be considered as heavy if the characteristic frequency of photons is much less than the electron mass!)

When a professional playwright mentions in the first act a gun hanging on the wall in the study of the main character, you can be sure that this gun will fire in the fourth act. Following this rule, I have hung in the pages of this book a small gun, which is going to fire *now*. I mean the effective chiral Lagrangian (11.51) that describes low-energy pion interactions in *massless* QCD.

We will now be interested not in pion physics, but only in mathematical aspects of this theory. As was already mentioned in Chap. 11, the effective chiral theory has many similarities with gravity. Like gravity, it involves a dimensionful coupling and is not renormalizable. The perturbative expansion (11.54) of the nonlinear Lagrangian (11.51) has the same structure as the expansion (16.2). In both cases, the quartic term involves two derivatives, so that the scattering amplitudes rapidly grow with energy,

---

[1] It was introduced into quantum field theories in 1971 by Kenneth Wilson.

and perturbative calculations become meaningless beyond some limit: $\sim 2\pi F_\pi$ in chiral theory and $\sim m_P$ in gravity.

Let us discuss chiral theory first and then use the lessons that we learn for gravity. Notice first that the Lagrangian (11.51) is the simplest structure invariant under $SU_L(2) \times SU_R(2)$ transformations, but not the only one. In fact, there are an *infinite number* of such structures with higher and higher numbers of derivatives and with higher and higher canonical dimensions d. The Lagrangian (11.51) has dimension $d = 2$. For convenience, we write it here again:

$$\mathcal{L}_0 = \frac{F_\pi^2}{4} \mathrm{Tr}\{\partial_\mu U \partial^\mu U^\dagger\}, \tag{16.4}$$

with $U = \exp\{i\pi^A \sigma^A / F_\pi\}$. There are two different structures at the level $d = 4$:

$$\mathcal{L}_1 = (\mathrm{Tr}\{\partial_\mu U \partial^\mu U^\dagger\})^2, \qquad \mathcal{L}_2 = \mathrm{Tr}\{\partial_\mu U \partial_\nu U^\dagger\} \mathrm{Tr}\{\partial^\mu U \partial^\nu U^\dagger\}. \tag{16.5}$$

Well, actually there are other invariant structures with four derivatives. But they are *not relevant* for us as they all vanish when the field U satisfies the equations of motion following from the basic chiral Lagrangian (16.4):

$$(\Box U) U^\dagger - U(\Box U^\dagger) = 0. \tag{16.6}$$

We will shortly see why and in what sense they are not relevant.

The tree pion scattering amplitude behaves as

$$M^{(0)}_{\pi\pi \to \pi\pi} \sim \frac{E^2}{F_\pi^2}. \tag{16.7}$$

Consider now the one-loop correction to the amplitude. It involves both power and logarithmic divergences and has the following structure (illustrated in Fig. 16.2):

$$M^{(1)}_{\pi\pi \to \pi\pi} = \frac{\alpha \Lambda^2}{F_\pi^2} M^{(0)}_{\pi\pi \to \pi\pi} + \frac{A(s,t)}{F_\pi^4} \ln \frac{\Lambda}{\mu} + \frac{B(s,t,\mu)}{F_\pi^4}, \tag{16.8}$$

with μ arbitrarily chosen within the range $E \ll \mu \ll \Lambda$. B logarithmically depends on μ in such a way that the whole sum is μ-independent. The last UV-finite term in (16.8) is nonlocal and complicated. But the first two ultraviolet divergent terms come from large loop momenta, where the loop shrinks effectively to a point, and their structure is much simpler. The quadratically divergent contribution is especially simple. It is proportional to the tree scattering amplitude and describes nothing but the renormalization of the bare coupling constant:

$$\frac{1}{F_\pi^2} \to \frac{1}{F_\pi^2}\left(1 + \frac{\alpha \Lambda^2}{F_\pi^2}\right) \quad \text{and hence} \quad F_\pi^2 \to F_\pi^2\left(1 - \frac{\alpha \Lambda^2}{F_\pi^2} + \ldots\right). \tag{16.9}$$

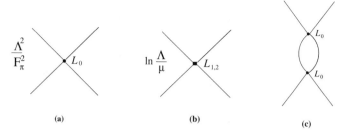

Fig. 16.2 One-loop contributions to the pion scattering amplitude. (**a**) renormalization of F_π^{-2}, (**b**) logarithmically divergent local contributions, (**c**) UV-finite nonlocal contributions.

The power divergences look more disturbing than the logarithmic ones. Equation (16.9) is only a first correction; there are also corrections $\sim\Lambda^4$, $\sim\Lambda^6$, which, in contrast to the logarithmic corrections to the effective charge in QED or in QCD, do not sum up into anything reasonable. Still, one can in principle *redefine* the coupling constant order by order and get rid of power divergences, expressing these contributions via the renormalized F_π. One can also remark here that there is a slightly artificial, but convenient way to regularize ultraviolet divergences (the so-called *dimensional regularization*) where power divergences do not appear at all.

Let us now discuss the second term in (16.8). It is both more interesting and in some sense more troublesome. In this case, the loop also effectively shrinks to a point, but this pointlike 4-point vertex is no longer the tree vertex determined by the expansion (11.54) of (16.4). It has a more complicated structure, including now four rather than two momentum factors.

The point is that this new vertex is a certain combination of the 4-point vertices associated with the higher-dimensional chiral-invariant structures in (16.5).

We understand now why we called "irrelevant" the structures that vanish when the field $U(x)$ satisfies the equations of motion. One can show that the vertices generated by such structures involve the factor p_i^2 for at least one of the external particles. But we are interested in the scattering of *asymptotic pion states* — the particles that appear after quantization of the free Lagrangian $\frac{1}{2}(\partial_\mu \pi^a)^2$. These particles are on *mass shell*, and the squares of their 4-momenta are zero for massless pions. Thus, those other structures do not contribute to on-mass-shell scattering amplitudes.

The higher-derivative structures (16.5) are usually called *counterterms*. In using this name, people mean the following. Suppose that the coupling constant multiplying the lowest dimensional structure in (16.4) in the *bare* Lagrangian is not $F_\pi^2/4$, but rather

$$\text{bare coupling} = \frac{F_\pi^2}{4} + \frac{\alpha\Lambda^2}{4} + \cdots \tag{16.10}$$

with the same α as in (16.8).

Suppose further that the bare Lagrangian involves in addition a certain combination of the structures (16.5) (and also the structures with six derivatives, etc.),

$$\mathcal{L}_{\text{bare}} = \mathcal{L}_0 \left(1 + \frac{\alpha \Lambda^2}{F_\pi^2} \right) + \beta \mathcal{L}_1 \ln \frac{\Lambda}{\sigma_1} + \gamma \mathcal{L}_2 \ln \frac{\Lambda}{\sigma_2} + \cdots . \qquad (16.11)$$

New terms in the Lagrangian give rise to new vertices and to new diagrams contributing to the $\pi\pi$ scattering amplitude. The point is that now the quadratic divergences in the sum of all contributions cancel out, and one can also assure the cancellation of the logarithmic divergences by choosing β and γ in an appropriate way. Thus, the terms $\sim\Lambda^2$ and $\sim \ln \Lambda$ in (16.11) represent a kind of "counterforce" that destroys the malignant ultraviolet divergences in the amplitude!

The coefficients β and γ can be fixed in this way, but the constants σ_1, σ_2 cannot. They are *new* coupling constants in our theory; one cannot calculate the loop corrections to the amplitude without specifying their values.[2] At the two-loop level, one can write down six different six-dimensional structures. This brings about six new coupling constants. The total number of coupling constants in a non-renormalizable theory is infinite.

To recapitulate. A non-renormalizable theory is mischievous not only due to wild power ultraviolet divergences that it possesses. One can in principle remove these divergences order by order by redefining the bare coupling constant. The logarithmic UV divergences can be removed by adding extra counterterms in the bare Lagrangian. But:

- Each counterterm brings about an extra dimensionful coupling constant σ_i. The total number of these constants is infinite.
- In a non-renormalizable theory, amplitudes grow rapidly with energy, and the perturbative expansion breaks down at a certain scale (the unitary limit). This limit is of the order of $G_F^{-1/2}$ in the Fermi theory, of the order of $2\pi F_\pi$ in chiral theory and of the order of m_P in gravity.

Now we go back to gravity. The structure of graviton scattering amplitudes is similar to the structure of pion scattering amplitudes discussed above, with one important distinction: in gravity, there are no *logarithmic* ultraviolet divergences at the 1-loop level.

Indeed, we have seen that the presence of logarithmic divergences is associated with the presence of counterterms of the appropriate dimension. At one loop, the relevant canonical dimension is $d = 4$. There are only two independent scalars involving four derivatives: R^2 and $R_{\mu\nu} R^{\mu\nu}$.[3] But both R^2 and $R_{\mu\nu} R^{\mu\nu}$ vanish on mass shell, when the metric satisfies Einstein's free equations of motion, $R_{\mu\nu} = 0$. People often say that "gravity is finite at one loop", but one has to bear in mind

---

[2] In real QCD chiral-symmetry applications, the coefficients $\sigma_{1,2}$ can be determined phenomenologically. Then loop calculations for not too large energies become possible. The corresponding technique was developed by Jürg Gasser and Heinrich Leutwyler in 1984.

[3] One can also write down the invariant $R_{\mu\nu\alpha\beta} R^{\mu\nu\alpha\beta}$, but it is not independent. One can show that

$$R_{\mu\nu\alpha\beta} R^{\mu\nu\alpha\beta} = 4 R_{\mu\nu} R^{\mu\nu} - R^2 + \text{total derivative} . \qquad (16.12)$$

This is the so-called *Gauss–Bonnet identity*.

that that concerns only on-mass-shell scattering amplitudes and only if the quadratic ultraviolet divergences are disregarded.

There exists, however, a nontrivial invariant of canonical dimension 6,

$$\mathcal{L}_{d=6} = R_{\mu\nu}{}^{\alpha\beta} R_{\alpha\beta}{}^{\gamma\delta} R_{\gamma\delta}{}^{\mu\nu}, \qquad (16.13)$$

which does not vanish on mass shell and can serve as a relevant counterterm for graviton scattering at the 2-loop level. There are also an infinite number of higher-dimensional counterterms.

We conclude that the physics of perturbative gravity is the same as the physics of chiral theory. Neither of them can be treated as a self-consistent fundamental theory.

16.1.3 Supergravity

As we learned in Chap. 14, certain troublesome divergent quantities, like the vacuum energy, may vanish in supersymmetric theories. A legitimate question is whether the divergencies in graviton scattering amplitudes may also cancel out in a super-symmetric extension of Einstein's gravity. We tell you right away that the probable answer to this question is negative, but before coming to this conclusion, let us give some details about these interesting theories.

The simplest supergravity theory (which was constructed by Dan Freedman, Sergio Ferrara and Peter van Nieuwenhuisen in 1976) involves, besides the metric field $g_{\mu\nu}(x)$, the anticommuting *Rarita–Schwinger* field $\psi_\mu(x)$. The latter is a 4-component Majorana field[4] carrying an extra vector index μ. It describes a neutral fermion of spin 3/2. The Lagrangian reads

$$\mathcal{L}_{\text{SUGRA}} = -\frac{m_P^2}{16\pi} R - \frac{1}{2\sqrt{-g}} \varepsilon^{\mu\nu\rho\sigma} \bar{\psi}_\mu \gamma^5 \gamma_\nu \mathcal{D}_\rho \psi_\sigma, \qquad (16.14)$$

where $\mathcal{D}_\mu$ is the gravitational covariant derivative.[5]

The spectrum of the theory includes the gravitons, carrying helicities $h = \pm 2$, and the gravitinos, carrying helicities $h = \pm 3/2$. A graviton of helicity $h = 2$ and a gravitino of helicity $h = 3/2$ carrying the same momentum form a supersymmetric doublet. There are also supersymmetric doublets formed by the graviton states with $h = -2$ and the gravitino states with $h = -3/2$.

---

[4]It is the bispinor (9.40) with $\xi \equiv \eta$.

[5]We managed to get along without defining $\mathcal{D}_\mu$ in the previous classical chapter. In principle we need to define it now, and not only when it acts on the vectors and tensors, but also for the spinors, explaining why the spin connection w_ρ^{ab} entering $\mathcal{D}_\rho \psi_\sigma$ depends in this case not only on the vierbein field e_μ^a, but also on ψ_μ (this brings about the term $\sim \psi^4$ in the Lagrangian). However, I shall conceal from the reader all these important details.

Table 16.1 The 256-plet of states of $\mathcal{N} = 8$ supergravity.

	Graviton	Gravitinos	Photons	Photinos	Scalars	Photinos	Photons	Gravitinos	Graviton
Helicity	-2	$-3/2$	-1	$-1/2$	0	$1/2$	1	$3/2$	2
Number of states	1	8	28	56	70	56	28	8	1

The Lagrangian (16.14) is invariant under general coordinate transformations and under *local* supersymmetry transformations, where the Grassmann transformation parameters are functions of x_μ. Local (gauged) supersymmetry has roughly the same relation to global supersymmetry as ordinary gauge symmetry to ordinary global symmetry.

We learned previously that gauge symmetry is not quite a symmetry in a certain sense. In particular, Goldstone's theorem, dictating the existence of massless bosons when a global symmetry is broken spontaneously, does not work for gauge symmetries. Likewise, certain theorems that are valid for theories with global supersymmetry are not valid for supergravity. In particular, one can no longer claim that the vacuum energy density is zero (the very notion of energy *density* loses meaning in gravity and in supergravity!). If local supersymmetry is spontaneously broken, a massless goldstino does not arise. Instead of that, this degree of freedom is eaten up by the gravitino, which becomes massive.

What is the perturbative structure of supergravity amplitudes? Well, the dimensionful coupling constant is still there and the theory is still not renormalizable. There is only one novel feature: logarithmic ultraviolet divergences in the on-mass-shell scattering amplitudes now cancel out at the 2-loop level. This is due to the fact that the 6-dimensional invariant (16.13) does not have a supersymmetric partner. Thus, a 6-dimensional counterterm is absent in supergravity. But at the 3-loop level and higher, such counterterms (of dimension 8 and higher) are present.[6] There are an infinite number of relevant counterterms, of extra arbitrary coupling constants, and the physical situation is the same as in gravity or in chiral theory.

The Lagrangian (16.14) describes the simplest $\mathcal{N} = 1$ supergravity. But there exist also *extended* supergravities (cf. the discussion of extended supersymmetric Yang–Mills theories in Sect. 14.3). The spectrum of $\mathcal{N} = 2$ supergravity, which is the next in complexity, includes a graviton, two different gravitinos and an Abelian gauge field. For a given momentum, there are two supermultiplets of states. One of them includes a graviton with $h = 2$, two gravitinos with $h = 3/2$ and a "photon" with $h = 1$. The other multiplet includes the states with negative helicities. The maximally extended $\mathcal{N} = 8$ supergravity has for each momentum only one 256-plet of states, which includes a graviton, eight gravitinos, twenty-eight different "photons", fifty-six massless Majorana particles of spin 1/2 — "photinos" and seventy scalars (Table 16.1). The numbers of bosonic and fermionic states are, of course, equal.

---

[6]That was confirmed by explicit calculations. One *does* see logarithmic divergences in the 3-loop scattering amplitudes in the theory (16.14).

The high symmetry of $\mathcal{N} = 8$ theory brings about a lot of nontrivial cancellations. Explicit calculations by the group of Zvi Bern displayed the absence of logarithmic divergences through four loops. This was later understood in the language of counterterms: one can show that the counterterms of canonical dimension 4, 6, 8, 10, 12, 14 all vanish on mass shell in this theory. This guarantees the absence of divergences throught six loops. Now one can write down a $\mathcal{N} = 8$ invariant of canonical dimension $d = 16$, which *might* mean that a logarithmic divergence and its associated new coupling constant arises at the 7-loop level. There are, however, certain heuristic arguments that this does not happen yet in 7 loops. But most experts believe that the divergences *must* appear at the 8-loop level[7] and higher — we simply do not see any reason for them to vanish.

Of course, to be *completely* sure about it, one would have to perform the actual explicit 8-loop calculation and show that the coefficient in front of $\ln \Lambda$ does not vanish. It is not clear to me that such a horrendous calculation has a chance of being carried through in this century. The question thus acquires a rather scholastic and academic flavour — similarly to the question of the consistency of conventional QED (see p. 65). A mathematician would comment that both are interesting as-yet-unresolved mathematical questions, and one should not say more until somebody *proves* something definite. But we physicists are impatient and, if the theorems are not yet proven, are often satisfied with plausible conjectures. The conjecture that even the maximally symmetric $\mathcal{N} = 8$ supergravity is not free from logarithmic divergences, bringing about an infinite number of unknown coupling constants, seems to me and to most my colleagues more plausible that the opposite one.

Starting from a certain order, perturbative series for physical observables are not defined in a non-renormalizable theory, no matter how highly symmetric or supersymmetric the latter might be.

16.2 Breaking of Causality □ Hawking Paradox

I must confess that I understand really well only two things in physics.

1. One of them is Newton's second law (4.1). It represents a second-order differential equation for the trajectory $x(t)$. If the force F is a known function of $\dot{x}$, x and t, and if the position and velocity are fixed at an initial time t_0, a Cauchy problem is formulated. It is intuitively clear (and mathematicians confirm it) that, if the function $F(\dot{x}, x, t)$ is not too wild, this problem always has a unique solution. The same applies to any dynamical mechanical system with an arbitrary number of variables. The equations of motion have a unique solution if the initial values for all generalized coordinates and velocities are fixed. And this applies also to field theories. In that case, we have to fix the initial values for all fields and their time derivatives at all spatial points.

---

[7]The corresponding counterterm of canonical dimension $d = 18$ was written back in 1981 by Renata Kallosh and independently by Paul Howe and Ulf Lindström.

2. The second thing that I understand is the Schrödinger equation (4.8). Again, by fixing the wave function $\Psi(x, t)$ at an initial moment t_0, we formulate a Cauchy problem that has a unique solution. The wave function can thus be found at any later time $t > t_0$. Knowing the wave function, we can determine the mean values of all observables. For field theories, one has to solve a similar Cauchy problem for the wave functional (4.34).

Unfortunately, as was already mentioned in Sect. 4.8, gravity cannot be formulated in these terms. The problem is that we do not have a notion of universal flat time any more.

Special relativity tells us that time is different in different reference systems, but once such a system is fixed, time is well defined, and the equations of motion in relativistic mechanics or in relativistic field theory represent a solvable Cauchy problem. But in gravity space-time is curved, so that the time and spatial coordinates are intertwined. As a result, Einstein's equations (15.18) cannot always be interpreted as the equations describing the time evolution of the metric.

To be more precise, if the deviation $h_{\mu\nu}$ of the metric from the flat Minkowski metric is small, this interpretation is still possible. It is not natural, and technically inconvenient, but $h_{\mu\nu}$ can in that case be represented as a tensor field in flat Minkowski space. In strong gravitational fields this does not work, however. In particular, Einstein's equations admit pathological Gödel solutions with geodesics representing closed time loops. This breaks causality.

In other words, not only quantum, but also classical gravity — general relativity — is not internally consistent. The theory works well for not too strong gravitational fields, but when the latter are strong, one meets difficulties of a fundamental nature. For example, the Schwarzschild solution (15.24) involves a horizon, so that the space-time region behind the horizon, the interior of a black hole, is not accessible to observations from the outside and hence is not too physical.

Things become worse in quantum theory. Staying at the classical level, one can simply disregard the troublesome solutions with closed time loops, saying that they are not physical and not realized in nature. Indeed, one can observe inter alia that these solutions are unstable — a small perturbation (like stepping aside from the prescribed path and killing a butterfly, as in a science-fiction story by Ray Bradbury) destroys them completely. But in quantum theory, there is no cheap way out. To find the amplitude of a process, one has to sum over all possible trajectories and over all virtual field configurations, including those with closed geodesics.[8] The amplitudes become non-causal.

The same applies to the problem of the horizon. In classical theory we may stay outside the black hole and simply not be interested in what happens within. But in quantum theory the interior of a black hole strikes back and brings about a paradox known as the *Hawking paradox*.

But to understand what it is, one has first to learn what is *Hawking radiation*.

---

[8]This is especially clear in the path integral approach, see Sect. 8.3.

In Sect. 4.7 [see p. 52], we mentioned that, if the charge of a nucleus slightly exceeded a certain critical value, $Z_{\text{crit}} \gtrsim 1/\alpha$, so that the energy of the lowest electron bound state dropped down below the threshold $E = -mc^2$, an electron-positron pair would be created, with the positron being emitted outside and the electron falling onto the nucleus and bringing its charge back to a subcritical value. For a nucleus whose charge substantially exceeds the limit Z_{crit}, a copious emission of positrons would be observed.

This is a threshold effect, because the electron is massive and creating an e^+e^- pair costs energy. Suppose, however, the existence of massless charged particles. Their pairs could and would be created without any cost, so that *any* charge would eventually be screened.

And this is exactly what happens in gravity. Photons are massless, but they carry energy and interact with gravity. Thus, photon pairs can be created spontaneously in any gravitational field, but of course, the stronger is the field, the more pronounced is this effect. In the presence of a horizon, this process acquires certain universal features unravelled by Steven Hawking in 1974. We will now describe these features, giving heuristic explanations.

Suppose that we have a black hole of mass M. The radius of its horizon is given by Eq. (15.25). The pairs of photons are mostly created in the region outside the black hole where the gravitational field is the strongest — in the vicinity of the horizon. [9] When charged particles are created by the Coulomb field, the net electric charge is conserved and the components of the created pair carry opposite charges. In gravity, it is the net energy (measured by a distant observer) that is conserved. Then one of the created photons should have positive energy and another one negative energy. Their fate is completely different. The one with negative energy falls down into the black hole and disappears there. Its positive comrade has more luck — it escapes the mortal clinch of the gravitational tentacles of the black hole and shoots its way out to infinity (see Fig. 16.3).

As for the black hole that swallowed its unfortunate negative brother, this black deed did not bring it any benefits. In fact, the black hole has not gained, but *lost* energy, carried away by the survivor. In other words, the process that we described boils down to spontaneous emission of photons by the black holes from the region around their horizon, associated with decreasing their mass, the *evaporation* of the black holes.

How to describe this radiation quantitavely? What is the energy spectrum and the intensity of this radiation? Hawking showed that the spectrum and the intensity are exactly the same as if the horizon of the black hole were *heated* up to the temperature

$$T = \frac{m_P^2}{8\pi M} \tag{16.15}$$

(in natural units). The spectrum of the radiation is the black body spectrum with the temperature (16.15). The full intensity P of the radiation is the surface of the

---

[9]The field is even stronger inside the hole, but whatever is created there cannot escape.

Fig. 16.3 Evaporation of black holes.

horizon,

$$\mathcal{A} = 4\pi r_g^2 = \frac{16\pi M^2}{m_P^4},$$

multiplied by σT^4, where $\sigma = \pi^2/60$ is the Stefan–Boltzmann constant. We obtain

$$P = \frac{m_P^4}{15 \cdot 2^{10}\pi M^2}. \tag{16.16}$$

The value (16.15) for the temperature of the horizon can be rewritten as $T = 1/(4\pi r_g)$. The estimate $T \sim 1/r_g$ looks natural if one notices that the maximum of the black body spectrum is of the order of the temperature and that the wavelength of the corresponding photons is of the same order as the size of horizon r_g.

We can now answer the question that our clever reader has without any doubt already formulated in his mind — why was I speaking only about photon radiation? Does the horizon of a black hole also emit other particles?

The answer is that other particles may indeed be emitted, provided that the size of the black hole is less than their Compton wavelength. The lightest massive particles are neutrinos. Bearing in mind the bound (12.69), we deduce that yes, all the types of neutrinos are emitted by black holes of size $\sim$1 μm or less.

For the black holes observed by astronomers, the radiation temperature and radiation intensity are negligibly small. If $M = M_\odot$, the temperature is only $\approx$10^{-7} K. But the smaller is the mass, the brighter the black hole shines. For example, the Hawking temperature for black hole with a mass equal to the mass of *Vesta* (one of the largest asteroids) would be $\approx$470 K. One could use it for cooking were it not for its minuscule size $\sim$0.4 μm.

A radiating black hole loses its mass, $\dot{M} = -P$. Bearing in mind (16.16), the time of the total evaporation is finite:

$$\tau_{\text{evaporation}} = 5 \cdot 2^{10} \pi \, \frac{M^3}{m_P^4} \, . \tag{16.17}$$

It happens to be of the same order as the age of the Universe for black holes of mass $M \sim 10^{12}$ kg.

I am not sure whether the process of evaporation of black holes is phenomenologically important. Probably not. But the fact that a black hole can disappear completely, emitting its mass in the form of thermal radiation enabled Hawking to formulate his famous paradox.

Suppose that a black hole is formed, and we throw into it the *Encyclopædia Britannica*. The *Encyclopædia Britannica* includes a lot of information, but once it crosses the horizon, this information is *completely* lost. The italics for "completely" are essential. If we burned the encyclopædia (especially, if we burned all existing copies of it) in a fireplace, the information it contained would also be lost for all practical purposes. But it would not be lost for a theorist! In classical physics, one can in principle follow the fate of all the atoms that constituted the volumes of encyclopædia and say that the information is still there, in the oxidized carbon atoms now hovering in thin air. It is then our problem that we are impotent to retrieve it.

In quantum mechanics, the information dwells in the wave function. As long as the evolution is unitary, it is also not lost and is in principle retrievable. But with a black hole it is different — whatever is gone, is gone without any hope of retrieval.

This is not a paradox yet for the following reason. Recall (see p. 300) that for a distant observer, objects falling into a black hole never cross the horizon. Thus, they are still theoretically accessible for observations.

Things become more worrysome if one includes the Hawking radiation into consideration. This radiation is of a stochastic thermal nature. And it knows nothing about the nature of the objects that fell down into the black hole. When the latter finally evaporates, the information about what preceded its formation seems to be lost forever.

But this unretrievable loss violates the fundamental principles of quantum mechanics. The Schrödinger equation describes a *unitary* evolution of wave function. That means that a *pure state* (a quantum state describable in terms of a wave function) can evolve only to another pure state, which contains in principle all the information that the initial state contained.

Currently, there is no agreement among experts on how serious this paradox is. A well-known and respected theorist Leonard Susskind published scientific papers and a popular book "*The Black Hole War*", in which he argued against Hawking's reasoning. I did not myself follow this discussion and cannot really remark on this issue, but my *suspicion* is that there is no simple way out...

16.3   Strings

Hopefully, I have now convinced the reader that Einstein's general relativity with the action (15.14) cannot be the fundamental theory of gravity. It has very serious, seemingly unsurmountable, difficulties both at the perturbative and non-perturbative level. Thus, most experts now agree that the theory (15.14) is an *effective* low-energy theory of gravity, similar to the way that the Fermi 4-fermion theory is an effective theory of weak interactions.

In fact, one may treat the electroweak sector of the Standard Model described in Chap. 12 as a nice and self-consistent way to provide ultraviolet regularization for the Fermi theory. But in contrast to QED and to QCD, regularizing the theory in the ultraviolet is not in this case a technical trick, and this regularization must not be eventually lifted. Λ_{EW} is a quite real physical parameter whose value is of the order of m_W and m_Z. When the energy approaches this limit, not only are the effective 4-fermion vertices modified, but also new physical degrees of freedom appear.

There should exist a similar physical ultraviolet regulator for the Einstein gravity. There is no consensus now about the nature of this regulator, but most theorists tend to believe that the underlying fundamental Theory of Everything (ToE) is a variant of string theory. I am not myself an expert in strings, have no intention to make my reader a string expert and will stay in this section at the popular heuristic level, not going into technical details.

The main idea is the following. When we talked about Feynman graphs in Chap. 10, we did not derive them accurately in the field theory framework, but used the analogy with scattering in quantum mechanics. These graphs were interpreted by saying that a particle (e.g. the electron) moves from infinity to the interaction point, then it scatters, then moves a short distance as a virtual particle, scatters again (or turns into another type of particle) and finally all the interacting particles move off to infinity, where they came from. Imagine now that we are dealing not with point-like particles, but with extended objects having an intrinsic size — strings. Finite size effects are irrelevant as long as the characteristic scattering momenta are much smaller than the inverse size. But when the momenta grow, these effects become important. In particular, the loop corrections for the amplitudes are not divergent in the ultraviolet any more — the finite string size provides a physical ultraviolet cutoff.

Two main types of strings are discussed — open and closed strings (Fig. 16.4). Open strings have the topology of an interval and closed strings have the topology of a circle. It is the closed strings that are relevant for gravity, and we will discuss only those here.

16.3.1   Bosonic Strings

As we learned in Sect. 8.3, quantum transition amplitudes can be calculated as path integrals over all possible trajectories connecting the initial and final points. This

Fig. 16.4 Strings.

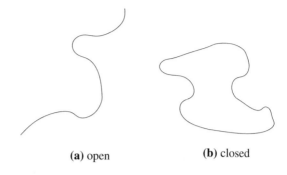

(a) open   **(b)** closed

also applies to *free* transition amplitudes, which are nothing but propagators, like (8.30) or (10.10), in the coordinate representation. To calculate path integrals, one needs to substitute there the expression for the action, $S[\text{path}]$. And we know that the action of a free relativistic particle is

$$S_{\text{free particle}} = -m \int ds = -m \int d\tau \sqrt{\frac{\partial x^\mu}{\partial \tau} \frac{\partial x_\mu}{\partial \tau}}, \qquad (16.18)$$

where τ is an arbitrary parameterization of the *world line* of the propagating particle.

A Lorentz-invariant action for a freely propagating string (the *Nambu–Goto action*) can be written by analogy,

$$S_{\text{free string}} = -T \int d\tau d\sigma \sqrt{\det\left(\frac{\partial x^\mu}{\partial \xi^\alpha} \frac{\partial x_\mu}{\partial \xi^\beta}\right)}, \qquad (16.19)$$

where $\xi^\alpha \equiv \{\tau, \sigma\}$ are the two coordinates on the *world sheet* of the propagating string.[10] The constant T carrying the dimension m^2 is called the *string tension*. It is akin to the tension (11.25) of the phenomenological QCD string.[11]

Consider now the scattering of two closed strings on each other. One can draw the diagrams displayed in Fig. 16.5.

The graph in Fig. 16.5a describes the tree string scattering amplitude. Topologically, it represents a 2-sphere, a surface of genus 0, from which four small discs are cut out. The graph in Fig. 16.5b describes a one-loop scattering amplitude. Topologically, it represents a surface of genus 1, a torus, from which four small discs are cut out. The analytic expressions corresponding to these graphs represent path integrals over all possible surfaces having four fixed crumpled circles as a border.

---

[10] ξ^α have nothing to do with x^μ. The latter are the coordinates in flat Minkowski space-time, where the string lives and which string theorists call usually the *bulk*.

[11] In fact, string theory was first formulated as one of the phenomenological models describing hadron physics and at that time did not attract so much attention. It was only after the *superstring revolution* of 1984–1985 that string (or rather superstring) theory began to be considered as a serious candidate for the role of the ToE.

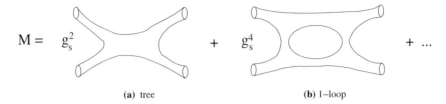

$$M = \quad g_s^2 \quad \text{(a) tree} \quad + \quad g_s^4 \quad \text{(b) 1-loop} \quad + \; \dots$$

Fig. 16.5 String scattering amplitudes.

These integrals can be calculated (especially in the limit where the circles are small, not so crumpled and far from one another), but we will not do so here. We only note that the integrals are mostly saturated by strings of characteristic size

$$l_s \; \sim \; \frac{1}{\sqrt{T}} \, . \tag{16.20}$$

The constant g_s entering the numerical factors in the amplitudes in Fig. 16.5 is called the *string coupling constant*. For the perturbative expansion to make sense, this constant should be small. The next term in the expansion involves the path integral over the surfaces of genus 2, the "pretzels". This contribution enters with the factor g_s^6. There are also higher-order contributions due to surfaces of higher genus.

"This is all very nice", you probably think, "but what is the relationship of these fancy string amplitudes to gravity?" To understand this central point, note that, for closed strings, $x^\mu(\tau, \sigma)$ are periodic functions in σ, $x^\mu(\tau, \sigma + 2\pi) = x^\mu(\tau, \sigma)$. Consider a Fourier expansion of $x^\mu(\sigma)$ describing the configuration of a propagating string at a given τ slice of the world sheet. We can write,

$$x^\mu(\sigma) \; = \; x_0^\mu + l_s \sum_{n \neq 0} \alpha_n^\mu e^{in\sigma} \tag{16.21}$$

with $\alpha_{-n}^\mu = (\alpha_n^\mu)^*$. We have thus described a nonlocal object, the string, in terms of an infinite set of local parameters: the position of the string centre x_0^μ and the Fourier coefficients $\{\alpha_n^\mu\}$.

I should now warn you that I was rather sloppy here. The reader can find more refined and correct reasoning in string textbooks. Instead of dealing with static string configurations, one needs to write down the Fourier expansion for the function $x^\mu(\tau, \sigma)$ satisfying the world sheet equations of motion $(\partial^2/\partial\tau^2 - \partial^2/\partial\sigma^2)x^\mu = 0$ and take into account both the right and left movers, $x_R^\mu(\tau - \sigma)$ and $x_L^\mu(\tau + \sigma)$. This allows one to perform an accurate quantization of the free string and to find the spectrum of its excitations.

Naturally, we cannot do all that in our book, but Eq. (16.21) may help us to acquire a heuristic understanding. To begin with, look at the first term in (16.21). It is the position of the string's centre of mass. If one disregards all other terms in the

expansion, our string effectively shrinks to a point and reduces to a pointlike scalar
particle.

And here we run into a serious impasse. The accurate quantization procedure[12]
mentioned before, but not spelled out, tells us that the mass of this particle is *imaginary*, $m^2 = -4\pi T$! This means that the effective Lagrangian corresponding to this
mode has the form

$$\mathcal{L} = \frac{1}{2}(\partial_\mu \phi)^2 + 2\pi T \phi^2 . \tag{16.22}$$

The potential energy does not grow with ϕ, but is negative and decreases. The vacuum
is thus unstable. Or better to say, there is no vacuum state. Pretty soon, at the end
of this chapter and of the book, we will argue that the absence of the vacuum is not
always a disaster. But in this particular case, it is. Thus, the Nambu–Goto action (see
p. 327) of the purely bosonic string does not describe a viable theory.

This malignant mode is isually called the *tachyon* for the following reason. If one
attempts to treat it as an ordinary particle and writes standard relativistic kinematical
relations, like $E = \sqrt{p^2 c^2 + m^2 c^4}$, one finds that the ratio pc^2/E, i.e. the velocity
of the particle, exceeds the speed of light. One should understand, however, that
the Lagrangian (16.22) does *not* describe any superluminal particles. It describes
nothing.

We will shortly see that the problem of the tachyon is solved in supersymmetric
string theory. But before talking about that, look again at Eq. (16.21) and consider
the term in the sum with $n = 1$. Geometrically, it describes a string of an ellipsoidal
form. An ellipse is characterized by a size, eccentricity and its orientation in the
bulk. When the string propagates, all these parameters oscillate. The oscillations
of the size are described by a scalar mode called the *dilaton*, with a clear enough
etymology. When the quantization is done, one finds that the mass of the dilaton
vanishes.

The oscillations of the eccentricity closely resemble a gravitational wave (see
Fig. 15.4). And indeed, the accurate analysis reveals the existence of a quadrupole
mode described by a symmetric tensor field, like the metric field. Moreover, the effective Lagrangian for this mode coincides with the Einstein–Hilbert action (15.14)!
Thus, the spectrum of excitations of the free bosonic closed string includes massless
gravitons.

Finally, there are oscillations of the orientation. They *roughly* correspond to the
presence in the spectrum of another massless mode, an antisymmetric tensor field
$B_{\mu\nu}$.[13]

Other modes in (16.21) with $n \geq 2$ all carry positive masses of the order of $\sqrt{T}$.
The larger is n, the greater is their mass and the higher is their spin. This is all
illustrated in Fig. 16.6.

---

[12] An additional important remark is that this accurate quantization cannot be carried out for every
bulk. Purely bosonic strings can only be embedded in 26-dimensional space-time. Otherwise, certain
bad anomalies do not cancel...

[13] It belongs to the adjoint representation of $SO(25, 1)$.

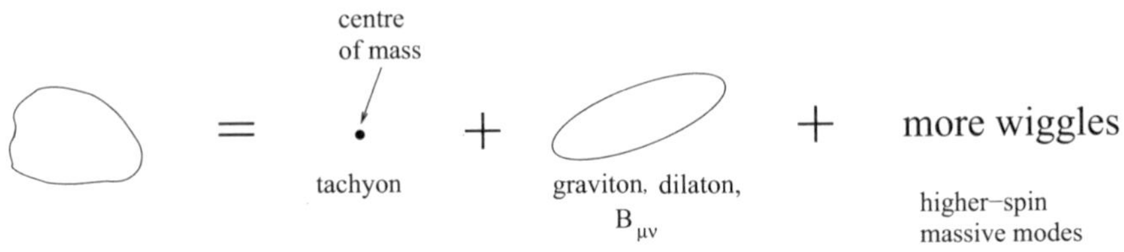

Fig. 16.6 Spectrum of closed bosonic string.

16.3.2 Superstrings

In superstring models, one adds, on top of the coordinates in the bulk $x^\mu(\tau, \sigma)$, their fermion superpartners $\psi^\mu(\tau, \sigma)$ (for each μ, ψ^μ represents a two-component Majorana spinor living in the string's two-dimensional world sheet). There are several variants of superstring theory, and a constellation of brilliant theorists[14] worked on their formulation and development.

The main good news about superstrings has already been announced — one can get rid of the tachyon in this case! This should not be so surprising. We learned in Chap. 14 that supersymmetric theories are characterized either by zero vacuum energy (if supersymmetry is not broken) or nonzero positive vacuum energy (if it is). One can thus expect the existence of a bottom in the spectrum, so that the tachyon instability as in (16.22) is ruled out.

The bad news (or at least an essential complication) is that, similar to bosonic string theory, superstring theory cannot be formulated in the four-dimensional bulk. For bosonic strings, the required dimension of the bulk was $D = 26$, while for superstrings it is $D = 10$.

For the most popular superstring model, the effective Lagrangian of the massless modes coincides with the Lagrangian of the so-called type IIA supergravity. This is a rather complicated theory, whose spectrum includes besides the 10-dimensional graviton and gravitino also the dilaton, dilatino, the antisymmetric tensor field $B_{\mu\nu}$ and some Abelian gauge fields of a special type. It possesses $\mathcal{N} = 2$ local supersymmetry. In ten dimensions this implies the presence of 32 real Grassmann transformation parameters.

Besides the massless modes, superstring theory has also massive higher-spin modes, their mass being of the order of $\sqrt{T}$. It is the string tension that serves as a physical ultraviolet regulator in all the string models, making the scattering amplitudes convergent. It is usually assumed that $\sqrt{T}$ is of the order of the Planck mass and hence the characteristic string size (16.20) is of the order of the Planck length. The effective ten-dimensional Newton's constant is estimated as

---

[14] André Neveu, John Schwarz, Pierre Ramond, Michael Green, Edward Witten, Alexander Polyakov and many others.

$$G_N^{D=10} \sim \frac{g_s^2}{T^4} \qquad (16.23)$$

(g_s is the string coupling constant, see Fig. 16.5). This estimate has exactly the same meaning as the expression (12.48) for the effective Fermi coupling constant in electroweak theory.

So far, so good, but how to reconcile the fact that our Universe is four-dimensional with that the superstring theory is formulated in ten, and not in four dimensions?

There are two possible answers to this question. One possibility is that our Universe represents a thin film, like a soap bubble embedded in the higher-dimensional bulk. We find it probable and will say few more words about this in the last section. The second conjecture (which most experts now prefer) is that the six extra dimensions are *compactified*.

Imagine an infinitely long water pipe. Topologically, it represents a direct product of an infinite straight line and a small circle. Likewise, our world may represent a direct product of the infinite four-dimensional Universe that we know and a small (of the order of the Planck length) six-dimensional compact manifold $\mathcal{M}^6$.

Accelerator energies are many orders of magnitude smaller than m_P, and the characteristic distances that we can probe in experiment are many orders of magnitude larger than the size of this compact manifold. And we simply cannot detect its presence.

What do we know about $\mathcal{M}^6$? The physical requirement is that we can still formulate superstring theory in the $R^4 \times \mathcal{M}^6$ bulk, as we can do so in R^{10}. This brings about certain constraints on $\mathcal{M}^6$. Not spelling them out, we just mention that $\mathcal{M}^6$ should be of the so-called *Calabi–Yau* type. The constraints are not restrictive enough, however. There are an infinite number of different Calabi–Yau manifolds, and there is no known way to give precedence to one of them.

16.3.3 New Complaints

I do not know whether the reader could appreciate the beauty of the whole string edifice. Maybe not — I did not give enough details. But it *is* very beautiful, and that is one of the most important reasons why it has been attracting so much attention from the theorists for the last 30 years. There is also a practical point — string theory seems to resolve (at least, in principle) the two main problems of gravity that we mentioned above:

1. It provides a physical cutoff for ultraviolet divergences.
2. If we do not talk about compactification, which is a separate story, string theory is formulated in a flat Minkowski bulk. And then there should be no problem with causality at the fundamental level.

But there are a lot of new questions that can be asked and new complaints that can be made.

1. We know how to calculate string scattering amplitudes at the tree level and at the
   one-loop level. But already two-loop string calculations represent a tremendously
   complicated problem. To the best of my knowledge, it has not been *completely*
   solved yet. One of the hindrances are kinematic singularities in the path integrals
   appearing from the string configurations when one of the handles of the "pretzel"
   becomes infinitely thin. I am not aware of any calculation at the 3-loop level or
   higher.
2. Suppose that string perturbation theory is well defined, and all the problems
   there are of a technical rather than fundamental character. Still, we do not have a
   nonperturbative formulation of string theory.
   In a field theory, the fundamental object is the action functional depending on the
   fields, which are ordinary functions of the space and time coordinates. In string
   theory, the role of these functions should be taken by *functionals* depending on the
   way a string is embedded into the bulk, $\Phi[\mathcal{C} \equiv x^\mu(\tau, \sigma)]$. And we do not know
   how a meta-functional of action $S\{\Phi[\mathcal{C}]\}$, depending on these contour functionals
   and describing both propagation of free string and string interactions, might look
   like. String theory is a beautiful edifice, but one can enjoy its beauty only above
   a certain floor; the first floors and the foundations are hid in mist.
3. More than 30 years have now passed since the superstring revolution of 1985,
   but string theory still cannot boast of phenomenological successes. It has not
   contributed much to our understanding of why the world we see is this and not
   that. In particular, the 26 free parameters of the Standard Model mentioned on
   pp. 237–238 are still free parameters; superstrings did not help to determine them.

One of the main reasons for this impasse is the necessity to go down from ten to
four dimensions. Everything depends on a particular compactification scenario,
and we do not have any guiding principle that could allow us to choose between
the myriad of different possibilities.

16.4   Author's Sweet Dreams

16.4.1   The Universe as a Soap Bubble

Personally, I am most worried about complaint no. 2. I admitted previously that I
only understand well the concepts in classical physics that reduce to Newton's laws
and the concepts in quantum physics that reduce to the Schrödinger equation. But
string theory does not. It is not in the cube of the physical theories in Fig. 4.5, which
we explored together with Sophie in Chap. 4.

Thus, I now invite the reader to think about the possibility that the ToE is not
string theory, but a conventional field theory formulated in flat space-time, which
can be placed in room G on the second floor of the theoretical physics building.

Fig. 16.7 A soap bubble model for the Universe blown by Sophie.

Of course, the first question that should be asked is how to handle gravity. Our 4-dimensional Universe is not flat, as we know. To answer that, I propose you to imagine a soap bubble (Fig. 16.7).

Soap bubbles are definitely curved. Their effective Hamiltonian is

$$H = \sigma \mathcal{A} = \sigma \int d^2 \xi \sqrt{g}, \qquad (16.24)$$

where σ is the surface tension. This Hamiltonian is invariant under reparameterizations and describes in fact a 2-dimensional gravity, with σ playing the role of the cosmological constant. But it is an effective rather than fundamental theory. Our thin bubble is embedded in flat 3-dimensional space, where the Fundamental Theory of Soap is formulated.

Similarly, our Universe may represent a thin (3+1) - dimensional film embedded in a flat higher-dimensional bulk. As I mentioned in the previous section, this is one of the possibilities that is now rather vividly discussed, along with compactification.

But in doing so, people usually keep having in mind that the ToE describing physics in the bulk is string theory. And we are not going to invoke strings in this section.

What is the reason for this Universal 4-dimensional film to appear? Who blew it?

I do not dare to pronounce myself on the second question, but the stable existence of such a film in the higher-dimensional bulk is something that is easy to imagine. Many field theories admit *solitons* — stable localized field configurations. Solitons were first discovered (by John Russel back in 1834) as solitary waves on water surface; such solitons represent solutions to nonlinear equations of hydrodynamics. But similar localized solutions also exist in many other theories. Not in QCD and not in electroweak theory, but, for example, in hypothetical Grand Unified Theories, such solutions are present.[15]

A soliton is not necessarily strictly localized. One can well imagine a soliton that is localized in *some* spatial directions, but is infinitely extended in the others. An example is the *Abrikosov string*, a magnetic vortex that represents a partly localized soliton in the effective Ginzburg–Landau theory. These vortices exist as physical objects in type II superconductors. Likewise, the Universe may represent a *3-brane* — a stable static solution of the fundamental field theory living in the bulk, which is extended in three spatial coordinates and localized in the other directions.

But what might this fundamental field Theory of Everything, defined in a higher-dimensional bulk, look like? It is not easy to answer this question. As a first try, we may write down a Lagrangian familiar from 4-dimensional physics, but extended into higher dimensions. For example, we can consider the 6-dimensional Yang–Mills Lagrangian

$$\mathcal{L}^{D=6} = \frac{1}{2f^2} \text{Tr}\{F_{MN} F^{MN}\}, \tag{16.25}$$

with $M, N = 0, 1, 2, 3, 4, 5$.

And here we meet a problem. To keep the action dimensionless, we should now attribute dimension to the gauge coupling constant, $[f] = m^{-1}$. But such a theory involves uncontrollable power and logarithmic ultraviolet divergences (indeed, we now integrate over $d^6 p$ rather than $d^4 p$ and the ultraviolet behavior is much worse) and becomes nonrenormalizable in exactly the same way as the Fermi theory or Einstein's gravity. As we have learned from previous discussions, non-renormalizable theories only have meaning as effective low-energy approximations of something more fundamental. They cannot be fundamental themselves.

As a second try, staying in six dimensions, we can write down gauge-invariant expressions of canonical dimension 6 — for example,

$$\mathcal{L}_1 = \alpha \text{Tr}\{F_{MN}\Box F^{MN}\} + \beta \text{Tr}\{F_M{}^N F_N{}^P F_P{}^M\}. \tag{16.26}$$

The coefficient α and β are dimensionless and the theory (16.26) is renormalizable. One can also be convinced of that by counting the divergences in Feynman graphs.

---

[15]They carry magnetic charge and are known as *'t Hooft–Polyakov monopoles*.

The extra box operator in the first term brings about an extra factor $1/p^2$ in the gauge field propagator, which eliminates power ultraviolet divergences in the integrals.

Can we take it as the fundamental theory? Unfortunately not. By replacing (16.25) with (16.26), we solved the problem of non-renormalizability, but we run into another serious problem. The Lagrangian (16.26) includes higher derivatives of dynamical variables. And such theories are known to include *ghosts*.

16.4.2 Living with Ghosts

Unfortunately, there is considerable confusion associated with this issue (the ghosts) in the literature. We make a digression now with an attempt to clarify this confusion and will return to our main topic (the ToE) at the end of the chapter. Our dedicated, but tired reader may go there directly, if s/he wishes.

One usually understands by "ghosts" those fields whose kinetic energy is negative. That corresponds to contributions like $-(\partial_\mu \phi)^2$ in relativistic Lagrangians. One can make the energy positive by attributing to the ghost quantum state a negative norm (so that unitarity is explicitly violated), but this is a confusing interpretation. It is much more physical to say that all the norms are positive, but neither classical nor quantum Hamiltonian has a ground state. Thus, a system with ghosts has similar properties (though not exactly the same) to a system with tachyons. The difference is that "ghost" means negative kinetic energy, while "tachyon" means negative potential energy.

Let us now explain why the existence of ghosts is a generic feature of higher-derivative systems. Consider first an ordinary mechanical system describing a particle moving in a central potential:

$$L = \frac{m\dot{r}^2}{2} - V(r).$$
(16.27)

The conserved energy is

$$E = \frac{m\dot{r}^2}{2} + V(r).$$
(16.28)

If the potential is bounded from below (as for the oscillator), so is the energy. For the attractive Coulomb potential, the classical energy is not bounded from below, but the spectrum of the quantum Hamiltonian is.

Consider now the simplest mechanical higher-derivative system, the so-called *Pais–Uhlenbeck oscillator* with the Lagrangian

$$L = \frac{1}{2}(\ddot{q} + \Omega_1^2 q)(\ddot{q} + \Omega_2^2 q),$$
(16.29)

which explicitly depends on the acceleration $\ddot{q}$, and not only on q or $\dot{q}$. The dynamical equations for this Lagrangian[16],

$$\frac{d^2}{dt^2}\frac{\delta L}{\delta \ddot{q}} - \frac{d}{dt}\frac{\delta L}{\delta \dot{q}} + \frac{\delta L}{\delta q} =$$

$$\left(\frac{d^2}{dt^2} + \Omega_1^2\right)\left(\frac{d^2}{dt^2} + \Omega_2^2\right)q = 0, \tag{16.30}$$

have an integral of motion — the energy

$$E = \left(\ddot{q} - \dot{q}\frac{d}{dt}\right)\frac{\delta L}{\delta \ddot{q}} + \dot{q}\frac{\delta L}{\delta \dot{q}} - L$$

$$= \frac{1}{2}\ddot{q}^2 - \dot{q}\left(q^{(3)} + \frac{\Omega_1^2 + \Omega_2^2}{2}\dot{q}\right) - \frac{\Omega_1^2\Omega_2^2 q^2}{2} \tag{16.31}$$

($q^{(3)}$ stands for the triple time derivative). One can observe that (16.31) includes terms of different sign and can be made as positive and as negative as one wishes. One can prove that it is an inherent property of all higher-derivative systems.

Moreover, in this case the quantum Hamiltonian also does not have a bottom. By a canonical transformation, the *Ostrogradsky Hamiltonian*[17] for the system (16.29) with $\Omega_1 \neq \Omega_2$ can be brought into the form[18]

$$H = \frac{1}{2}(P_1^2 + \Omega_1^2 Q_1^2) - \frac{1}{2}(P_2^2 + \Omega_2^2 Q_2^2). \tag{16.32}$$

In other words, the system (16.29) is equivalent to two ordinary oscillators, but the Hamiltonian represents the *difference* of two oscillator Hamiltonians, rather than its sum. The spectrum of the quantum problem is

$$E_{nm} = \hbar\Omega_1\left(\frac{1}{2} + n\right) - \hbar\Omega_2\left(\frac{1}{2} + m\right) \tag{16.33}$$

---

[16]The first line in (16.30) represents the left-hand side of the generalized Lagrange equation for a generic function $L(q, \dot{q}, \ddot{q})$. In our particular case, $\delta L/\delta \dot{q} = 0$.

[17]Mikhail Ostrogradsky showed how to derive Hamiltonians for higher-derivative Lagrangian systems back in 1850. He did so in a remarkable paper published in *Memoires de l'Académie Impériale des Sciences de St. Pétersbourg*, where he in fact derived the Hamiltonian formalism independently of William Hamilton. He applied this method to generic Lagrangians $L(q_i, \dot{q}_i, \ddot{q}_i, \ldots)$.

This paper did not attract much attention in his time (it was not written in an optimal pedagogical way), but is often cited nowadays. However, the observation that the Hamiltonian functions for higher-derivative systems are not bounded either from below or from above, what is called by many people "Ostrogradsky instability", belongs not to Ostrogradsky (he only derived general formulas and did not study particular dynamical systems), but to Abraham Pais and George Uhlenbeck, who studied the classical and quantum dynamics of the system (16.29) in a paper published exactly a hundred years after Ostrogradsky's paper.

[18]It is more tricky if $\Omega_1 = \Omega_2$ — the case that we will not consider here.

with integer n, m. If the ratio Ω_1 / Ω_2 is irrational, the spectrum is *everywhere dense*[19] and is what mathematicians call *pure point*, i.e. every eigenstate has a normalizable wave function. This is, of course, unusual, but is by no means a disaster. Both the classical and quantum problems are well defined, the wave functions of all eigenstates can be written explicitly, and the evolution operator (8.33) is unitary. In fact, for a *free* oscillator, including the Pais–Uhlenbeck oscillator, energy does not have much dynamical meaning and serves only for bookkeeping.

The situation changes if one switches on interactions. Heuristically, an interacting system whose spectrum is not bounded from below "tries" to transfer energy from the degrees of freedom giving positive contributions to the spectrum to the ghost sector with negative energies. And this may lead to *collapse*.

To exhibit collapse, a system need not necessarily involve higher derivatives. The simplest collapsing system has the Lagrangian (16.27) with the attractive potential

$$V(r) \; = \; -\frac{\alpha}{r^2} \,. \tag{16.34}$$

It is known that for strong enough attraction,

$$\alpha \; > \; \frac{\hbar^2}{8m} \,, \tag{16.35}$$

this system exhibits a *fall to the centre*, to the singularity at $r = 0$. Not only is the quantum Hamiltonian not bounded from below, but also the singularity at the origin does not in this case allow one to correctly pose the Schrödinger problem and determine the spectrum. Heuristically, the evolution is not unitary, because the probability "leaks away" through the singularity. The problem can only be made benign if the singularity is regularized. For example, one can assume that the potential has the form (16.34) only for $r > r_0$, while $U(r \leq r_0)$ is a constant. The spectrum in such a system is bounded, but the values of the energy levels depend essentially on r_0.

Note that, for any α, there are *classical* trajectories running into the singularity in finite time. On the other hand, for small enough α not exceeding the limit (16.35), the quantum spectrum is bounded and benign. This teaches us that quantum fluctuations try to prevent the system from falling down to the singularity and "leaking through". If the attraction is not too strong, they manage to do so. Saying it differently, a system undergoing collapse in quantum mechanics also has classical collapsing trajectories, but the inverse is not necessarily true. On the other hand, *if a classical system is benign and never runs into a singularity, its quantum counterpart is also benign.*

An interacting PU oscillator can be obtained by adding to the Lagrangian (16.29) a quartic term like $\sim \alpha q^4$. Not going into detail,[20] we only say that in that case there *are* classical trajectories running in finite time to the singularity at $q = \infty$. In all

---

[19] For any energy E and for any $\epsilon > 0$, one can find an eigenstate $|nm\rangle$ with $|E - E_{nm}| < \epsilon$.

[20] An interested reader can consult *arXiv:1710.11538 [hep-th]* and references therein.

Fig. 16.8 Benign
oscillations in a benign
ghost-ridden system.

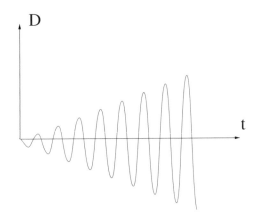

probability, the quantum problem is also not well defined in this case — unitarity is
violated due to "malignant" ghosts.

Consider, however, the following nonlinear Hamiltonian:

$$H = pP + D(\Omega^2 x + \lambda x^3), \tag{16.36}$$

where (p, P) are the canonical momenta for (x, D). The kinetic part of the Hamiltonian (16.36) is not positive definite and hence the Hamiltonian is ghost-ridden: its
spectrum is not bounded. But the ghosts are not malignant, they are *benign* in this
case!

The system (16.36) has a nice property: it involves, besides the Hamiltonian,
another integral of motion:

$$N = \frac{1}{2}p^2 + \frac{\Omega^2 x^2}{2} + \frac{\lambda x^4}{4}, \tag{16.37}$$

so that both classical and quantum problems are exactly solvable.

The classical trajectories are expressed via elliptic functions. Both $x(t)$ and $D(t)$
oscillate. $x(t)$ stays bounded, while the amplitude of the oscillations of $D(t)$ grows
linearly with time, as in Fig. 16.8. The linear growth is not a collapse, however —
the system does not run into infinity in finite time.

In this case one can also explicitly find the spectrum and the eigenstates of the
quantum Hamiltonian. The spectrum is mostly continuous, with the eigenvalues lying
in the two symmetric intervals, $-\infty < E \leq -\Omega$ and $\Omega \leq E < \infty$. Besides, there
are an infinite number of states with $E = 0$. The evolution operator is unitary.

Another example of a Hamiltonian with benign ghosts is

$$H = \frac{1}{2}(P_1^2 + \Omega_1^2 Q_1^2) - \frac{1}{2}(P_2^2 + \Omega_2^2 Q_2^2) + \lambda(Q_1 - Q_2)(Q_1 + Q_2)^3. \tag{16.38}$$

This is the PU oscillator with a very specially chosen interaction. The system (16.38) is not analytically solvable, but the classical solutions can easily be found numerically. A reader having Mathematica or Maple on his/her computer can switch it on and be convinced that both $Q_1(t)$ and $Q_2(t)$ represent oscillating functions with rather slowly growing amplitudes. There is no classical collapse and hence, as we previously argued, there is no quantum collapse.

There are also other similar quantum mechanical examples. In addition, we have at our disposal today an example of a nontrivial field theory involving ghosts that are benign and do not bring about collapse and the associated loss of unitarity. It represents a natural field-theoretical extension of the system (16.36). The Lagrangian reads

$$\mathcal{L} = \partial_\mu \phi \partial^\mu D - D(\Omega^2 \phi + \lambda \phi^3). \tag{16.39}$$

This nonlinear dynamical system has an infinite number of variables $\phi(x)$, $D(x)$ and only two integrals of motion [the energy and the field theory counterpart of (16.37)]. Thus, it is not exactly solvable. Solutions to the field equations of motion can, however, be found numerically. We looked at these solutions for the theory (16.39) in the simplest case of only one spatial dimension and found no trace of collapse.

The analysis above leads to the following conclusions:

1. Higher-derivative systems always include ghost degrees of freedom, which means that the spectra of their Hamiltonians are not bounded from below.
2. In most cases, this leads to quantum collapse and loss of unitarity.
3. There are also some special quantum mechanical and field theory systems where ghosts are present, but they are benign and do not induce collapse. Such systems are unusual, but not sick — the Hamiltonian of such a system is Hermitian and the evolution operator is unitary.

We can finally formulate our sweet dream:

The ToE is a higher-derivative field theory with benign ghosts living in a higher-dimensional flat space-time. Our Universe represents a 3-brane — a solitonic solution extended in three spatial and the time directions and localized in the extra dimensions. Gravity arises as effective theory on the world volume of this brane.

Unfortunately, we do not have the slightest idea today of what this Holy Grail theory might be.

Part VII
Coffee

Chapter 17
Recommended Reading

The literature on quantum field theory and related subjects is vast. I am bound to make a subjective selection and mention only a few books. I have already done so in the main text, but here I collect the most relevant references together, accompanied with personal clarifying comments.

- *R.P. Feynman, R.B. Leighton and Matthew Sands, The Feynman Lectures on Physics, Addison-Wesley, 1963.*

  This course on general physics is recommended to every physics student. We assumed that the reader was familiar with it. This acquaintance is sufficient to enjoy the *Aperitif* and *Hors d'Oeuvres* in our dinner.

- Chapter 5 involves some historical comments. A reader who wants to learn more about the fascinating story of how QED was created can consult Feynman's Nobel lecture,

  (www.nobelprize.org/nobel_prizes/physics/laureates/1965/feynman-lecture.html)

  and also read a very good book

  S.S. Schweber, QED and the Men Who Made It: Dyson, Feynman, Schwinger and Tomonaga, Princeton University Press, 1994.

- Chapter 6 represents a crash course in group theory. For a more detailed and comprehensive course addressed to physicists, see

  H.F. Jones, Groups, Representations and Physics, Taylor & Francis Group, 1998.

- Similarly, Chap. 7 represents a crash course in analytical mechanics. For more details, please look into

  L.D. Landau and E.M. Lifshitz, Mechanics, Pergamon Press, 1969.

  (Volume I of the Landau Course) and the first four chapters of Volume II:

© Springer International Publishing AG 2017
A. Smilga, *Digestible Quantum Field Theory*,
DOI 10.1007/978-3-319-59922-9_17

L.D. Landau and E.M. Lifshitz, The Classical Theory of Fields, Pergamon Press, 1971.

- Chapter 8 is devoted to scattering theory in quantum mechanics. We recommend reading

E. Fermi, R.A. Schluter, Notes on Quantum Mechanics, Univ. of Chicago Press, 1995.

Scattering theory is discussed there in Sects. 23 and 33, but these brilliant lecture notes of Fermi deserve to be read completely.

More details on scattering theory can be found in Chapter VIII of the textbook

C. Cohen-Tannoudji, B. Diu and F. Laloë, Quantum Mechanics, Wiley & Sons, 1991.

And the advanced reading consists of Chaps. XVII and XVIII of Volume III of the Landau Course,

L.D. Landau and E.M. Lifshitz, Quantum Mechanics, Pergamon Press, 1965.

The last section of Chap. 8 discusses path integrals, and the classical reference is

R.P. Feynman and A.R Hibbs, Quantum Mechanics and Path Integrals, Emended Edition, McGraw-Hill, 2005.

- For those who wish to learn quantum field theory at the professional level, I recommend two textbooks:

M.E. Peskin and D.V. Schroeder, An Introduction to Quantum Field Theory, Addison-Wesley, 1995.

and

A. Zee, Quantum Field Theory in a Nutshell, Princeton University Press, 2003.

The second book is more modern and discusses, besides the Standard Model, other aspects of QFT, but it involves fewer calculational details.

- A lot of calculations for different QED processes are brought together in Volume IV of the Landau Course:

V. Berestetsky, E.M. Lifshitz, L.P. Pitaevsky, Quantum Electrodynamics, Butterworth-Heinemann, 1982.

As further reading on QCD, I dare to suggest my own book:

A.V. Smilga, Lectures on Quantum Chromodynamics, World Scientific, 2001.

To learn more about electroweak theory, I recommend

L.B. Okun, Leptons and Quarks, North Holland, 1985.

- The first section of our *Trou Normand* chapter was based on Feynman's memoir book,

  R.P. Feynman, Surely You're Joking, Mr Feynman!, Norton & Company, 1997.

  There are several books telling about Landau and his school. I mention two such books:

  I.M. Khalatnikov, ed., Landau, the Physicist and the Man: Recollections of L.D. Landau, Pergamon Press, 1989.

  and

  M.A. Shifman, ed., Under the Spell of Landau: When Theoretical Physics Was Shaping Destinies, World Scientific, 2013.

- The standard classical textbook on supersymmetry is

  J. Wess and J. Bagger, Supersymmetry and Supergravity, Princeton Series in Physics, 1983.

- To study general relativity, I recommend the five last chapters of *The Classical Theory of Fields* by Landau and Lifshitz and also the all-embracing

  S. Weinberg, Gravitation and Cosmology, Wiley & Sons, 1972.

- It is difficult to give a particular reference on quantum gravity — this is an emerging science and the time for textbooks has not yet come. There are several serious monographs on string theory, but it is not easy reading. A little more user-friendly is

  B. Zwieback, A First Course in String Theory, Cambridge University Press, 2004.

Subject Index

This index is complementary to the table of contents. We have tried, whenever possible, to avoid "double counting".

A

Aether, 32, 281
Anomaly
    chiral, 153, 206, 207, 225
    conformal, 207
Asymptotic
    freedom, 72, 73, 139, 240
    states, 219, 281
ATLAS collaboration (at CERN), 233, 234

B

Berezin integral, 106
Big
    Bang, 17, 18, 50, 200, 302, 303
    Crunch, 49, 303, 304, 306
    Freeze, 306
Bjorken scaling, 71, 72
Black holes, 21, 22, 298, 300–301, 319–321
    evaporation of, *see* Hawking radiation
Bohr
    correspondence principle, 37
    radius, 23
    spectrum, 39
Bootstrap, 69, 260
Born–Oppenheimer approximation, 82
Bottomonium, 190
Brookhaven National Laboratory (BNL), 185–187
Bulk (hypothetical multidimensional space where the Theory of Everything might be formulated), 323, 325–330

C

Cabbibo angle, 212, 215
Calabi–Yau manifolds, 327
Canonical
    dimension, 144, 212, 312, 314, 315, 317, 330
    Hamiltonian, 114, 117, 122, 145
    momentum, 110, 112, 114, 117, 122, 134, 149, 273, 275, 334
    transformation, 332
Casimir operator, 183
Cauchy
    problem, 32, 35, 55, 317, 318
    theorem (of complex analysis), 131
Causality (violation of), 55, 318
CERN (European Centre for Nuclear Research), 70, 78, 85, 86, 186, 219, 230, 231, 265, 285
Charmed mesons, 190
Charmonium, 187–190
Chirality, 155
CKM mixing matrix, 215, 229, 238
Clifford algebra, 13, 150, 156
Closed time loops, *see* Gödel solutions
CMS collaboration (at CERN), 233
Compton length, 10, 11, 41, 67, 83, 121, 216, 320
Condensate
    Bose, 251
    gluon, 208
    quark, 203–207
Conduction and valence bands (for electrons in crystals), 147
Conservative systems, 108
Constant
    cosmological, 48, 50, 282, 296, 301, 306, 329

© Springer International Publishing AG 2017
A. Smilga, *Digestible Quantum Field Theory*,
DOI 10.1007/978-3-319-59922-9

Name Index

And this index is complementary to the table of contents and to the subject index. (Thus, we did not make an index entry for Feynman each time a Feynman diagram was drawn.)

A

Abrikosov, Alexei, 56, 66, 251, 254
Akhiezer, Alexander, 254
Akulov, Vladimir, 284
Anderson, Carl, 148
Archimedes, 136, 278
Aristotle, 57

B

Bagger, Jonathan, 341
Barish, Barry, 298
Barkov, Lev, 82
Becquerel, Henri, 27
Berestetsky, Vladimir, 61, 160, 341
Berezin, Felix, 149
Bern, Zvi, 317
Bethe, Hans, 254
Blinnikov, Sergei, 298
Blaizot, Jean-Paul, 7
Bohr, Niels, 76, 250, 252, 253, 260
Born, Max, 252
Bose, Satyendranath, 250
Bradbury, Ray, 318
Bronstein, Matvei, 252, 255
Brout, Robert, 219
Brunelleschi, Filippo, 7
Bryusov, Valery, 131

C

Chadwick, James, 76
Chaichian, Masud, 7
Cockroft, John, 257
Cohen-Tannoudji, Claude, 123, 340
Copernicus, Nicolaus, 71
Cowan, Clyde, 77

D

Davis, Raymond, 235–237
De Broglie, Louis, 250
De Tocqueville, Alexis, 57
Dirac, Paul, 42, 250, 252, 253
Diu, Bernard, 340
Durell, Clement, 3, 4
Dyson, Freeman, 65, 134
Dzyaloshinsky, Igor, 56

E

Eddington, Arthur, 294
Einstein, Albert, 4, 10, 18, 46, 250, 266, 276, 282, 287, 288, 292, 300
Eletsky, Tatiana, 7
Englert, François, 219
Euclid, 276, 294
Ezhov, Nikolai, 256

F

Fermi, Enrico , 47, 58, 123, 340
Ferrara, Sergio, 315
Feynman, Michelle, 249

© Springer International Publishing AG 2017
A. Smilga, *Digestible Quantum Field Theory*,
DOI 10.1007/978-3-319-59922-9

Printed in the United States
By Bookmasters